There is a major effort throughout the world to develop genetically improved trees like this magnificent specimen of *Pinus caribaea* var. *hondurensis* from Venezuela; but the effort will fall short if the wood produced by the selected tree is not acceptable. Therefore, it is essential to include wood properties in forest tree breeding programs. Most wood properties have strong genetic control

Springer Series in Wood Science

Editor: T.E. Timell

M.H. Zimmermann
Xylem Structure and the Ascent of Sap (1983)

J.F. Siau
Transport Processes in Wood (1984)

R.R. Archer
Growth Stresses and Strains in Trees (1986)

W.E. Hillis
Heartwood and Tree Exudates (1987)

S. Carlquist
Comparative Wood Anatomy (1988)

L.W. Roberts/P.B. Gahan/R. Aloni
Vascular Differentiation and Plant Growth Regulators (1988)

C. Skaar
Wood-Water Relations (1988)

J.M. Harris
Spiral Grain and Wave Phenomena in Wood Formation (1989)

B.J. Zobel/J.P. van Buijtenen
Wood Variation (1989)

P. Hakkila
Utilization of Residual Forest Biomass (1989)

J.W. Rowe (Ed.)
Natural Products of Woody Plants (1989)

K.-E.L. Eriksson/R.A. Blanchette/P. Ander
Microbial and Enzymatic Degradation of Wood and Wood Components (1990)

R.A. Blanchette/A.R. Biggs (Eds.)
Defense Mechanisms of Woody Plants Against Fungi (1992)

S.Y. Lin/C.W. Dence (Eds.)
Methods in Lignin Chemistry (1992)

G. Torgovnikov
Dielectric Properties of Wood and Wood-Based Materials (1993)

F.H. Schweingruber
Trees and Wood in Dendrochronology (1993)

P.R. Larson
The Vascular Cambium: Development and Structure (1994)

M.-S. Ilvessalo-Pfäffli
Fiber Atlas: Identification of Papermaking Fiber (1995)

B.J. Zobel/J.B. Jett
Genetics of Wood Production (1995)

Bruce J. Zobel Jackson B. Jett

Genetics of Wood Production

With 79 Figures

Springer-Verlag
Berlin Heidelberg New York
London Paris Tokyo
Hong Kong Barcelona
Budapest

Prof. Dr. BRUCE J. ZOBEL
Professor Emeritus
College of Forest Resources
North Carolina State University
P.O. Box 8002
Raleigh, NC 27695–8002, USA
and
President
Zobel Forestry Associates
P.O. Box 37398
Raleigh, NC 27627, USA

Prof. Dr. JACKSON B. JETT
College of Forest Resources
North Carolina State University
P.O. Box 8001
Raleigh, NC 27695–8001, USA

Series Editor:

T.E. TIMELL
State University of New York
College of Environmental Science and Forestry
Syracuse, NY 13210, USA

ISBN-13: 978-3-642-79516-9 e-ISBN-13: 978-3-642-79514-5
DOI: 10.1007/978-3-642-79514-5

Library of Congress Cataloging-in-Publication Data. Zobel, Bruce, 1920– . Genetics of wood production/Bruce J. Zobel, Jackson B. Jett. p. cm. – (Springer series in wood science). Includes bibliographical references (p.) and index. 1. Wood – Variation. 2. Trees – Breeding. 3. Trees – Genetics. I. Jett, Jackson B., 1943– . II. Title. III. Series. SD535.7.Z624 1995 634.9'56 – dc20 94-48497

Softcover reprint of the hardcover 1st edition 1995

Typesetting: Macmillan India Ltd., Bangalore 25

SPIN: 10043351 31/3130/SPS – 5 4 3 2 1 0 – Printed on acid-free paper

This book is dedicated to Dr. Tore Erik Timell

It is rare indeed to find an individual who is thoroughly knowledgeable about, and who has contributed to, a number of disciplines at the same time. One such person is Dr. Tore Timell. His contributions have been many in teaching, in research, and in the literature related to wood. Truly, as he says, "trees have been at the center of my interests all my life". He graduated from the Royal Institute of Technology in Stockholm in 1950 and then joined the Pulp and Paper Research Institute of Canada in Montreal. In 1962, he moved to the College of Forestry in Syracuse, where he still works. In later years, he wrote a monograph on compression wood and has been very active in establishing and editing the Springer Series in Wood Science. Although Dr. Timell is not a specialist in the genetics of wood, he has a broad understanding of the subject, strengthened by his in-depth knowledge of many aspects of wood and forestry. We are, therefore, highly honored to dedicate our book to someone who has made such a significant contribution to forestry and the field of wood science.

Preface

Over the past years, a great deal has been learned about variation in wood properties. Genetic control is a major source of variation in most wood properties. Wood is controlled genetically both directly in the developmental or internal processes of wood formation and indirectly by the control of tree form and growth patterns. Emphasis in this book will be on the internal control of wood production by genetics although there will be two chapters dealing with the indirect genetic control of wood, which was covered in detail in the previous book by Zobel and van Buijtenen (1989).

The literature on the genetics of wood is very variable, some quite superficial, on which little reliance can be placed, and some from well-designed and correctly executed research. When suitable, near the end of each chapter, there will be a summary with the authors' interpretation of the most important information in the chapter. The literature on the genetics of wood can be quite controversial. This is to be expected, since both the environment and its interaction with the genotype of the tree can have a major effect on wood properties, especially when trees of similar genotypes are grown under widely divergent conditions. Adding to the confusion, studies frequently have been designed and analyzed quite differently, resulting in conflicting assessments of results.

It is essential to remember the differences in wood properties between the hardwoods and the conifers. Their different types of wood sometimes respond similarly to genetic control but sometimes they differ quite strikingly. General trends will be emphasized in this book but when the trends differ, they will be noted.

This book is partially a sequel to an earlier one in the Springer-Verlag wood series (Zobel and van Buijtenen 1989, *Wood Variation, Its Causes and Control*) and frequent references will be made to the earlier book. For example, growth rate can have an effect on wood properties; but growth rate is a very complex characteristic where genetics and environmental factors interact in numerous ways. There is a chapter in the above book on growth rate and wood properties but it is not possible to say how much of the relationship is related to genetics. Because of the strong effect of environment and growth on wood properties, we may miss some of the genetic effects on wood and, in a few instances, may give credit to genetics when environmental differences are the real cause.

It is important to know that a considerable amount of wood variation is under genetic control, and this control is usually strong enough to allow breeding programs to effectively change wood. The need for wood for specialty purposes is increasing rapidly as is the use of species with wood now considered to be of marginal utility. In the past, not enough application has been made of the genetic manipulation of wood. This most valuable tool has often been ignored, with tree breeders emphasizing only the growth and form. Wood quality and uniformity must be a primary objective of a forest tree improvement program. Genetic manipulation is a major tool to help obtain this objective.

The numerous reports regarding genetic control of wood properties are widely scattered throughout the literature. It is the objective of this book to bring these references together and summarize them for use by the reader. The book deals primarily with those wood properties having a strong enough genetic component that a meaningful change in wood is possible by the use of genetic manipulation. Additionally, ideas and concepts gained by the authors through many years of working intensively with wood will be combined with information from studies reported in the literature to suggest how wood properties can be improved using genetic manipulation.

Raleigh, North Carolina
Spring 1995

Bruce J. Zobel
Jackson B. Jett

Acknowledgments. A book such as this cannot possibly be done well by one or two persons and requires help from others. The authors are most grateful to those persons who made suggestions and corrections, edited chapters, supplied publications as well as data, graphs, and photographs. We wish to thank the College of Forestry of the North Carolina State University for supplying the facilities and much of the support that made the preparation of this book possible.

Following is an alphabetical listing of those who contributed to the development of the book:

Name	*Organization*	*Location*
Tom Adams	Oregon State University	Corvallis, Oregon
Roger Blair	Environmental Protection Agency	Corvallis, Oregon
Floyd Bridgwater	US Forest Service	Raleigh, North Carolina
Kevin Harding	Queensland Forest Service	Brisbane, Australia
John Harris	Timber Consultant	Christchurch, New Zealand
Mike Kane	Zobel Forestry Associates	Raleigh, North Carolina
Robert Kellison	NC State University	Raleigh, North Carolina
Bob Kellogg	Consultant	Vancouver, B.C., Canada

Name	*Organization*	*Location*
Clark Lantz	US Forest Service	Atlanta, Georgia
William Lowe	Texas A&M University	College Station, Texas
George Lowerts	Union Camp Corporation	Savannah, Georgia
Steve McKeand	NC State University	Raleigh, North Carolina
Hubert Polge	INRA	Nancy, France
Robert Purnell	International Paper Co.	Bainbridge, Georgia
Beatriz Vera Pozzi Redko	Companhia Florestal Monte Dourado	Monte Dourado, Brazil
Donald Rockwood	University of Florida	Gainesville, Florida
Fred Taylor	Mississippi State Univ.	Mississippi State, Mississippi
J. van Buijtenen	Professor Emeritus	College Station, Texas
Elisabeth Wheeler	NC State University	Raleigh, North Carolina
Jeffrey Wright	Carton de Colombia	Cali; Colombia
Marvin Zoerb	Union Camp Corporation	Rincon, Georgia

Special Acknowledgment. In developing a book, a tremendous amount of detailed and repetitive activity is necessary as the book is finished and revisions are made. The authors are indebted to Jean Pittman, who so capably attended to the details of typing and making revisions, and whose efforts contributed significantly to making this book possible.

Contents

Chapter 1

The Role of Genetics in Wood Production – General Concepts

1.1 Background Information[1]

For many years most tree improvement programs included growth, form, adaptability, and pest resistance in their assessments but did *not* include wood properties as such. It was recognized that tree form, growth rate, and pest tolerance could all affect wood in several ways, as was described by Zobel (1971) (Fig. 1.1). However, the potential for improving wood by direct application of genetics was not generally appreciated, although persons like Harris (1983) strongly emphasized the value of employing genetics to improve wood. Many people felt that if growth, form, and adaptability were central to a genetics program, there would be little opportunity left for altering wood, even if there were strong genetic control. However, as noted by Zobel (1972): "Addition of a wood property will enable moderate changes in wood while still being able to maintain the desired form, growth, and adaptability." As early as 1935, Schreiner recognized the possibilities for genetic manipulation of wood and published an article on how pulping characteristics might be improved by breeding; these ideas were later expanded in his 1958 paper. The economic impact of changes in wood quality were recently outlined by Cubbage (1990). However, little proof was available regarding the inheritance of wood properties except for studies like those of Pawsey (1965) in Australia, who tested the wood of clones of radiata pine (*Pinus radiata*) and found large differences among clones but great similarity among ramets within a clone. The general possibility of improving wood quality through breeding was covered by Lahiri (1959) and an assessment of the wood qualities to use in tree breeding was made by Dadswell et al. (1963) in *Pinus radiata* and for Douglas-fir (*Pseudotsuga menziesii*) by Kellogg (1990).

In a unique study to determine the inheritance of wood, Zobel et al. (1962) assessed the wood of 33 grafts of slash pine (*Pinus elliottii*) using samples taken just above and just below the graft union. The two wood samples from each tree were unrelated (Fig. 1.2), with $r = -0.122$ for density and $r = 0.002$ for cell length, even though the wood was all produced in the same environment, by the same root system, and the same crown. The cambial control of wood production was strong, producing greatly different specific gravities and tracheid lengths above and below the graft union. This was one of the earliest studies

[1] This first chapter is introductory, mentioning numerous subjects covered in detail in later chapters. Its objective is to provide a general overview of the part that genetics plays in the determination of wood properties.

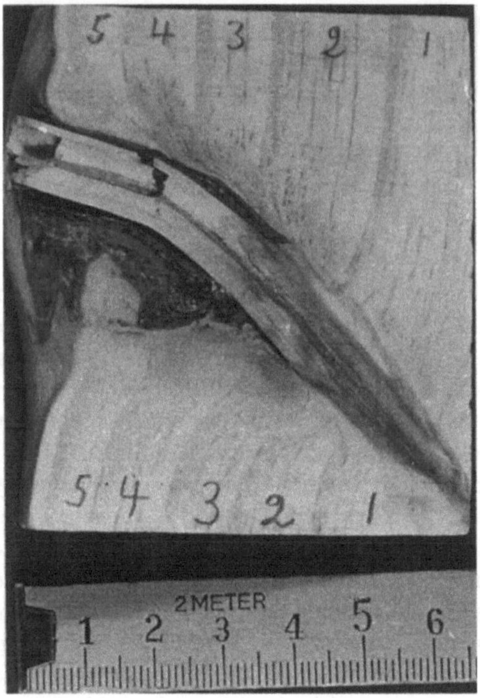

Fig. 1.1. All types of adverse wood properties are associated with limbs. In this young tree, note the ingrown bark, the resin deposits, and the abnormal wood in addition to the knot volume itself which is associated with large knots and steep-angled limbs. As shown, branches arise from the center of the tree and influence a considerable volume of wood around them

specifically designed to assess genetic control of wood properties, and led to the statement: "These heritabilities of wood density and tracheid length add further data to the growing results indicating intensity of inheritance of wood properties."

Once the genetic control of wood was appreciated, the question was then raised as to whether it should be a primary or secondary trait in a tree improvement program. Because of the relative importance of growth rate and form, and also because of the high heritabilities for most wood properties, Zobel (1963) suggested that wood be considered a secondary trait in a tree improvement program. His suggestions were similar to those of Currò (1972), who advocated selecting first for form, growth habit, pest resistance, volume, and finally for wood density. The statement made by Zobel in 1971 still holds today that: "Breeding for wood usually is a secondary effort, within the primary effort of getting faster growing, better formed, well-adapted, pest-resistant trees. It can be superimposed on a regular breeding program with the major extra effort required being a greater number of parent trees to satisfy the wood requirements." In a series of studies, like the one reported by van Buijtenen et al. (1975), it was found that volume and wood density are the two most important tree characteristics in a breeding program to produce pulp for linerboard (Fig. 1.3).

As will be described in the following chapters, tests on genetic control of wood have now been made on a number of both conifer and hardwood species.

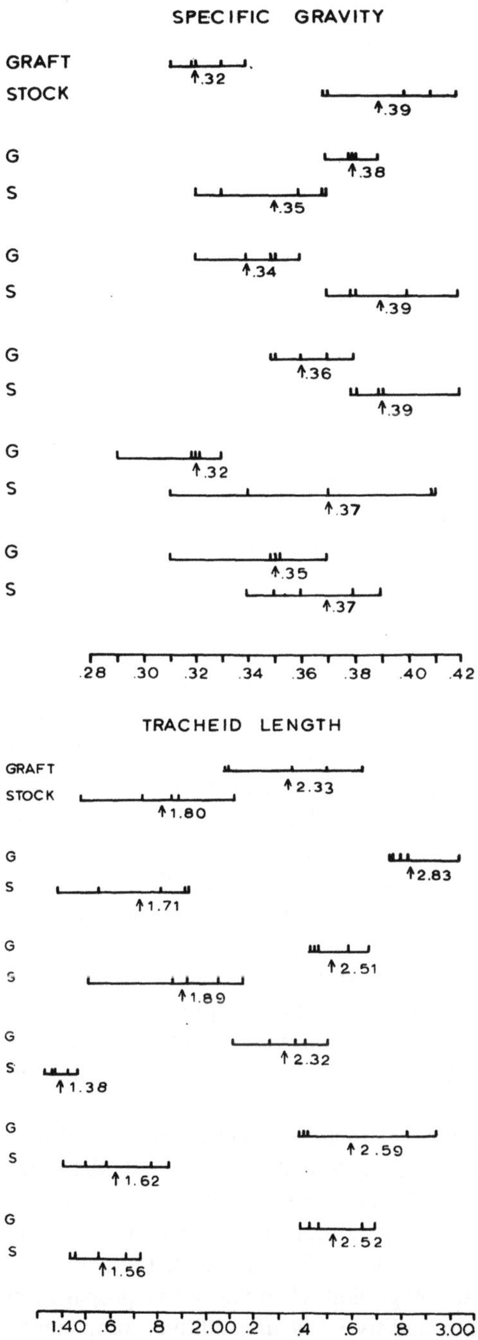

Fig. 1.2. One of the earliest studies to "prove" the genetic control of wood was made on grafts where wood samples were obtained 15 cm above the graft and 15 cm below the graft. (The trees had the same root systems and the same tops.) Note the differences in specific gravities (*above*) and tracheid lengths (*below*) between the scion (graft) and the stock from the same plant. This indicates that the genetic control of wood properties is regulated by the cambium, since the roots and tops of the plants were identical (Zobel 1967)

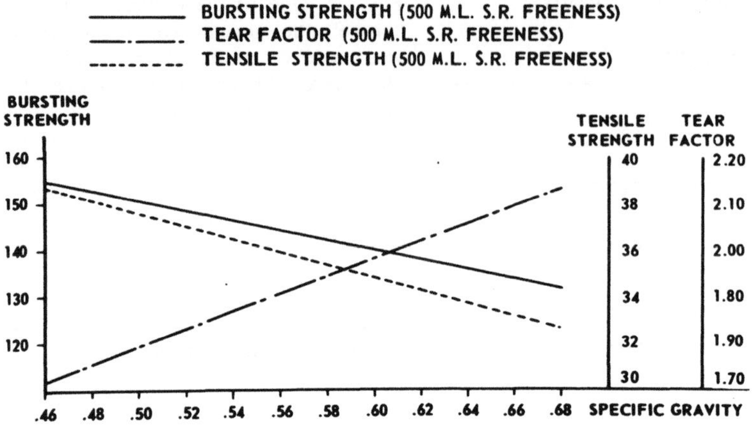

Fig. 1.3. Breeding for certain kinds of wood properties can greatly affect the properties of the final product. Shown is how tear, tensile, and bursting strengths will change as specific gravity is increased. Because of the sometimes opposing effects, as shown, great care is needed in making the decision as to which wood properties are emphasized in a breeding program. (After Einspahr et al. 1969)

The recognition of the potential for inclusion of wood properties in tree improvement programs was evident by studies such as those by Zobel (1961, 1963). The prevalent inheritance patterns in wood were shown in papers such as the one by Khalil (1985) on *Picea mariana* (black spruce), where he found that all characters except fiber wall thickness had a heritability greater than 0.30. These findings generated considerable interest and resulted in the inclusion of wood in more tree improvement programs. The current importance of the genetics of wood is indicated by projects such as the one by the Western Tree Improvement Research Cooperative and the Forest Products Department of Oregon State University which are described later.

Wood improvement is most needed in pines grown as exotics in the tropics. These trees often have poor form and because they grow rapidly, have a high proportion of juvenile wood when large enough to harvest. Real interest in improving trees grown in the tropics was generated by extensive studies summarized by Palmer in 1977.

This book is not the place for a detailed treatment of genetic principles. However, a few genetic terms are used which the reader must understand; these are more thoroughly covered in Chapter 2.1.1. A very simple explanation of genetic principles was given in the book by Zobel and Talbert (1984). Although genetic variability of wood properties can be complex, it can be manipulated rather easily to obtain significant gains. These gains are magnified when vegetative propagation is used. Genetic variation is divided into additive and nonadditive components, and the two combined make up the total genetic variance which is obtained using vegetative propagation. Additive variance results from the cumulative effects of alleles at all gene loci (see Zobel and Talbert 1984 for details). The nonadditive genetic component consists of *dominance* variance resulting from the interaction of alleles at a gene locus and *epistatic* variance that is due to interactions between loci.

The additive genetic portion of the variation can be used in a standard selection and breeding program. Its strength is indicated by narrow sense heritability (h^2) which is the ratio of additive variance to the total variance (genetic plus environmental). The broad sense heritability (H^2) is the additive plus nonadditive variance divided by total variance. It can be utilized by making specific crosses or when using vegetative propagation. In general, most wood properties have high heritabilities, which makes breeding successful. There are, however, exceptions, such as for cellulose yield, which is largely controlled by nonadditive genetic variance, and gains in cellulose yield will be best when using high cellulose producing clones in a rooted cutting program.

In this introductory chapter, only a few examples will be used to illustrate that wood properties are usually genetically controlled strongly enough to permit gains from selection and breeding. One example is for the hybrid poplars, where Hamilton and Wendel (1967) state: "Decisions concerning the choice of clones for planting might profitably include information concerning wood characteristics in addition to that for survival and growth." Numerous other authors have concluded that inclusion of wood as a breeding objective is warranted, including Harris (1970), who summed up his broad assessment as: "Many important wood properties can undoubtedly be improved by tree breeding." This was emphasized for *Pinus radiata* by Bannister (1970), who stated that wood density, as well as volume, must be included in an assessment of productivity. Earlier, Dadswell (1957) concluded that in all tree breeding work the wood of the parent trees should be assessed along with other characteristics.

The importance of controlling wood quality was emphasized by van Buijtenen (1967a). This was also expressed by Gonzalez and Kellogg (1978) for western US conifers when they stated: "Increased volume production is always a main goal of any tree improvement program. However, wood quality factors, such as tracheid length, cell wall thickness or specific gravity generally are not included in the objectives of many such programs."

There have been many objections to including wood properties in forest tree breeding programs. The most common of these was: who knows what will be the desired wood 30 years from now? The question is certainly pertinent and has to be answered as best possible. However, there is one wood property, then and now, that will always be desired. It is *uniformity*, both within and among trees (see Chap. 11.3 for a more complete coverage).

Wood, as it occurs in natural stands or unimproved plantations of a species, is usually highly heterogeneous and difficult to manufacture into uniform products. As an example, the authors found that the specific gravity of 1000 loblolly pines (*P. taeda*) of the same age, growing on the same site in coastal North Carolina, varied from 0.45 to 0.65. When such varied woods are pulped together there is no standard cook that will give uniform pulp and paper. The mechanical strength of boards made from these trees will vary substantially in strength. Regardless of the other desired wood properties, uniformity of wood, both among and within trees, can be improved by genetic manipulation.

Wood properties have always been considered of more importance when vegetative propagation is used as the method of reproduction because of the clearcut differences in wood among clones and the gains in quality and wood uniformity obtained. The magnitude and opportunities for improving wood employing rooted cuttings have been outlined by Campinhos (1980) and Ikemori et al. (1986).

Lack of inclusion of wood in the early breeding programs resulted from several reasons in addition to the uncertainty of what would be desired in the future. One was that many foresters believed that the wood produced by a tree was primarily the result of the environment in which the tree was grown (see Chap. 1.5). Additionally, obtaining representative samples of the wood was difficult and methods for sampling and testing wood samples had not been well worked out (see Chap. 3). The patterns of within-tree development of wood were poorly known, and some nonexplainable results were obtained when things such as the age of the annual ring were not considered. Methods to compare wood from young trees with those from old trees had not been well determined, nor had the best location from which to obtain a wood sample from a tree.

There are numerous current reports about a strong inheritance of wood properties in forest trees – these are summarized throughout the book. When a number of wood properties are included for a number of species, the inheritance relationships tend to become clouded, although summary papers like the one by van Buijtenen (1965) on North American conifers are most helpful in examining the inheritance relationships of multiple wood properties. A few of his generalizations were: (1) tracheid diameter, wall thickness, and fibril angle are less strongly inherited than tracheid length. There is good evidence of strong environmental control of tracheid diameter and wall thickness; (2) percent latewood appears to be inherited as strongly, or more strongly, than wood specific gravity; (3) there is a close relationship among cell wall thickness, tracheid diameter, and wood specific gravity.

A strong inheritance pattern apparently holds for trees from the tropics as well as those from the temperate climates. Wilcox (1977) emphasized for tropical trees: "Wood density is highly heritable and, because of its influence on pulp yield, paper quality and timber quality, should always be considered in a breeding program." At the same time, Palmer (1977) discussed the inheritance of wood properties in the tropical pines and the potential to reduce the large amount of variation and to help improve the relatively poor wood qualities associated with the pine species and short rotations used in the tropics. Less positive results have been reported for the hardwoods, with the possible exception of the poplars and the eucalypts.

There can be no genetic gain unless there is variation. In general, there is a huge amount of variation in wood properties. This was shown by the recently recognized tropical species of importance (*Pinus tecunumanii*) in their annotated bibliography by Dvorak and Kellison (1991). Among the species that have been well studied, only a few have shown a small extent of variation. In the pines,

Pinus resinosa (red pine) is noted for a low degree of variability, and yellow birch (*Betula alleghaniensis*) is reputed to have few wood differences. There may be others with uniform woods, especially among the tropical hardwoods which have not been well studied. The pines, Douglas-fir, eucalypts and, in fact, most species studied have a substantial amount of tree-to-tree variation in wood properties. Sources of variation include provenances, individual tree differences and even, for most wood properties, within-tree variability; but one must be careful not to assume that just because there is variation, there will be strong genetic control. Variation is usually categorized as being under *environmental* or *genetic control*, and it is ultimately determined by the interaction between the two. If the genetic component is large, the characteristic is usually referred to as being under genetic control. If the genetic control influences only a small part of the total variation, it is assumed that differences in the environment cause the variation. This breakdown into genetic and environmental influences is a rough categorization but is useful. It is safe to say that **anything that affects the physiology and growth of the tree can alter its wood properties.**

According to van Buijtenen (1967a), there are two general goals to genetic improvement of wood. One is to directly modify the fiber; for example, to increase its strength, flexibility, or length. The other objective of the geneticist is to develop trees with more uniform wood properties, resulting in increased efficiency in manufacturing. The uniformity can be obtained by production of trees that have consistent fiber properties. This was alluded to by Zobel (1978), who said: "Although many wood characteristics respond to genetic manipulation, emphasis in applied programs has also been on indirectly influencing wood quality by improving tree bole straightness and limb form and by reducing degrade from pests ... ."

Wood almost always shows a great deal of within-tree variability. This was stressed by van Buijtenen (1967a), who stated: "... we will always be faced with very high variability within a tree ... the within-tree variability is higher than the between-tree and the difference between the top log and butt log in any one tree is usually greater than the largest difference between any two butt logs of trees of the same species growing under similar conditions ... ." In an attempt to reduce some of this variation, van Buijtenen suggested that the application of genetics can make a 10% improvement in the population average of wood properties with one cycle of selection; the magnitude of improvement can be increased over several cycles of selection.

Genetic studies have recently been made on many different species; some have been excellent and well designed, but many of them were done without taking into account either, or both, the tree-to-tree or within-tree wood variation. For example, a number of studies consisted only of an in-depth, detailed study of one or two trees of a species, with the results then being applied to the whole species. In a very few instances, the results were even applied to *all* conifers when the particular tree that had been examined was a pine.

There was, and still is, a lack of understanding about the importance of sampling enough trees for a given assessment if the results are to be reasonably reliable, accurate, and statistically analyzable. This number is about 30 trees for open-pollinated progenies (see Zobel and van Buijtenen 1989). Many researchers have used too few samples (often five trees) to assess differences in wood characteristics for a population. Inadequate sampling precludes detecting statistical differences among populations because of the magnitude of both the genetically and environmentally caused variation in wood (see Chap. 3).

1.2 Categorization of Wood and Trees

Foresters usually divide trees into hardwoods and conifers (often referred to as softwoods). In fact, the categories of hardwood and softwood are not good because some of the trees with the softest, lowest density wood belong to the hardwood group – examples are the members of the genus *Populus*. Conversely, some softwoods have very hard wood such as some individuals of the southern pines of the USA. Despite these inconsistencies, the categories of *conifers* and *hardwoods* will be used throughout the book because of their common usage, even though the terms gymnosperms and dicot angiosperms would be more technically correct.

The woods of the conifers and hardwoods differ greatly, as does their variability and response to genetic manipulation. The relatively simple cellular structure of the conifers, consisting primarily of tracheids and ray cells, makes the wood easy to use, to change, or to create more uniformity for the final product (Fig. 1.4). According to Wakeley (1969), the simplicity of ray and tracheid development of the conifers is a result of early evolution. The wood of hardwoods, consisting of vessels, rays, fibers, fiber tracheids, and other cell types is more difficult to manipulate genetically (Taylor 1977b; Fig. 1.5). In the hardwoods, the fibers, rays, and vessels all make a major contribution to wood quality but their differences and importance (and sometimes interaction) make manipulation of the wood for a high quality final product difficult. Added to this is the different distribution pattern of the vessels which creates the so called ring-porous and diffuse-porous hardwoods (Fig. 1.6). The differential responses to genetic manipulation are frequently greater between the ring-porous and diffuse-porous hardwoods than exist between the conifers and the diffuse-porous hardwoods. Much more is known about the genetics of diffuse-porous hardwoods than about the ring-porous ones. The genetics of vessel and ray production and distribution in the ring-porous hardwoods is very poorly understood.

The following terminology misusage was pointed out in Chapter 1 of Zobel and van Buijtenen (1989) and it applies equally well here. The term *fiber* is commonly used for both the true wood fibers of hardwoods and the tracheids of conifers (MacDonald and Franklin 1969). Although this is botanically incorrect, the general use of the term *fiber* must be recognized, since numerous publications refer to the fiber characteristics of the conifers as well as to the real fibers of

Fig. 1.4. The wood of conifers is relatively simple, being primarily made up of tracheids. Shown is a 27× magnification of the wood of *Pinus caribaea* var. *hondurensis* illustrating the relative simplicity and uniformity of the wood of the conifers. Many tracheid characteristics can be changed by breeding. (Courtesy of Companhia Florestal, Monte Dourado, Brazil)

hardwoods. For example, one frequently reads or hears reference to "pine-fiber-length". Statements in this book will attempt to avoid the confusion between fibers and tracheids, but it will not be possible to do so when citing published material.

The woods produced early in the year and late in the growing season are categorized as earlywood (springwood) and latewood (summerwood). The terms *earlywood* and *latewood* have now become standard and will be used in this book. However, the terms springwood and summerwood were commonly employed in the past, and are still sometimes seen in the literature, especially in the United States. They will be used in this book only when a direct quotation or citation is being made.

Differences in, and between, the earlywood and latewood are of key importance. Generally, it is the manipulation of latewood that most affects overall wood quality though sometimes earlywood can be of equal importance. In studies where earlywood and latewood have been compared both separately and in

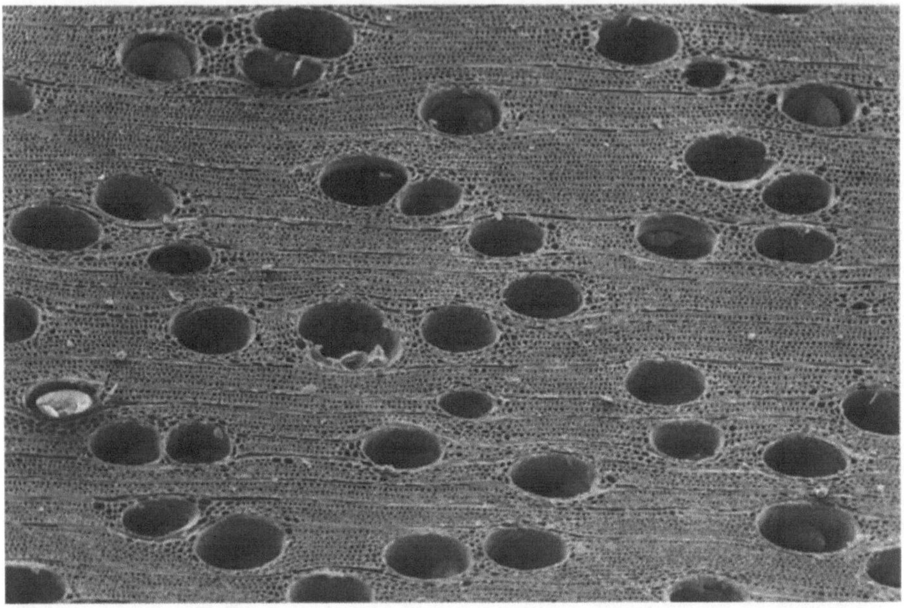

Fig. 1.5. The hardwoods have a complex wood structure. Shown is a diffuse-porous wood with the vessels scattered among the fibers within the annual ring, as in this *Eucalyptus deglupta* with 31× magnification. The number and size of vessels appear to be reasonable strongly inherited genetically. (Courtesy of Companhia Florestal, Monte Dourado, Brazil)

mixtures, Watson and Dadswell (1962) found that earlywood yielded the best burst, tensile and folding endurance and the pulp developed strength rapidly[2]. Latewood tracheids were more important for tearing strength in conifers but latewood fibers were less important in the hardwoods. In fact, in hardwoods, the presence of thick-walled fibers caused a reduction in all strength properties. In a pulping study of earlywood and latewood, Gladstone et al. (1970) found that latewood yielded 2 to 7% more pulp per unit wood weight on an oven-dry basis; this was due to a greater resistance of latewood to degradative pulping reactions. Thus, genetic manipulation that will affect specific gravity of a tree can do so through altering either or both earlywood and latewood. These two are among the easiest to change, both genetically and silviculturally.

1.3 Wood Properties of Importance

A description of important wood properties in general was given by Zobel and van Buijtenen (1989) (pp. 6–32). An excellent characterization of important wood

[2] Measures of various physical properties of paper.

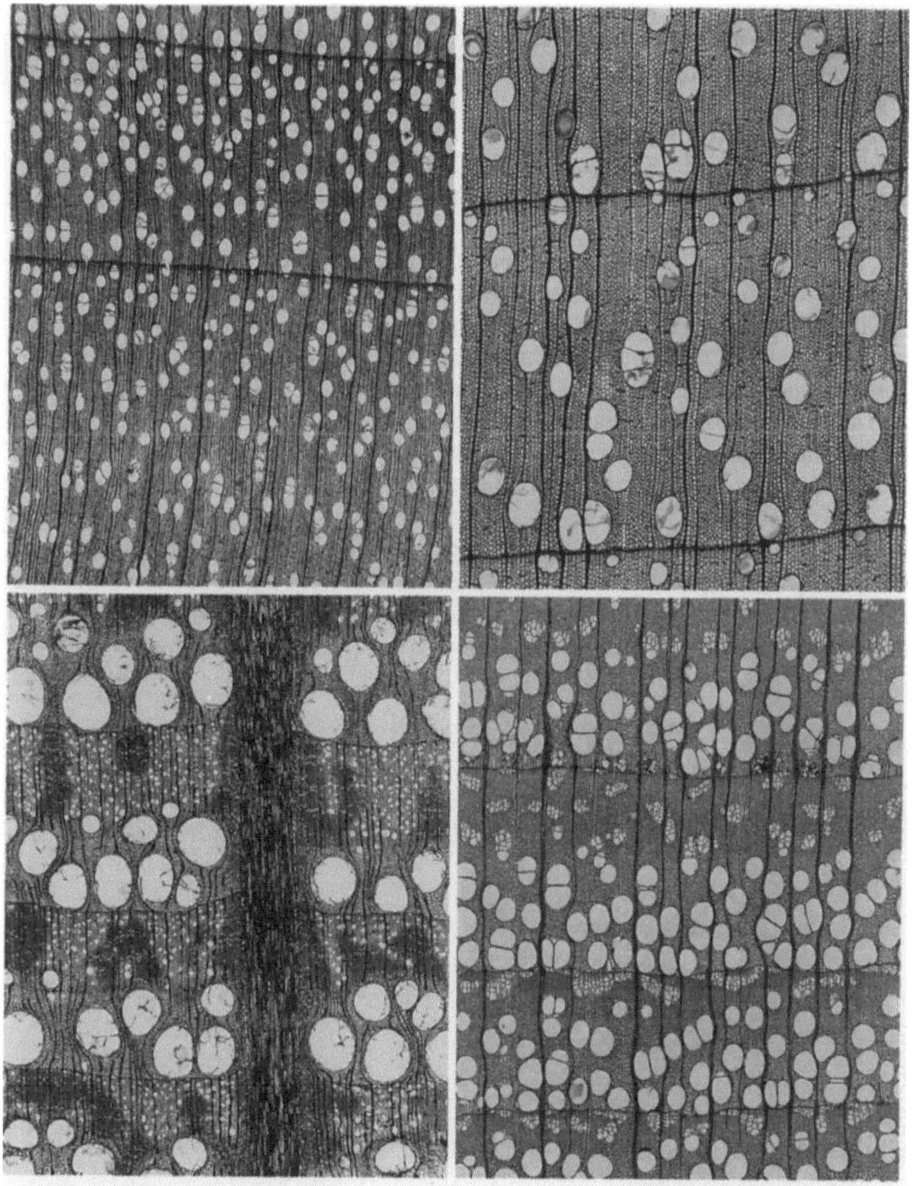

Fig. 1.6. Hardwoods have complex wood properties that result in difficulty in breeding programs. For example, the diffuse porous hardwoods *above left* and *right* (shown at 35×) respond quite differently to genetic improvement than do the ring porous hardwoods *below left* and *right* (also shown 35×). The complexity and location in the annual ring of vessels, fibers and rays determine the value of the wood for the final product. (Courtesy of Elisabeth Wheeler, NC State University)

properties of *Pinus radiata* can be found in Bamber and Burley (1983). Although their book specifically deals with radiata pine, the wood property descriptions apply well to other conifers. There are numerous textbooks on wood technology that can be consulted for characteristics of wood properties; examples are Brown and Panshin (1940), Koch (1972), Chudnoff (1980), Panshin and de Zeeuw (1980), and Baas (1982). Therefore, the characteristics of wood properties will not be detailed here.

The emphasis in this book will be on cell morphology (anatomy) rather than on wood chemistry, although the latter is covered. There are two reasons for this: (1) there is much more known about the genetics of cell morphology; (2) for both wood and paper, the majority of the control of the quality of the final product depends upon cell morphology. This was stated clearly by Barefoot et al. (1964): "... at least 93% of the variation of all of the above (kraft) paper properties could be accounted for in terms of fiber morphology." Latewood cell wall thickness was the most important property, accounting for 74% of the variation in paper quality.

The question constantly arises as to the relative effect that improving growth rate and tree form will have on wood properties. The concepts are developed more completely in Chapter 10.2 and in Chapter 8.2. From the growth studies, it is clear that the largest gains can be obtained through volume improvement (Fig. 1.7). In that sense, wood properties are secondary but it is possible to improve volume

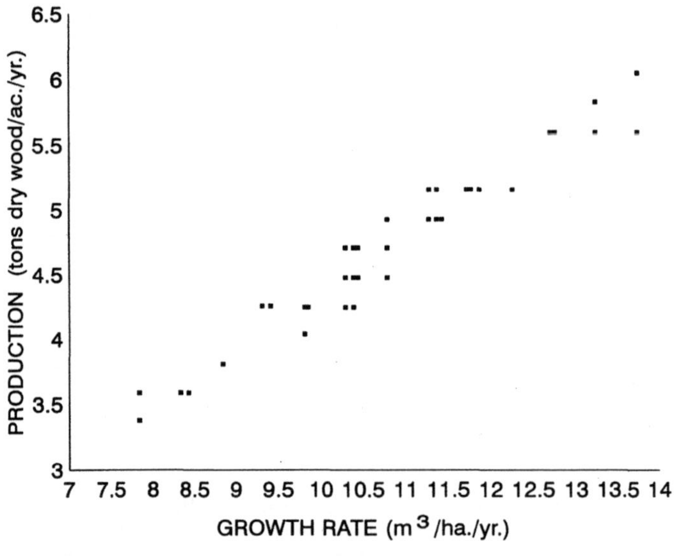

Fig. 1.7. Growth rate change by genetics which results in more wood is important. There are some questions about growth rate and related wood quality (see Chap. 10.2). The response of increasing volume production in tons of average dry wood is shown. It is now understood that growth and wood properties can be improved by genetics at the same time. The tonnages shown would differ depending on whether high or low density wood was produced

along with tree form, which results in better wood. Wood properties themselves can also be altered at the same time because they often have only a weak, or no, genetic correlation with growth and form and often also with each other.

1.3.1 Wood Density (Specific Gravity)

Wood density (Specific gravity) is the term used to express how much wood substance is present in a given volume of wood (Zobel and van Buijtenen 1989). All aspects of specific gravity were covered in detail by Simpson (1993). Wood density is usually simply expressed as the ratio of the oven-dry weight of wood to its green volume and is measured in units such as kg/m^3 or lbs/ft^3. Wood specific gravity essentially expresses the same thing as wood density but is the ratio of the dry weight in grams of a given volume of green wood in ml (at 4 °C). Since it is the ratio of grams of dry wood to milliliters of water, (1 ml of water weighs 1 g), specific gravity is a unitless measure and is expressed in values such as 0.45. The two ratios (wood density and specific gravity) can be calculated from one another. Specific gravity times 62.4 (the weight in pounds of 1 ft^3 of water) equals pounds per cubic foot. Thus a tree with a specific gravity of 0.45 has a wood density of approximately 28 lbs/ft^3 (Dadswell 1972). Density in the metric system is divided by 1000 to calculate specific gravity; wood with a density of 450 kg/m^3 has a specific gravity of 0.450. Density and specific gravity are more fully described and defined by Tsoumis (1991).

As will be thoroughly covered throughout this book, wood density has been the most studied and is the most important of wood characteristics (see Chap. 5 for details). The concept was summarized by Zobel (1974) as: "The most consistently responsive characteristic has been wood specific gravity." However, a number of other wood properties with proven genetic variation and heritability and of importance to the wood utilization industry will be dealt with.

1.3.1.1 The Genetics of Wood Density in Conifers – General Introduction

Breeding to change the wood density of conifers appears to be relatively easy, compared to the same activity in the wood of hardwoods. Even then it is not simple. Wood density is, in fact, not a single wood property but a combination of wood properties (latewood percent, wall thickness, cell size, and others). Each of these wood elements has its own genetic control, most of them quite high, so each can be manipulated genetically. Despite this apparent genetic independence of characteristics that make up wood density, the total of their effects (which is specific gravity or wood density) responds to genetic manipulation very well, usually with a high inheritance factor. Details of the complexity of wood density are covered in Chapter 9 of Zobel and van Buijtenen (1989) and Chapter 12 of Zobel and Talbert (1984). It was also stressed by van Buijtenen (1964, 1965) and Harris (1965b) that specific gravity is determined by a combination of cell wall thickness, cell diameter, and latewood percent. Despite its complexity, wood density reacts genetically as though it were a single, simple characteristic.

In this book, specific gravity (wood density) will be handled as a single independent characteristic, even though it is known to be controlled by a complex of characteristics. If the complexity of specific gravity had been known initially, there would have been many fewer genetic studies done on specific gravity per se in the early days although studies on cell morphology, such as wall thickness, would have been made.

1.3.1.2 The Genetics of Wood Density in Hardwoods – General Introduction

Just as for the conifers, the general statement can be made that inheritance patterns for wood density in hardwoods are strong. This has been quite well proven even though much less is known about the genetics of other wood properties of the hardwoods. The complex wood of the hardwoods makes practical genetic improvement difficult. For example, if one alters the vessels, then the accompanying changes in rays and fibers cause confusion. This situation was reported by Huber and Bongarten (1981), who found that the specific gravity of sycamore (*Platanus occidentalis*) increased as the ray and fiber contents increased, but decreased as the vessel volume increased.

In addition to the complexity of the wood, the economic value of the various wood properties of the hardwoods is not well known. The positioning of the differing cell types within the annual ring is of key importance. It is possible to have two hardwoods with identical wood densities, which, when handled similarly during manufacture, produce greatly different final products. Also, differing vessel volumes in ring-porous hardwoods will have a much greater effect on the final product than will the same vessel volumes in diffuse-porous hardwoods. Thus, specific gravity is a less definitive, and thus less useful characteristic in the genetical improvement of the hardwoods than in the conifers.

There are active breeding programs (including hybridization) to improve wood properties, especially in the lower density, diffuse-porous hardwoods. For the poplars, Smith and Rumma (1971) advocated: "... intensive selection and breeding with North American cottonwoods might develop even better hybrids combining fast growth with higher specific gravity and longer fibers." This methodology of improving wood properties is known to work well but it is seldom used.

Although some temperate zone wood characteristics of the hardwoods are fairly well understood, most of the tropical hardwoods are still a mystery as far as genetic manipulation and even utilization is concerned. A knowledge of density, fiber length, fiber wall thickness, or ray and vessel volume may be of only limited value in predicting the kind of paper a tropical hardwood produces or whether it will make acceptable furniture. Of course, some characteristics, such as the presence of silica in tropical wood, obviously impact its usefulness. Especially in tropical hardwoods, such things as internal log stresses cause deformation of the wood, which can result in its' being of limited use for lumber or plywood (Turnbull 1965). Proof of genetic control of the varied characteristics of the wood of hardwoods is scarce but the few available results show a strong inheritance pattern.

1.3.2 Other Wood Properties

There are numerous wood properties for which genetic control has been demonstrated. Some of these, like cell length, have been intensively investigated and the results are clear. Fibril angle, which is of major importance to product quality, has not been well studied. Conversely, spiral grain, which is primarily of concern only to solid wood users, has been the subject of a considerable amount of genetic assessment and is under moderate to strong genetic control. Spiral grain is becoming of greater importance as rotation ages become shorter (Harris 1984). Other characteristics, mostly related to cell anatomy, such as wall thickness, latewood percent, and cell size, have been studied to some degree. These three are closely related to wood density, and their inheritance patterns are generally quite similar to those of wood density.

There are special wood characteristics, such as heartwood formation, wood discoloration, and decorative grains that can be manipulated genetically. It has been found that inheritance of these wood properties is reasonably strong, as covered in Chapter 8.4.3, Chapter 11.2.1.1, and in Chapter 7.4.

One "wood property" that can be of real importance is the moisture content of the wood. This trait shows a strong genetic component (Zobel et al. 1968a). It is often closely, but negatively, related to wood density within a species but not always among species. Moisture content is difficult to work with because it changes with tree age, methods of handling the wood specimen, and time of year. Yet it can be of vital importance, especially when wood is bought by green weight, as is the case in several countries throughout the world.

Most wood users consider uniformity in wood from tree to tree, the most important wood property (Harris 1969). Uniformity was the theme of Zobel (1984), in which the need for less variable wood was strongly stressed, and in Zobel (1972): "Breeding for wood should bring greater uniformity as well as move the average in the desired direction." A detailed discussion on uniformity is found in Chapter 11.3.

There are two other types of wood uniformity in addition to tree-to-tree, namely within the annual ring and from ring to ring. Both of these can be partially controlled by silviculture and to some extent by genetics. Silvicultural actions, such as fertilization regimes have proven to be effective in improving uniformity, usually increasing the wood density of the earlywood and decreasing that of the latewood. The use of silvicultural methods to produce more uniform wood has been quite well studied (Gladstone and Gray 1972, Hsu and Walters

1975, Blair and Olson 1984). The lowered latewood percentage produced trees with wood with less internal variability.

Occasional trees are found with similar juvenile and mature wood and there are also trees with an earlywood of higher density than the latewood. Manipulation of the genetic factor relative to the density of earlywood and latewood is possible but not greatly studied (Zobel et al. 1978). Nicholls (1967a) concluded that reduced within-ring variation is less important than tree-to-tree uniformity.

The adverse effects of variability were emphasized by Kanô (1957) when he stated about the use of Todo fir (*Abies sachalinensis*) in Japan: "Variation in wood quality reduces the reliability of wood material and is a fundamental problem in the utilization of (woody) materials ...." When products are made from wood that has been reduced to a chip or cellular form, either mechanically or chemically, the inherent variability in the wood can have a major effect on product quality and manufacturing efficiency. Uniformity of wood is a major demand by the manufacturers of all wood products (Zobel et al. 1983). Much more breeding for wood uniformity is needed.

Wood chemical properties are difficult to work with genetically. Some studies, such as the one by Byrd et al. (1965), showed that even though considerable chemical differences are present, they have a relatively small impact in papermaking compared to cell morphology. The few genetic studies on cellulose inheritance have shown a strong nonadditive trend (Zobel et al. 1966, Jett et al. 1977).

In addition to specific wood properties per se, there are numerous cell ratios that are of importance in pulping such as cell length to cell width. In a few instances, genetic analyses have been made on such ratios and, surprisingly, some show a genetic trend.

Although there are exceptions, it is safe to say that most wood properties are under moderate to strong genetic control and can be manipulated through sexual breeding. Those that are not, can often be handled through the use of vegetative propagation. An example is cellulose yield; little gain can be made in cellulose production per se through a seed regeneration program, but good gains are possible through the use of rooted cuttings (Jett et al. 1977). There are numerous papers where the authors concluded, as did Lewark (1982a), that: "The calculated responses to selection show that all characteristics studied can be improved by clonal selection" or the paper by McElwee (1963), that lists wood and pulping properties that respond to genetic manipulation.

1.4 The Causes and Types of Wood Variation

Causes of variability in wood are many and were recently summarized by Zobel and van Buijtenen (1989). In general, causes of variation are categorized as being the result of specific environmental factors or internally controlled genetic factors. Although this is a usable way of categorizing variation, it is simplistic. There is *always* an interaction between the genetic potential of a tree to produce a certain kind of wood and the influence exerted by the environment. One key issue is to determine *how much*, or *how easily*, these two different forces affect wood. Anything that causes a tree to change its growth pattern will probably have some effect on the wood produced.

The control of wood properties by changing tree growth, tree form, and response to extreme environments through genetic manipulation is now fairly well known and appreciated. Also quite well known are the genetics of family-to-family and tree-to-tree wood variation. However, less well understood is the genetic control of the wood properties within a tree. Additionally, anatomical variation within a tree can be related to unusual internal wood structure. One example was given by Smith (1977), who emphasized that: "Defects such as defective knots, resin and bark pockets, pitch streaks and heart shakes can adversely affect ... value recovery appreciably more than intrinsic wood properties, especially for solid wood products." This could also be emphatically said for bole straightness, which in itself has a relatively strong genetic control. This trait affects total recovery and value of solid wood products as well as pulpwood properties.

The way in which genetics affects wood properties is through an alteration of the physiological functions in the tree. Similarly, environmental factors can also affect the physiology within the tree and thus the wood produced. The latter aspect was partially covered by Richardson (1960) in his article related to the role of physiology in forest tree improvement. He points out, for example, that cell wall thickness is a function of the balance between photosynthesis and respiration while tracheid size seems to be linked to the activity within the terminal meristem. The importance of physiological control relative to wood production is discussed a number of times in this book. The topic is covered in some detail by van Buijtenen (1955), Dietrichson (1963), Larson (1973), and several others.

1.4.1 Importance and Magnitude of Wood Variation

Wood is a remarkable substance, whose variability and flexibility make it useful for many purposes (Keating and Bolza 1982). Regardless of whether one considers it an advantage or disadvantage, the variability present in wood makes it a versatile material suitable for a large number of uses. The same variability sometimes makes wood inferior for certain products, and often more wood is used than is really needed in a building because construction must make allowances for the weaker boards.

One advantage of wood is that it is a renewable resource, as was strongly stressed by Bethel (1977). Wood is relatively easy to work with and it is known as a useful material by the general public. It is most important that a large overall amount of variation be maintained. Without variation, usefulness for multiple products, as well as genetic gains, would not be possible. Yet the manufacturer of a specific product prefers a uniform wood. The expressed desire of most industries is to have more uniformity in the wood raw material they are using; the more uniform the basic wood used, the better will be the quality of the desired specific product. It also will be less expensive to manufacture, and will make a more uniform product. Thus, *foresters have the dilemma of keeping overall variability in wood for long-term development but developing uniformity for specific products from designated plantations for short-term gains.*

There is one basic and irrefutable fact – one cannot exploit either the variability of wood or create greater uniformity without a knowledge of both the cause and control of variation. Thus, there is a need to know more about genetic control of wood. A recognition of the necessity of becoming involved in breeding for wood properties was emphasized many years ago by Mitchell (1955), who stated: "I sincerely believe ... that industry's support of such research is not anywhere in line with either the current or long-term benefits it stands to gain." When Mitchell made the above comment, very few tree improvement programs included wood in their operations and this is still largely the case today.

Wood has been so little altered genetically by man's interference, and its genetic control is so strong, that large gains are usually possible through selection and breeding; but care must be taken to assess the value of possible gains. For example, a relatively small increase in growth rate will often produce more tons of wood per unit area of forest than will a large increase in wood density. It is necessary to assess the relative contributions of improving growth rate and wood density; one cannot assume that a concentration on the characteristic with the strongest inheritance value is the way to achieve the highest desired yield improvements or economic gains. This was stated well by Parrot (1960): "... for species such as Norway spruce (*Picea abies*) where growth rate is inversely related to specific gravity ... rapid growth should be the principal criterion and high density secondary ...."

Frequently, the genetic improvement of wood is considered as the "cream on the milk" in a tree improvement program. The tree breeder does everything possible to efficiently produce large volumes of high quality trees. The second concern is to ensure that the trees have wood with desired properties. When combined, the result is more, as well as better, wood. Such genetic improvement would not be feasible if there were not a high inheritance in the wood properties concerned, because the selections for improved wood qualities will be made in a population of individuals considerably smaller than the one initially selected for fast growth and form.

A common error in tree improvement is to use volume production as the sole criterion for success. This was successful when Geary (1967) worked with 4-year-old *Pinus kesiya* and found that diameter measurements alone were sufficient to rank families for dry-wood weight yields. Normally, however, wood properties should also be considered and, as stated by Bannister (1970), who studied wood density of radiata pine in New Zealand: "... clearly, this reinforces earlier beliefs that assessments of genetical variation in productivity ... should consider the density as well as the volume of wood." It is clear that dependence on volume alone for an estimate of productivity or reliance on wood density assessments alone are rarely sufficient to estimate dry wood productivity since volume differences contribute to dry weight and density contributes to dry weight and quality. In his summary, Klem (1965) emphasized the need to consider both volume production and specific gravity when assessing the value of a species for pulp production. A similar emphasis was made by Zobel (1973): "A well rounded tree improvement program must include wood qualities."

The methods of obtaining better wood genetically, either by improving growth, form, or disease resistance, or by breeding directly for morphological or chemical characteristics, are covered in detail in this book, and briefly summarized below. The effect that disease can have on the final product is shown in Fig. 1.8.

The bulk of the work on genetic manipulation of wood relates to wood properties of special importance for paper production but there are also great, perhaps even greater, benefits possible in improving wood for sawtimber or plywood (Karki 1978). Benefits are especially evident in growing specialty woods, like curly grained birch (Heikinheimo 1951). Although not yet proven, it seems evident that this will also be true for many tropical hardwoods that frequently are used for specialty products.

The interesting thing in working with wood is that the "normal" variation patterns do not always hold for individual trees. Many examples could be cited but

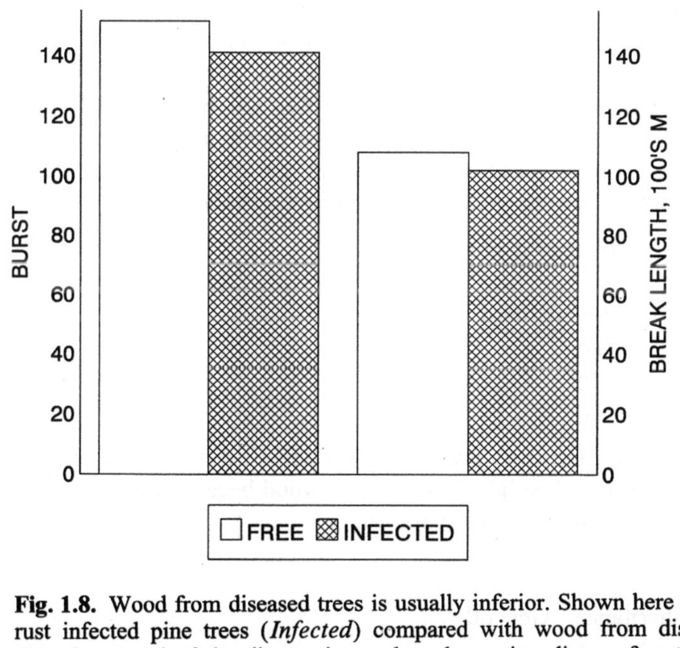

Fig. 1.8. Wood from diseased trees is usually inferior. Shown here is the effect of fusiform rust infected pine trees (*Infected*) compared with wood from disease-free trees (*Free*). Genetic control of the disease is good, and growing disease-free trees will result in considerable improvement in the properties of the final product, here shown as burst and breaking length

the most notorious is radiata pine C55 in New Zealand. Its wood is so different, and produces products so abnormal, that Kibblewhite and Bawden (1992) label it a "genetic misfit". Such trees are usually of only academic interest, but when they are of excellent form, have fast growth, and have a major effect on product quality, as C55 does, then they will be extensively used in breeding programs, especially when vegetative propagation is employed.

1.4.2 Assessing Genetic Improvements

Assessing improvements from genetic manipulation of wood must be done cautiously. The most flagrant mistake is to use percentage gain as the measure of success. This can be very misleading, since percentage varies by the size of the base upon which it is calculated, which, in turn, varies with the age of the tree. Erroneous predictions for time of harvest are frequently obtained by using percentages developed at the juvenile stages of the tree, (Zobel 1964).

The percentage gain also must be related to the portion of the characteristic that is being considered. For example, assume the total range in specific gravity of a provenance is 0.4 to 0.6, with 0.5 average. If one improves average specific gravity from 0.50 to 0.55, the *percentage of the possible gain* from genetic manipulation is 0.05 or 25%. However, the actual gain in wood weight is 0.05 or only 5%. It is this value that is of interest to those who utilize the wood. Thus, percentage gains in actual wood weight from changing wood density are always small. Given the biological constraints to potential gains, percentage improvements in wood density are almost always less than 10%. The best way to report gains is in actual wood weight or dollars rather than in percentage. Although the data are somewhat hypothetical because of the young age of trees, Stonecypher et al. (1964) showed the following yields from improved specific gravity.

Improvement in specific gravity	Increased green weight: Pounds/cord	Increased green weight: kg/m^3
0.008	80	18
0.016	160	36
0.030	300	69

A given gain in wood weight from juvenile wood will be different percentage-wise than the same gain in wood weight from mature wood because the respective basic wood weights of the two kinds of wood differ.

1.4.2.1 Strength of Inheritance – General

For nearly all tree characteristics, most species have a large range in variation of inheritance patterns. Although many characteristics are inherited strongly, the strength of genetic control varies from species to species. A good example is red pine, which produces quite uniform wood throughout its range. This species is much less variable in inheritance of wood properties than loblolly pine (Fowler and Morris 1977). The strength of inheritance for each wood property is discussed in greater detail in the following chapters.

Improvement of certain characteristics by genetic manipulation is sometimes marginal in forest tree breeding (Zobel and Talbert 1984), but for most wood properties it is positive and often profitable. There are exceptions where, although genetic control may be very high, changing a particular wood property is not worthwhile because of its small influence on the final product. Examples include fiber length in hardwoods and tracheid length in conifers. While the inheritance of both fiber and tracheid length is high (Wheeler et al. 1965), there is so little variation to work with in the hardwoods that absolute changes in length frequently are of little economic value because of a very small effect on the final product. Similarly, an increase in tracheid length of mature wood in conifers also has little effect on the product.

The key to the use of genetics in wood improvement is how similar the wood of the progeny is to the parent. The calculation of heritabilities is a good but still theoretical method to estimate the gains that will be obtained. Parent-progeny correlations give better indications of what can be obtained by breeding. These are described in Chapter 2.1.1. Generally, there is a close relationship between the wood of the parent tree and the wood of derived rooted cuttings, but this is not always so. In a study of radiata pine, Nicholls et al. (1976) found that the cuttings had longer tracheids, larger spiral grain angles, and less dense wood than the parent, but they were overall quite similar. Most researchers find the woods of the cutting and donor tree to be almost identical when woods of the same age are compared (see Chap. 5.4).

1.5 Environmental vs. Genetic Influence on Wood

There are strong differences of opinion and sometimes senseless arguments relative to the influence of environment versus genetics on wood properties. Some of these were covered by Koeler (1939) in his discussion of heridity versus environment to improve wood. As previously mentioned, at one time, most foresters believed that wood properties produced by a tree were primarily the direct result of the environment in which the tree grew, including its growth rate. Because of this, certain early workers in tree improvement were scoffed at because they wanted to use genetics to try to improve wood properties. Unfortunately, this attitude is still too prevalent. In actuality, those believing in environmental determination of wood properties were correct in the sense that conditions under which a tree grows, reacting with the genetic potential of the tree (this is sometimes called genotype × environment interaction), determine wood properties (see Chap. 2.1.1.4).

A good example of the response of wood to environmental differences is in radiata pine (Cown et al. 1991a). The wood of this species was found to be highly dependent on the site where the trees were grown. There was a trend for wood density to decrease with increasing latitude and elevation. Mean annual temperature was found to be the environmental factor that was best correlated with wood density. Tracheid length also showed large regional differences but could not be closely correlated with any environmental factor.

Numerous articles could be cited about environmental control of wood properties, as was discussed by Zobel and van Buijtenen (1989). Many studies have been made of the reaction of the wood of specific species to silvicultural activities. One was for a 24-year-old Douglas-fir stand by Jozsa and Brix (1989). They found that fertilization reduced ring density about 16% for the 3- to 4-year period after treatment, but not thereafter. A decrease in the density of the latewood, as well as a decrease in the percentage of latewood, occurred after fertilization. Thinning resulted in a small gain in wood density. It is apparent that *both* genetics and environment are important. The dual and interactive control of each is so large that a determination of the genetic control of wood properties cannot be made without specifying the environment in which the tree is grown. Conversely, the effects of environment and growth differences cannot be assessed unless the parentage of the trees on which the study is being made is known to be reasonably uniform.

1.6 Literature on the Inheritance of Wood

The literature on the genetics and breeding of wood is voluminous and scattered. There are a few books or in-depth articles, but many of the genetic studies are mentioned only incidentally in papers dealing with broader aspects of wood assessment. Some excellent studies were never published because of proprietary reasons or because the researchers were working with organizations that did not emphasize the value of publishing.

Several attempts have been made in the past to summarize what is known about the genetics of wood. They all contain valuable information but fall short of providing a satisfactory coverage of the subject. The attempt in this book is to simplify what is known and reported about the genetics of wood properties in a form for easy understanding without sacrificing technical accuracy, and to provide a better and more complete coverage than is currently available. It has combined the pertinent parts from broad-based publications on the genetics of wood (an example is Burley and Nikles 1973) and from the many hundreds of scattered publications related to wood as a breeding objective in tree improvement.

A caution must, however, be voiced here. It is impossible in a review of the literature such as attempted in this book to indicate the reliability of the genetic methods used or the soundness of the study design, including number of trees, replications, or other. A weak study appears, in print, as important as a strong one. We, as authors, cannot pick and choose reliability of study results, although, if glaring weaknesses such as too small sample size are known, this will be pointed out. However, each reader must take the responsibility to consult the original reference indicated if he or she is going to use the information as a basis for major decisions and actions.

There is hardly any single statement related to inheritance of wood that cannot be challenged or for which contradictory statements cannot be found in the literature. However, at the end of all but one chapter in this book, summaries will be made of the apparent trends. Additionally, this book will present procedures for sampling wood variation in different types of studies that will result in accurate and reliable results.

Perhaps the largest portion of good research that is "buried" in publications is in conjunction with association meetings, short courses, and in other difficult to find places. These are frequently made available to participants of the meeting but are never formally published. They do not fall in the category of "refereed" publications and thus are not included in the computer or card file lists of many libraries. They are either never filed, or if filed, cannot be relocated. A glance at the **Literature Cited** section of this book will show a number of such articles which really are difficult to categorize as publications; some could only be referred to as "mimeographed reports." Some of these contain first-class information of significant value relative to the genetics of wood. It is evident that this book does not contain all publications relative to genetic control of wood production, but it does bring together a very large number of reports previously not easily available.

Reading the English literature, one can obtain the impression that the bulk of the work on the genetics of wood has been done in the English-speaking countries. This certainly is not true. For example, in his two-volume review of forest genetics translated from Japanese to English, Toda (1970) cites many studies dating back several hundred years; some of these dealt with wood. Many publications not in English, like the one in Portuguese on the eucalypts by Navarro de Andrade (1961), have valuable information on wood which would be of interest to those working in tree improvement. Even though the literature on the genetics of wood has been summarized as far as possible in this book, certainly some references have not been recognized. To make it easier for the English reader, the English title of the article is cited when available. If it is not available, the title in the original language will be used.

There have been some previous attempts to summarize the literature related to the genetics of wood in tree improvement. One example is the book by Keith and Kellogg (1985) for the Forintek Canada Corporation. An early publication (Anonymous 1962) was the TAPPI Monograph Series Number 24 which contained 883 annotated citations relating the effects of both environment and genetics on wood properties. A similar publication today would require several thousand citations. The 1163-page publication on the use of the southern pines (Koch 1972) contains many references related to the effects of changing wood properties.

1.7 Summary

In the past, tree improvement programs often ignored wood as an objective of the genetics program. Currently, there is much interest and research related to the genetics of wood. It is the objective of this book to simplify and summarize what is known about inheritance of wood properties and how this information can be used. It is now known that many important wood properties are inherited strongly enough to make meaningful gains possible in operational programs.

The place that wood should occupy in a genetics program is debated. We feel that it should be a secondary trait to be used on the fastest-growing, best-formed, and well-adapted genotypes. This is possible because most wood properties have a sufficiently strong genetic control, combined with large variation. Most wood properties are inherited independently of growth, form, and adaptability, making it possible to tailor-make desired trees that also have the desired wood.

Wood improvement is needed most for exotic species planted in the tropics, such as for *Pinus caribaea*. Growth conditions are such that the trees are harvestable while still relatively young and have a high proportion of juvenile wood. Much effort is going into improvement of juvenile wood.

The benefits from, and objectives for, using wood in genetic programs are discussed. Maximum gains in wood will be realized when vegetative propagation becomes more generally used.

The relative importance of genetics and environment, plus their interactions, is a major source of inquiry. Answers are not clearcut and both genetics and environment always play a part in determining wood quality. It can be summarized that anything that changes the growth pattern of a tree can cause variation in its wood.

Although the separation is somewhat misleading, trees are categorized as conifers and hardwoods. The hardwoods are further divided into ring-porous (vessels laid down in the spring) and diffuse-porous (vessels scattered throughout the annual ring). Much is known about the genetics of wood in some pines, Douglas-fir and spruce. Considerable information is also available on the diffuse-porous eucalypts and poplars. Much needs yet to be determined for other groups, especially the ring-porous species and the diffuse-porous tropical hardwoods.

The importance of various wood properties was mentioned but it is clear that specific gravity (wood density) has by far the greatest value. Wood density and other wood properties are introduced in this chapter but are covered in detail in later ones. The importance of wood uniformity for all species and all products is evident. It should be a number one breeding objective. However, for long-term breeding programs, wood diversity is needed. These opposing objectives create a real problem for the tree breeder. Wood chemical properties are relatively unimportant relative to cell anatomy or cell morphology.

It has often been difficult to assess the value of improving wood, but this is becoming much clearer now. Part of the problem is how to express gain. The error of using percentage gain can be very large, especially if it is based on young trees. Genetic control of wood frequently changes with age of the cambium.

The literature relative to the genetics of wood is voluminous but it is scattered. Many excellent research results have been unused because they were presented in obscure places, often in non-refereed journals, and so are not available in normal library sources. Also much valuable information is considered to be proprietary. This book brings the information available for a completeness on the genetics of wood not available anywhere else.

Chapter 2

Genetic Controls in Wood Formation

2.1 Controls Influencing Wood Development

An explanation has already been given in Chapter 1 about how external factors (environmental), combined with internal factors, control the formation and qualities of wood. Both external and internal factors influence the physiology of the tree which, in turn, determines the kind of wood formed. Thus, without a good knowledge of physiology, one can only hypothesize what causes the production of differing types of wood. That is a major problem because, unfortunately, *too little is known about the physiological processes and cell differentiation that control formation of wood with differing properties* (Fig. 2.1).

Current knowledge about the physiology of wood development and control is well covered in other books in the Springer Series in Wood Science and in articles such as by Larson (1960). It is only necessary to clarify here that, as stated earlier in Chapter 1: "... anything that affects the physiology and growth of the tree can change wood properties" (Zobel 1978). The speed and extent of the contribution of genetics to change wood characteristics depends upon the ability of the forester to predict wood values at a young age (see Chap. 3.3.1.1). Good predictability of wood properties from young trees enables a rapid turnover of generations which is the key to obtaining large genetic gains. Such predictability can be quite good, especially for wood density. The mean progeny values of specific gravity at early ages can serve as predictors for specific gravity at older ages (Stonecypher and Zobel 1966). Therefore, the age at which the degree of inheritance is estimated becomes all important. When age is not known, the genetic values must be suspect because inheritance patterns usually change with the age of the cambium.

As will be emphasized throughout this book, wood properties can be affected both by changes in external tree characteristics – the tree morphology – and by internal changes through genetic control. Both morphology and physiology show genetic control but these can vary from very strong, as they influence wood density, to essentially zero for cellulose yield (Fig. 2.2). Based upon current knowledge and techniques, both the additive variance and nonadditive variance patterns are strong enough to yield real and meaningful changes in wood properties if handled correctly. Tree form, growth rate, and internal genetic controls and, most importantly, their interactions determine the kind of wood that will be produced (Fig. 2.3).

Writing about larch (*Larix decidua*), Balatinecz and Kennedy (1968) covered the importance of physiology in wood production well when they stated:

Fig. 2.1. Tree-to-tree variation is usually great, and proper parentage is very important. It is difficult to illustrate this in wood but it is shown above by growth differences within families. Just as for growth, some families have all trees with similar woods and good uniformity while others have very variable wood. The physiological reactions that determine this must be known. (After Zobel and Talbert 1984)

"These results are compatible with the hypothesis that endogenous growth substances in forest-grown trees play a key role in the physiological mechanisms of earlywood-latewood differentiation." The authors emphasized that xylem differentiation was related to changes in growth substance levels that directly control the physiological processes related to wood properties.

In his articles on the physiological basis for wood specific gravity, Larson (1960) stated that cell diameter is determined by hormones, usually auxins originating in the bud or growing shoot. A high auxin content in the apical meristem will result in the production of large diameter cells; anything that reduces apical activity will result in smaller diameter cells. Generally, when cell diameter decreases, wall thickness will increase. Larson felt that cell diameter and wall deposition are related to nutrition.

Strong genetic control is evident in phenological development. This was assessed for Norway spruce (*Picea abies*) by Worrall (1970). There are "early and late flushers" and undoubtedly the physiology of time of flushing is part of the genetic control of wood density. This was reported for 150-year-old Norway spruce (*Picea abies*) where late-flushing trees had a lower wood specific gravity

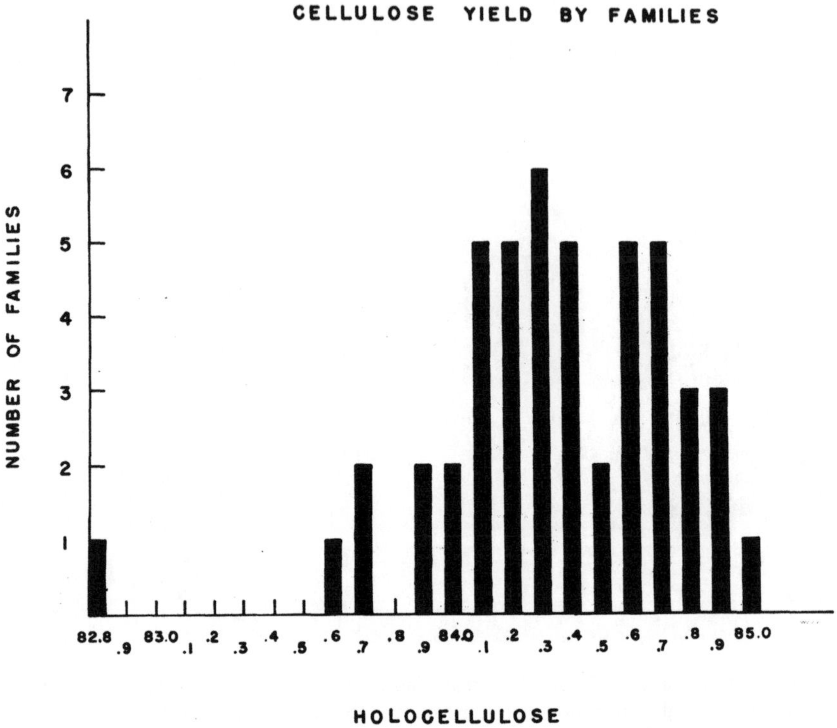

Fig. 2.2. Variation in cellulose yield is large, both by family and by individual tree. However, the genetic differences result from nonadditive genetic factors so a standard selection and breeding program will not work. Good gains can be made with the large variability in cellulose from the use of rooted cuttings. (After Zobel and van Buijtenen 1989)

than those that flushed early. This was true for both populations and individuals. It was found that Douglas-fir (*Pseudotsuga menziesii*) trees that produced latewood earlier in the season and grew for a longer period had denser wood. Tree-to-tree differences for this pattern were considerable. Family heritability for this characteristic was 0.30, indicating that some positive response could be obtained by selecting for time of conversion from earlywood to latewood (Anonymous 1990). Thus it was felt that selection for increased wood density would be associated with both an earlier transition to latewood and with extended duration of latewood formation. In his study of the formation of latewood, and thus wood density in Douglas-fir, Vargas-Hernandez (1990) found that differences in wood density among families were related to the date of latewood transition; families with higher wood density normally had an earlier date of late-wood transition, increasing the period of latewood production at the expense of the earlywood formation. Vargas-Hernandez found that wood density was also positively correlated with the duration of the cambial growth period, caused

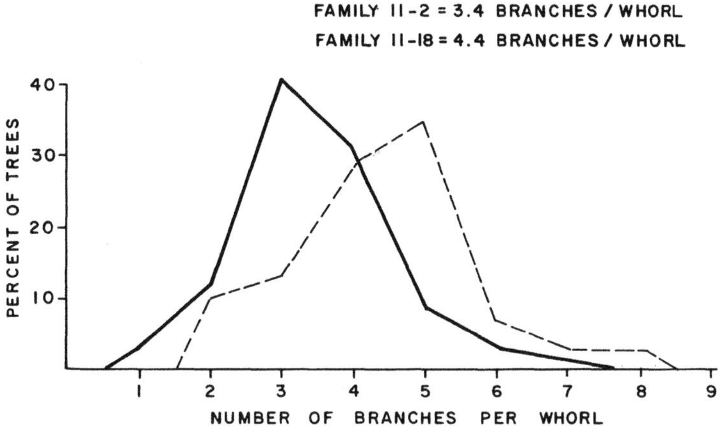

Fig. 2.3. There are many differences in tree form that are under reasonably strong genetic control. Among these is the number of branches, which vary greatly among individuals and which respond to breeding programs. The number and size of branches have a major effect on wood properties. Note the variation in the number of branches per whorl for the young progeny of two loblolly pines of the same age grown under identical conditions

by an earlier initiation and a later cessation of cambial growth. Unlike cambial phenology traits, date of bud burst was not correlated with wood density traits. In an expansion and more detailed assessment of the study, Vargas-Hernandez and Adams (1994) found that wood formation traits had a heritability of $h^2 = 0.20$. They reported: "Overall core density was negatively correlated with the dates of cambial growth initiation ($r = -0.41$) and latewood transition ($r = -0.62$) and positively correlated with the date of cambial growth cessation ($r = 0.40$)." They concluded that higher wood density was associated with a longer duration of cambial growth ($r = 0.67$) and a slower rate of wood formation ($r = -0.37$). The above comments are in agreement with observations of the present authors for both loblolly (*P. taeda*) and slash (*P. elliottii*) pines.

It is apparent that most persons concerned with wood development are convinced that the initiation of latewood cells is related to cessation of height growth via development of a resting terminal bud and its control of auxins. In an attempt to clarify this concept, Mitchell (1961) studied several species of conifers from British Columbia, Canada. Mitchell anticipated that species with an abrupt transition from earlywood to latewood would abruptly stop growing in height, while those with a gradual transition would slowly cease terminal growth. He found this to be essentially true. In Douglas-fir saplings there was a close correlation between cessation of height growth and the formation of latewood. According to Mitchell (1961), there is possibly a similar trend in grand fir (*Abies grandis*) trees. However, he found that saplings of western white pine (*P. monticola*) and lodgepole pine (*P. contorta*) did not exhibit this relationship. Results with seedlings of all species were inconclusive. Our wide experience, and studies such as van Buijtenen (1955), indicate that formation of a true resting bud usually is necessary before latewood formation is initiated in mature trees. Since time of bud formation and growth cessation tend to be under strong genetic control, as reported by many workers (Kozlowski 1963, for example) the bud-auxin-latewood formation relationship is obviously related to genetic control of wood production. In Douglas-fir, McKimmy (1966) found that beyond the juvenile wood zone a very strong correlation ($r = 0.82$) existed between time of bud burst and specific gravity.

For jack pine (*P. banksiana*), the development of mature xylem lagged behind leader growth (Kennedy 1969). The first mature latewood was produced 17 days after height growth had ceased. Kennedy found that latewood initiation began almost at the same time as the cessation of height growth and also noted that it usually took about 20 days for completion from differentiation by the cambial initials to complete lignification. Thus, provenances originating in milder climates, which should be adapted to more growing days, should have a later completion of height growth, resulting in lower wood density. A correlation of $r = -0.69$ was found for jack pine between specific gravity and number of growing degree-days. In Norway spruce, Wodzicki and Witkowska (1961) studied the effect of bud inactivation or removal on cell wall thickness and found that only when the whole plant was influenced by short-day conditions would thick-walled tracheids be induced.

The importance of needle development to auxin production has been emphasized by Larson (1973). He hypothesized that cell radial diameter is regulated by a growth hormone produced by the needles while they were actively elongating. The newly maturing needles also produce more photosynthetic products, which affect wall thickness. Needle development and auxin production take place at nearly the same time but are physiologically independent. Larson felt that where there were gradients along the stem, the characteristics of the tracheids produced varied both with position in the stem and with time of formation in the growth ring. When these physiological gradients overlap, or interact, transition wood is produced, consisting of tracheids that cannot be classified as either true earlywood or latewood. Larson (1973) concluded that variations in wood specific gravity within trees, found with increasing size and age and with environmental differences, are related to changes in tracheid dimensions, which vary relative to needle development. Not everyone agrees totally with this simplified explanation but it certainly is a good start in explaining the physiological control over the development of wood.

The subject of control of wood formation by chemical substances such as auxins is very complex and controversial. For the nonspecialist it is only necessary to know that this control mechanism exists, and that there are numerous interactive forces that *affect* both external and internal mechanisms. Wood formation is complex and any generalizations about this phenomenon will have exceptions. The reader is also warned that improving wood by genetic manipulation can be difficult, although, in fact with such complex wood properties as wood density or cell length, genetic gains can occasionally be quick and sometimes dramatic.

The objective of this book is to try to make the relationships involved in internal wood production understandable.

If one is to understand the genetics of wood formation, it is essential to have some knowledge of the stages involved in the development of wood, which were clearly described by Franklin (1979). It essentially consists of an understanding that a stand of trees goes through phases: a juvenile genotypic phase, a mature genotypic phase, and a codominance-suppression phase. The heritabilities are initially high, as the result of differences in seed size and germination rate, but decrease as these effects wear off. When competition sets in (the mature genotypic phase), heritabilities rise again as inherent differences in competitive ability are expressed. Heritabilities drop again as the stand begins to approach the carrying capacity of the site (the codominance-suppression phase). This theory is still controversial but it is clear that, if early selection for wood properties is to be effective, meaningful heritabilities must exist at the age at which the selection takes place so that results are evident at the age at which the trees are harvested.

A different method for control of wood formation was given by Schniewind (1962), who related changes in wood density with age to the support function of the stem wood. He showed how higher density wood is needed at the periphery of the stem rather than near its center, and related the amount of void volume in the cells with crown structure and growth rate.

A very thorough knowledge of the physiological basis of wood formation is needed. Arguments frequently occur as to which is more important in wood formation, physiology or genetics. Such arguments are pointless because the only way genetic control can work is by affecting physiology, and physiological processes are in turn influenced and controlled by genetics. It is important to remember that genetics is only one way to influence physiology and that even when genetics is a major reason for a change in physiology, the final results are almost always influenced by interaction with environmental factors.

2.1.1 The Kinds and Strength of Genetic Control In Wood

It should already be obvious from the introductory comments in Chapter 1, that genetic control of most wood properties is moderate to very strong. The original broad variability in wood properties present in indigenous stands is still there, essentially unaffected by human interference. The statements by Hamrick et al. (1979) and Conkle (1980) that trees have more than double the amount of variability when compared to other organisms, and that conifers have much more variation than other plants, applies to wood as well as to other characteristics of forest trees. The book *Wood Variation – its Causes and Control* by Zobel and van Buijtenen (1989) throughout emphasizes the large variability in wood. It is the delight of the geneticist to have such great variation; without it, potential gains would be limited. However, having variation does not assure that gain can be made using genetics.

The variation patterns in wood are sometimes the result of genetic differences. This aspect will be dealt with in detail in parts of several chapters, but the importance of *large genetic* variability must be recognized when dealing with changing wood through genetic manipulation. As an example, Dodd and Power (1989) emphasized that there were substantial differences among eucalypt trees in specific gravity and fiber length; this finding is repeated essentially in every report dealing with wood variation in nearly all species of forest trees.

To determine the potential value of any genetic program (including wood) and how much effort should be expended on it, the following questions need to be addressed:

1. *How much variation is present*? This question has already been answered—for most wood properties, a large amount of variation exists. In fact, there is so much variation in properties like wood density that it is sometimes hard to believe. As examples, species such as *Eucalyptus grandis* have wood specific gravities of individual trees that range from 0.35 to 0.85. Woods with such different densities in no way resemble each other and are not suitable for the same end uses even though they are from the same species. Loblolly pine is another example where there is huge variation within a stand of trees of a given age, with the specific gravity of some trees as low as 0.40 and others as high as 0.75.
2. *What causes the variation*? As mentioned in Chapter 1, if the wood differs largely because of varied environmental conditions, a genetics program will not be warranted. For example, provenances within some species have greatly differing woods, but when the seeds of the provenances are planted on one site, the wood can be very similar by provenance source (Zobel and van Buijtenen 1989). Some species are site-sensitive; when radiata pine (*Pinus radiata*) of a given source is planted on different sites, the wood can be quite different (Harris 1965b).
3. *What kind of genetic control is present*? This subject is complex and can become very detailed; but an over-simplification is to divide the genetic control into additive and nonadditive components.
   a) The additive variance is the easiest to work with because results are predictable. For example, if a tree with high wood density is crossed with one with low density, the progeny produced will have wood in the mid-range of density. Characteristics primarily and strongly controlled by the additive genetic pattern respond well to selection. For example, in a pine plantation where the individual tree specific gravities varied from 0.40 to 0.60 (average 0.50), if one selected only the trees of 0.55 and better (whose average might be 0.57) and let them cross among themselves, their progeny would average 0.54 (instead of 0.50 in the original plantation), if the heritability were 0.6. This value is determined by the rough approximation of the gain formula ($G = SD \times h^2$) where (SD) selection differential is the difference between the average of the selected and original populations. (Although not technically accurate, one can obtain a good estimate of gain using this approximation formula).

The relative magnitude of variance components due to general combining ability and specific combining ability is the way to show the importance of additive and nonadditive genetic variances. General combining ability was found to be predominant in wood density, tracheid length, cell diameter, and moisture content for black spruce (*Picea mariania*) (Boyle et al. 1987). These respond well to selection, and this has been true for most species. However, characteristics such as cellulose yield have a low combining ability and cannot be easily improved by standard selection (Jett et al. 1977).

b) For nonadditive variance, selection and crossing are *not* profitable and of little value, since the nonadditive variance is mostly dependent on gene and allele associations that occur randomly when different genotypes are crossed. The ways to use this nonadditive genetic variance are by vegetative propagation, in which both additive and nonadditive variance are transferred by means of the cutting from the donor parent to the new plant, or by making specific crosses that have shown the genetic combinations to produce the desired wood property. The best current example of this is the cellulose content of wood (Jett et al. 1977). Little gain in cellulose yield per dry weight of wood is possible by selection but real improvements are obtained using a rooted cutting program or specific crosses. This has been well shown for *Eucalyptus* (Brandáo 1984).

4. *Are the gains that can be achieved worth the effort and cost*? This sometimes is a difficult question to answer, and is discussed more fully in Chapter 2.2 and in Chapter 13. For example, the genetic changes possible in the length of fibers of hardwoods are very large from a percentage standpoint, but the actual physical increase obtained in length often has very little effect on the value of the final product. Therefore, the effort to increase fiber length in some hardwoods is not justified even though the genetic manipulation may be very successful (Zobel and van Buijtenen 1989). An estimate of the value in the final product to be gained from breeding for wood properties should be ascertained so as to prevent a great waste of money, time, and effort. In an attempt to determine the relative value of changing cell characteristics by genetics, silviculture, or mill operations, a huge study was made on pine linerboard by TAPPI in the Southern USA (van Buijtenen et al. 1974). The results showed the great importance of wood density and volume production but indicated that tracheid length was of secondary importance.
5. *What is the relative size of the heritability value*? A heritability value is only an approximate estimate to indicate relative differences, even though h^2 is given as a definite, (and often assumed accurate) numerical value. This warning holds especially when gain predictions are calculated, using a formula such as $G = SD \times h^2$. Gain values can only be relative and any decision made on minor differences stands a good chance of being incorrect.

All the analyses and questions mentioned above cannot be handled successfully without first determining the proposed end-use of the wood. The most common mistake, which can result in huge losses, is to grow wood without a clear concept of what the wood will be used for because the wood properties most suitable for specific products (lumber, linerboard, newsprint, tissues, energy, etc.) can be different. There is no such thing as a "best" wood – always the definition must include "best for what?" Within each product requirement is the need for uniformity. Quite frequently, uniformity of wood is of much greater value than specific wood properties. This need is rarely emphasized even though some uniformity can be obtained through genetic manipulation (see Chap. 11.3).

2.1.1.1 The Measurement of Genetic Control

Only a portion of the wood differences is due to genetics. The remainder is affected by the environment. The wood we see, or measure, is termed the phenotype, which is, in turn, the result of the interaction between the genotype and the environment (expressed as $P = G + E$). The phenotype is what can be seen or measured while the genotype (the actual genetic potential of the tree) can only be determined by testing.

In order to make an effective assessment, there must be a way to express the degree of genetic control. Statements such as "inheritance is strong" can be quite meaningless and frequently misleading. In the final assessment, the inheritance must be categorized by the gain to be obtained from application of genetics but there is still an initial need to assess strength of inheritance. There are numerous books dealing with these concepts and only a few of the simplest but most important ones will be briefly covered. Chapter 4 in Zobel and Talbert (1984) presents a more detailed but still simplified discussion of the concepts of the genetics involved.

Although there are several methods of describing the importance of genetics, the one most used is termed *heritability*. Heritability is generally used relative to genetic differences among individuals within a given population of a species, but it is also occasionally applied to populations, such as to families, to provenances, and, rarely, even to species differences. Therefore, it is always of importance to express what kind of heritability (individual, family) one is dealing with. For example, just by the method of calculation, the family heritability value is larger than that for individual trees, but the two kinds are applied to different kinds of selection differentials when calculating gain. Heritability is nothing more than a ratio that expresses the degree to which a characteristic is passed from parents to offspring.

Heritability is used as a portion of the gain formula to give a concrete estimate of what genetics can produce. There is an accurate formula ($G = i\,h^2\sigma_p$, where G = gain, (or response to selection), i = intensity of selection, h^2 = heritability, σ_p = phenotypic standard deviation). We often use the previously described, simplified version of $G = SD \times h^2$, where SD is the selection differential. The above is applicable when individuals or families are selected based only upon their phenotypic characteristics without information about relatives.

Heritability is of two types.

a) *Broad sense heritability* (H^2). This is the ratio of total genetic variation (additive + nonadditive) to the total variation (additive + nonadditive + environmental). Broad sense heritability is expressed as

$$H^2 = \frac{A + NA}{A + NA + E} \quad \text{or} \quad \frac{G}{P}.$$

(Here $G =$ total genetic variance, $P =$ phenotypic variance, $A =$ additive variance, NA = nonadditive variance and E = environmental variance). The concept of broad sense heritability has had a limited application, usable only where both the additive and nonadditive variation can be transferred as is the case for vegetative propagation or controlled crosses. At present, with a strong trend toward use of rooted cuttings and more intensive breeding programs, the use and importance of H^2 is increasing.

b) *Narrow sense heritability* (h^2). This refers to the ratio of the additive genetic variation to the total variation. It can be expressed as:

$$h^2 = \frac{A}{A + NA + E}.$$

When there is no genetic control, heritability equals 0; if there is complete genetic control, it equals 1. The usual measure of heritability seen in the literature is expressed as h^2. It will vary with the proportion of additive to nonadditive + environmental variance, with values from 0 to 1. Narrow sense heritability can never be greater than broad sense heritability, although as the proportion of additive variance increases, the two come closer together. In a study on slash pine, Einspahr et al. (1964) reported that for several important wood properties "... where both narrow and broad-sense values are available ... narrow sense values run approximately 60 percent of broad sense heritability." In a recent paper on the crown characteristics of noble fir (*Abies procera*), Doede (1993) found that family heritabilities were twice as large as individual tree heritabilities. This, of course, changes for species, conditions, wood property, and age but it does give an indication of the relationship between the two kinds of heritabilities.

c) *Parent-progeny and ortet-ramet correlations.* When the correlation between the ramet and ortet, or the parent and its progeny are known, an estimate of heritability can be obtained by squaring the coefficient of correlation between the two (Zobel and Talbert 1984). This method is not used very much but in some ways it is best in that it indicates the *actual* results in the young trees when using older parents or ortets with differing wood properties. An example was reported by Einspahr et al. (1964) in slash pine. The ortet-ramet correlation was squared and an $h^2 = 0.25$ for specific gravity was then obtained. This is lower than the $H^2 = 0.44$ to 0.49 they obtained by using the standard formulation for broad sense heritability; such a result would be expected. Another example of the use and small size of the parent-progeny heritability for wood density was reported by Lima (1987), who obtained a typical $h^2 = 0.52$ on an individual tree basis but only $h^2 = 0.23$ on the basis of the parent-progeny correlation. In *Pinus pinaster*, Keller (1973) reported that the h^2 calculated from parent-progeny correlations is sometimes considerably lower than that obtained from the regression method on the progeny.

The parent-progeny correlation does change, sometimes considerably, as the progeny age and wood properties change with age. A parent-progeny correlation thus entails comparing old wood with young wood. Old wood often has chemical (resin and polyphenol) deposits in the wood, hard to extract. The young trees are often mostly juvenile wood and have a high proportion of reaction wood and other defects. This results in low correlations between the two; but it does show what the wood of the progeny will be like from older selected parents.

The easiest way to understand heritability was summarized by Zobel and Talbert (1984) as follows: "The basic and key point about heritability is that it is a ratio between genetic and phenotypic variances; thus, it is not a fixed value for a given characteristic or a given species. Estimates of heritability are not made without error; therefore the ratios obtained are only a relative indication of genetic control and should not be interpreted as absolute or invariant values. ... An important but often overlooked aspect of heritability estimates is that they apply only to a particular population at a particular point in time. For example, estimates of heritability for a group of trees grown in a greenhouse would not be appropriate for the same trees growing in a field environment ... estimates of h^2 obtained in the greenhouse will usually be higher than those from the field, because there is less environmental variation in the greenhouse. ... The heritability values of a given characteristic in a population often change ... as the trees mature."

The genetic value of parents is frequently expressed in terms of their *combining abilities*. Throughout this book the term *good general combiners* is used and occasionally also *specific combiners*.

a) *General combining ability* is defined as the average performance for a given characteristic (in this case wood) of the progeny of an individual when it is mated to a number of other individuals in the population (Falconer 1960). A good general combiner for a specific wood property is a parent tree that produces progeny with wood partially independent of the quality of the wood of the second parent.
b) *Specific combining ability* refers to the average performance of a cross between two parents that is different from what would be expected based upon the general combining ability of the parents. It always refers to a specific cross, never to a particular parent by itself. Specific combining ability has little use in wood assessment. Even though the cross may be a good specific combiner because it is better than what was expected, it still may be inferior to the population average (see Zobel and Talbert 1984, Chap. 4, for a short explanation).

Since general combining ability refers to the performance of a parent when used in many crosses, it is a reflection of the additive genetic value of the parents. Parents that transmit a large amount of genetic control for a given trait are called good general combiners; but this does not always infer that the combining ability is desirable. For example, if one wanted high wood density, a given parent that is a good general combiner for low wood density would not be desirable for wood production even though it is classified as a good general combiner for wood density. Since general combining ability can be predicted because it is additive, it is frequently referred to as the dependable portion of the tree's genetic constitution. It can be utilized in seed orchards based upon selection of parents. The improved wood of the planting stock (from the seed orchard) results from the accumulation of favorable alleles that have an additive genetic effect on the wood of the trees grown from seed from the seed orchard.

The ease with which a wood characteristic can be controlled genetically depends upon its complexity, which is related to the number of genes and alleles involved and their interaction with one another and with the environment. Unfortunately for the tree breeder, genetic control of most wood properties is complex (illustrated later for wood density). Experience has shown that most important traits are controlled by several genes (or alleles), each with small effects rather than by one or two major genes. This makes necessary an in-depth use of statistical methods to study the genetic differences. Changes are rather small by breeding so they take several generations to become established. However, the variation is generally so large that meaningful improvements will result by use of simple, standard breeding methods (Zobel and Talbert 1984). Once established, the genetic gains achieved are slow to disappear or to be lost.

2.1.1.2 The Change of Genetic Control with Tree Age – Juvenile and Mature Wood

The age (ring number) when heritabilities of wood are taken is important, because heritabilities change with the age of the cambium. The effect of tree age on growth was expressed by Franklin (1979) as he described the three phases of development: a juvenile, a mature, and a suppression phase (see Sect. 2.1). Heritabilities are usually high in the juvenile phase, decrease in the mature phase as competition sets in, increase when competitive abilities are expressed, and drop again in the suppression phase. There has been much discussion whether the same parttern holds for aging of the cambium. It appears that it does, but the age of the cambium differs from base to top of the tree and therefore is not directly related to the age of the tree. It is determined by the number of annual rings from tree center. The inheritance pattern may vary because of cambium age, which appears often to be the case. An example is *Pinus pinaster*, where Keller (1973) reported $h^2 = 0.38$ for wood density at breast height for 11-year-old trees and only 0.13 for 30-year-old trees.

Relative to the genetics of wood, there are two major concepts related to age. The first is a change in genetic parameters of given wood properties as the wood is laid down by cambia as they age, already described above. This refers to changes from pith to bark. Differing zones of wood are formed which are referred to as juvenile and mature wood. These have been covered in detail in Chapter 3 of Zobel and van Buijtenen (1989) and in Chapter 12 of Zobel and Talbert (1984) and are briefly described in Chapter 8.3 of this book. The concepts of juvenile and mature wood are generally understood and widely used. It has been found that parent trees with wood of high or low wood density produce progeny with juvenile wood similar to that of the parents (Zobel 1973) (Fig. 2.4). Juvenile wood occurs at any given height in a tree.

The number of rings from the pith when the cambium changes from producing juvenile to producing mature wood is one key to wood quality. Generally, longer-lived species produce juvenile wood for a long period of time while shorter-lived species form juvenile wood for a relatively short period of time. The question is whether the transition age is genetically controlled and whether it can be changed by breeding (Fig. 2.5). This aspect is more fully discussed in Chapter 8 and Chapter 3.3.1.1, where various possible methods to alter the characteristics and duration of the juvenile period, including use of genetics, are covered.

The other concept related to age is the predictability of the wood properties of an old tree from the wood of a young tree. Expressed in another way, it is

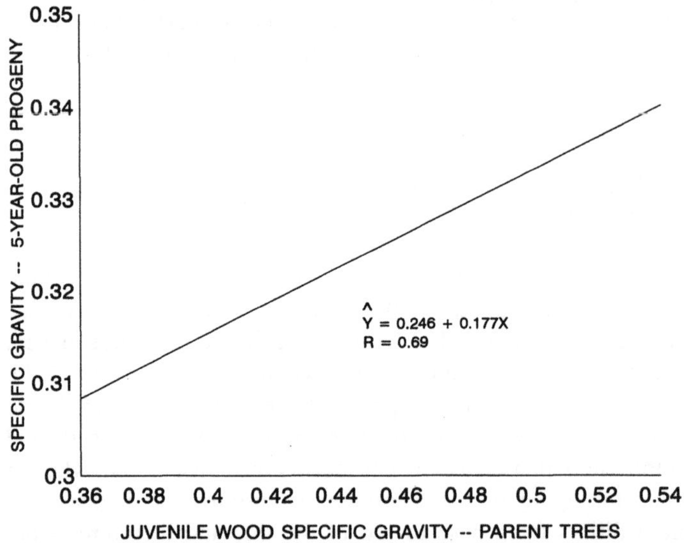

Fig. 2.4. There is always a hurry to obtain results when breeding for improved wood properties. Trees were produced as progeny from parents with different density juvenile woods. As shown on the graph, some improvement is possible by selective breeding, but the increase in specific gravity of juvenile wood from using parents with high specific gravity wood is only moderate

Fig. 2.5. Trees within a species, as well as species themselves, differ in the extent and severity of juvenile wood production. It is affected by the environment as well as by genetics. An extreme example of juvenile wood production is this *Pinus caribaea* grown in the Amazon region of Brazil. The tree is 6 years old, with obvious juvenile wood produced for four rings and mature wood later. (After Zobel and Talbert 1984)

the correlation between juvenile and mature wood within a tree. The pattern of radial change determines the juvenile-mature relationships. Usually the two are closely correlated, as reported for 20-year-old radiata pine where the juvenile to mature wood had $r = 0.79$ (Burdon and Young 1991a). In white spruce (*Picea glauca*), Corriveau et al. (1987) found a positive link between juvenile and mature wood densities strong enough for improvement of wood density to be based on selections based on juvenile wood. Differences between juvenile and mature wood are usually large in the conifers but may be (usually are) small in the temperate hardwoods. This is indicated for fiber length of five hardwood species in Table 2.1 and is illustrated in Fig. 2.6.

The wood density profile from the center of the tree outwards varies with species, with provenance, and with individual trees at various heights. For example, Loo-Dinkins and Gonzalez (1991) reported that the relative density profile from pith to bark in Douglas-fir was different at different heights in the tree but these differences disappeared with increasing cambial age. Often, the characteristics of juvenile wood are closely related to mature wood (Fig. 2.7).

Table 2.1. Juvenile and mature wood of most temperate hardwoods are similar, as shown for fiber length[a] of the five species represented.

Species	Age segment 1–15	(no. of rings) 16+
Liquidambar styraciflua	1.90	2.13
Platanus occidentalis	1.93	1.98
Fraxinus pennsylvanica	1.24	1.41
Acer rubrum	0.93	1.02
Quercus nigra	1.41	1.60

[a] Length in mm at different distances from tree center.

Considerable work has been done to understand the relationship between juvenile and mature wood or how many rings of wood from the pith are required to give a good estimate of the wood formed nearer the bark. Results vary widely, depending on species or circumstances. For example, McKimmy (1966) concluded that wood density of Douglas-fir could not be reliably evaluated until the trees were 25 years of age. For the same species in the same area, Gonzales and Richards (1988) found that reliable selection for wood density required 12 to 15 growth rings from the pith at breast height. In the southern pines it is 6

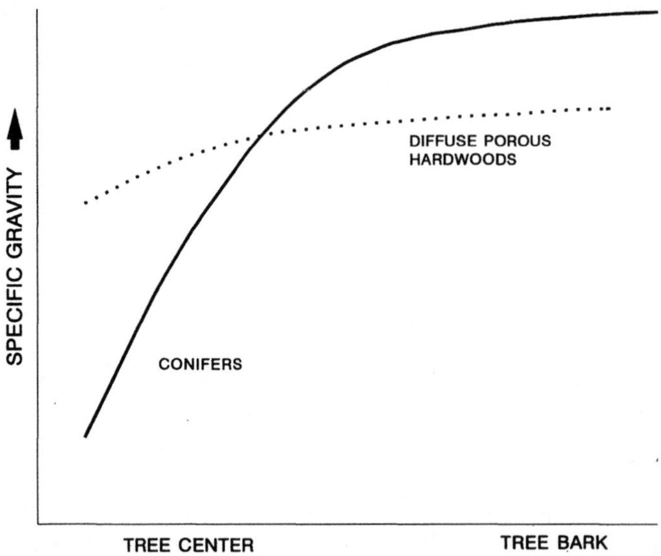

Fig. 2.6. The change in wood characteristics radially from the pith varies with species. This is illustrated schematically for conifers and diffuse-porous hardwoods. As a result, juvenile wood of conifers is more different from mature wood than it is in the diffuse-porous woods

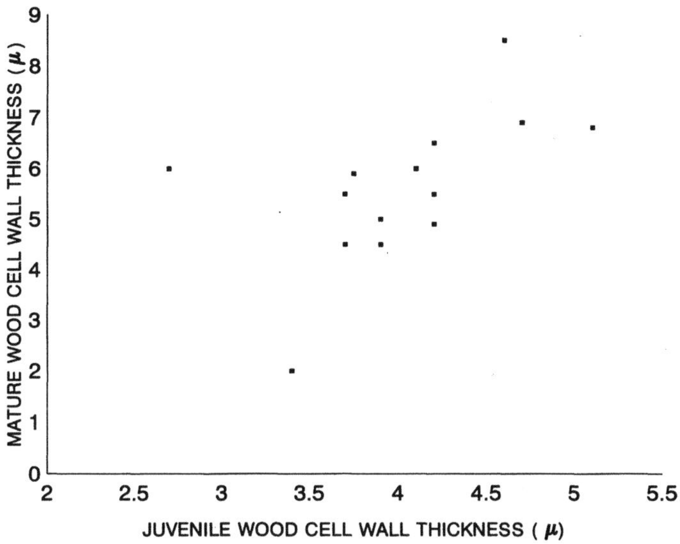

Fig. 2.7. One frequently finds the wood properties of the juvenile wood reasonably well related to those of the mature wood. This is illustrated for wall thickness

to 11 years, depending on the individual tree (Zobel et al. 1972) and in some tropical pines it is less than that.

As the cambium ages, inheritance patterns as well as wood properties usually change. In the conifers, the wood property change can be large and abrupt. In most diffuse-porous temperate hardwoods, the change with cambial age is usually slight (see Table 2.1) although a juvenile zone is definitely present in otherwise uniform woods such as *Gmelina arborea* (Ku et al. 1991). Usually, wood density changes with ring number in the ring-porous hardwoods because of the percentage of vessels. Zhang and Zhong (1991) concluded for *Quercus liaotungensis* that age was of key importance in controlling wood specific gravity. As the tree shifts from the production of juvenile to mature wood, differing genetic controls often occur. Thus one cannot safely report that wood of a species (or tree) has one overall heritability. It is necessary to assess the genetic changes for the different categories of wood found within a tree.

Perhaps the most common error in estimating gains in wood properties through genetics is to make the assumption that the heritability value remains constant with increasing distance from the pith. This error certainly is understandable – forest researchers usually cannot delay assessing studies of inheritance until near harvest age. The data are needed *now* so they can be applied to the current breeding program and values for juvenile wood are therefore used as a predictor. Yet, under some conditions and for some wood characteristics, the inheritance patterns (heritabilities) change, as determined by the genetic control and the environmental conditions under which the wood is formed. Thus it usually happens that the intensity of genetic control is assessed earlier than ideally in order to save time and to obtain useful information early, but usually the values obtained at the early cambium age cannot be accurately applied to wood produced at the time of harvest.

Although this phenomenon of change in heritability has often been observed, it has rarely been recorded. One group of researchers (Dadswell et al. 1961) did report a greater environmental control of percentage latewood in radiata pine in the rings near the pith than for those rings closer to the bark, where the genetic influence dominated environmental influences. Often, tree growth and wood properties also change as the trees grow because of greater competition with their neighbors. The genetic control of wood density and cell characteristics is often much different in the wood of seedling or sapling trees than it is in mature ones.

The above circumstances related to heritabilities dictate that a WARNING must be given. It is that the *magnitude* of the *measurements* of *inheritance*, such as *heritability, should be viewed only in a relative sense*, not as an absolute and accurate value. At the least, the environment will change as the stand of trees matures. Since the heritability value calculation includes the environmental variance, it is inevitable that the heritability value will also change with age, even if the strength of genetic control should be similar. However, if the genetic variance also changes, for which there seems to be good evidence, then a heritability value from young trees is certain to be different from that of the same tree when it is older.

Although too early assessment of h^2 is constantly raised as a danger, valid research data showing the inheritance of wood properties at harvest age are scarce indeed. To obtain such data requires a long-term, expensive effort based on the proper design and suitable parents. The wood of 25-year-old trees of *P. taeda* is now being assessed but results are not yet available. All that can bc said with certainty is that a knowledge of plant developmental patterns and the formulations to predict gain indicate that there will be a change of genetic control with the age of the cambium, but how large this change will be and how well early assessments will predict gains in wood properties at harvest time is really not well known. Changes of inheritance with time are known to be of importance for growth, pest tolerance, and adaptability but these cannot be directly applied to wood. For example, our extensive observations on pine show a rapid increase in the size of h^2 of wood density as trees grow to a competitive size and then it levels off, but this has not been generally reported in the literature.

A few publications can be cited that show a change of heritabilities of wood properties with ring number from the tree center as previously cited for Keller (1973) in *Pinus pinaster* (maritime pine). In radiata pine, Nicholls (1965) found a systematic change of heritability with rings from the pith, for grain inclination, wood density, and tracheid length in 25-year-old rooted cuttings. He warned that care must be taken as to where genetic information on wood is obtained. This was also shown for radiata pine by Dadswell et al. (1961), who reported for wood density a $h^2 = 0.50$ for rings 2 to 10, but a heritability of 0.70 for the outer 9 rings in 20-year-old trees. Also for *Pinus radiata*, Nicholls (1966) states that: "... it is clear that heritability values for fibre length are likewise subject to patterned change with increasing age, in that *heritability* increases from the pith outward until a broad maximum is reached at about the fifth to the ninth ring from the pith and then declines again. ... The trend for the heritability of basic density ... is one of a decrease from the pith outward until a minimum is reached at about the ninth growth ring from the pith followed by an increase in heritability with further increase in age." Nicholls noted that this reported specific gravity trend was not found in a second plantation tested. Resins were removed from wood samples from the area where heritabilities of wood density changed but not from wood from the other area where heritabilities did not change. In a following paper (Nicholls 1967b) says of wood density: "Genetic control of this characteristic appears to be a maximum in the early life of the tree and therefore maximum gains from selection can be obtained in the first-formed wood." It is of interest that this finding is diametrically opposite to that of the authors of this book, who found the lowest heritabilities from the youngest rings. For example, a significant change of heritabilities of specific gravity with age was found by Zobel (1964) for loblolly pine when h^2 increased rapidly from young ages to 15 years (h^2 of 0.25 to 0.50 and above). Tracheid length shows the same trend for increase to 15 years of age. In contrast, spiral grain has shown a decrease in heritability with age of annual ring from tree center. Similarly, a recent study on *Pinus elliottii* by Hodge et al. (1992) showed a definite trend in the heritability of ring density from the center outward, with the older rings having a somewhat higher heritability than those near the tree center (Fig. 2.8).

2.1.1.3 The Environmental Control

It has been stated many times that wood variation results from environmental differences, genetic differences, and their interactions. Numerous publications could be cited to show the importance of environmental control. As an example, one study by McKimmy and Nicholas (1971) was made on 50-year-old Douglas-fir for which they reported: "This analysis indicates that the environmental components ... and genetic components ... are both significant factors affecting specific gravity." Contrary to most other investigators, Rink and Thor (1973) found weak genetic control of Virginia pine (*Pinus virginiana*) specific gravity relative to the environmental effect. They reported that specific gravity in this species appeared to be strongly affected by the environment, with different populations reacting differently to changed environments. Similarly, Nicholson (1967) found that for Douglas-fir the variances and means for tracheid characteristics were greatly influenced by photoperiod and associated environmental factors.

Conversely, only a small environmental effect on wood was reported by Ericson (1960a) for Scots pine (*Pinus sylvestris*) and Norway spruce, where they found the wood of grafts to be nearly the same as that of the parent trees even when the grafts were grown in greatly differing environments. This indicates that the differing basic densities of the original trees were not greatly determined by

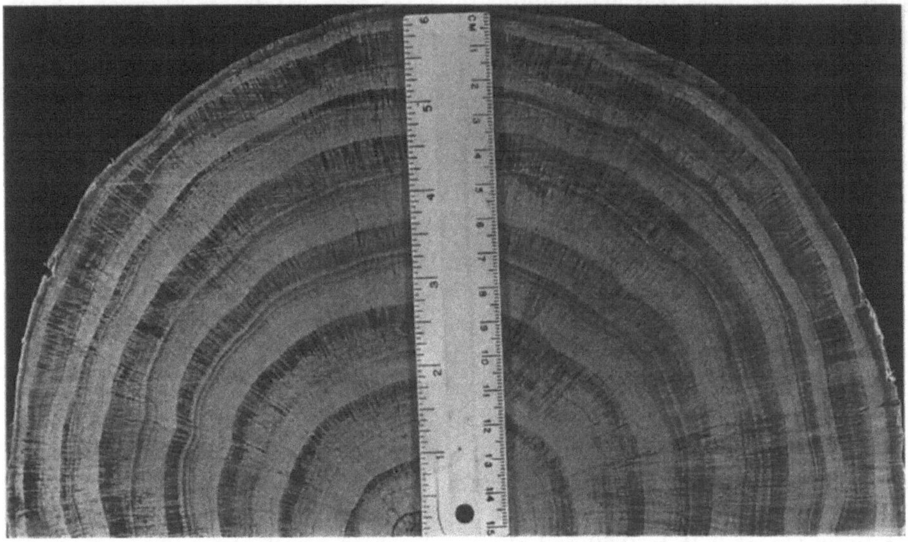

Fig. 2.8. The characteristics of juvenile wood can be altered by breeding. Shown is a fast-growth tree in which the juvenile wood has a relatively high specific gravity. Such trees are sought when high density wood is desired at short rotations. Trees with increased juvenile wood density can be developed. (After Zobel and van Buijtenen 1989)

the environment. Ericson also found that fiber length and other fiber properties of the parental tress agreed with those of young trees grown from seeds. He concluded that the wood properties largely depended on genotypical differences between the trees and were not seriously affected by the environment. In contrast, Harris (1966) reported that latewood density in *P. radiata* in New Zealand was closely correlated with mean annual temperature ($r = 0.94$); a basic density of 0.40 was produced with a temperature mean of 49 °F while it was 0.50 for a mean of 57 °F. This appeared to be nearly independent of soil type and rainfall regimes. Such direct and clearcut relationships between one aspect of the environment and wood properties are rare.

A series of studies have been made on how wood changes as the environment changes. Most of these have been based upon gross environments. Some have been related to specific environments and precisely measured; an example of this is van Buijtenen (1956). A summary of a few of his results in loblolly pine seedlings are: (1) an increase in temperature caused a reduction in tracheid wall thickness, without influencing cell diameter; (2) high soil moisture resulted in large tracheid diameters, but wall thickness was not affected; (3) long photoperiods did not affect wood properties. In older trees, continuous illumination and irrigation did not affect the type of wood formed. Based on the type of physiological responses obtained by altering the environment and the frequent lack of changes in wood, van Buijtenen concluded that *initiation of latewood formation* in loblolly pine must be from a *change* in the *condition* of the *tree itself rather than from external environmental stimuli.*

Many environmental factors affect wood but the one most commonly cited is the moisture regime. This was shown by Nicholls (1967a), who worked with *Pinus pinaster* in drier climates. He emphasized that wood quality improvement can only be wholly successful if proper attention is also given to growing conditions, particularly to those factors that are associated with moisture availability to the mature tree.

Apparently there is strong environmental control of wood in mahogany (*Swietenia macrophylla*) (Chudnoff and Geary 1973). The basic densities of the progenies, regardless of parental specific gravity or Puerto Rican growth conditions, were almost identical, with all being 0.53. Here, the environment appears to be the major control of wood density, with genetics playing only a minor part.

Environmental control of wood properties is not the subject of this book, but it is clear that all factors of the environment must always be considered when assessing the effect of genetics of wood.

2.1.1.4 Reaction with the Environment – Genotype × Environment Interaction

The effect of different sites on wood production has previously been discussed and will not be covered here except for a reminder that the wood produced often is quite different, depending on the site on which the trees grow, especially when the sites differ a great deal. As mentioned in Chapter 1, it is not possible to simply state that wood properties are controlled by either genetics or environment. There is always some interaction between them ($G \times E$), which is a change in the performance ranking of given genotypes when grown in different environments. The question is whether the interaction is large enough to have any practical significance. When environments differ only slightly, $G \times E$ is usually small. Even when the environments differ considerably, the $G \times E$ interaction for wood density of a number of provenances of loblolly pine was found to be very little (van Buijtenen 1978).

Although not too many well-designed $G \times E$ interaction studies for wood properties have been made, essentially all have shown a relatively small interaction unless the differences among the environments are very large. The term sometimes used to estimate how much a genotype changes with differing environments is "stability-variance" (Shukla 1972). If the stability-variance for a genotype is significant, the genotype is considered to be unstable and has a meaningful $G \times E$ interaction.

$G \times E$ interactions are of particular importance in clonal forestry, where they can be utilized to great advantage. $G \times E$ can be evident for clone × environment, clone × fertilizer, clone × site preparation and many other combinations. Often clone × wood interaction with differing environments is unimportant. Environmental effects on wood vary by species; a species that seems to be especially site-responsive is radiata pine. There are numerous references to this response; they are well summarized by Cown et al. (1991a), who stated that wood density differences in radiata pine associated with site are important and well documented. Also for radiata pine, Burdon and Harris (1973) found large site effects on wood properties, especially on phosphate-deficient sites where a wood density increase was found. However, they observed that the $G \times E$ was very small because the clones × site interaction for wood was minor.

The Mountain Pine Ridge source of *Pinus oocarpa* (this population is now called *P. tecunumanii*, instead of *P. oocarpa*), showed that 85% of the families had stable wood densities based on all sites tested (Lima 1987). When 29 families were grown on each of three very different sites in South America, (Colombia and two different sites in Brazil), the wood specific gravities of only five families exhibited significant $G \times E$ interaction (Lima et al. 1990). These few families showed extreme changes in ranking from site to site although, in general, reasonably uniform wood was produced over the environments tested. It is usual for only a few families in each test to show significant $G \times E$ interaction. In another tropical pine (*P. caribaea*) Wright (1990a) reported that the site × provenance interaction was not significant, so provenances were consistently ranked similarly wherever tested.

In their study of 10-year-old control pollinated loblolly pine, Talbert et al. (1982a) also found genotype × environment interaction to be of little consequence for specific gravity of juvenile wood. When 18 families of loblolly pine were grown in seven different environments, only four families of 12-year-old trees showed a large $G \times E$ (Jett et al. 1991). They summarized: "These four families accounted for 49.5% of the genotype × environment interaction sum of squares. However, the loss in potential gain in a breeding program for specific gravity due to the presence of a significant genotype × environment interaction was estimated to be only 1%." Also for loblolly pine in Texas, McKinley et al. (1982) found family by location interaction to be small on seven open-pollinated progeny tests consisting of three types of progenies.

A $G \times E$ stability also occurs in the wood of hardwoods, as shown by Ferrari and Scaramuzzi (1980) for hybrid poplars, and Farmer (1970) for cottonwood (*Populus deltoides*). Farmer (1970) found little change in clonal ranking in stress vs. favorable environments for either wood density or fiber length.

Sometimes, $G \times E$ is reported for unexpected situations. For example, for a wood strength test Waugh (1972) found the value for poplar clones grown in Australia had "... a highly significant interaction between clone and locality ... not all clones will react in the same way to a particular change in environments". He reported that the ramets within a clone responded almost identically for the strain strength values of the wood.

2.2 The Value of Genetic Differences in Wood

Regardless of whether wood can be changed environmentally or by genetic manipulation, it is important to know the worth of a unit change in a wood property to the final product value. Once this is determined, one must know how many units of change can be achieved through genetics or other means and thus whether it is worthwhile to try to obtain a change in a given wood property (see Chap. 13 for a more complete coverage of this subject).

As pointed out earlier, just because one can achieve a good percentage improvement does not necessarily mean that a breeding program should be developed for a given wood property. As explained in Section 2.1.1, the heritabilities are high, relative variations are large, and percentage improvements are large for the fiber length of the short-fibered temperate hardwoods. Yet the maximum change possible will have only a minor effect on the final product and a genetic program to increase fiber length of most hardwoods cannot be economically justified.

Therefore ONE OF THE MOST IMPORTANT DECISIONS TO MAKE IN A GENETICS PROGRAM WITH WOOD IS *WHICH PROPERTIES SHOULD BE IMPROVED*. This decision is not only important, it is extremely difficult, and is perhaps the poorest understood area in tree improvement. There are dozens of publications that deal with this subject, the majority related to pulp and paper. This book is not the place to cover in detail the wood property to product quality relationship. That subject is large and important enough for a whole book. However, *if one is to breed for wood properties intelligently, it is necessary to know the value of the property*, the *genetic control* of the *property* and thus *how much time* and *effort* should be expended on it.

A few examples have been selected throughout this book to illustrate the value of wood properties in pulp and paper, in solid wood, and in other products with the knowledge that only samples are cited and the subject is in no way complete. It is included in this book as a WARNING to anyone contemplating improving wood genetically, how complex the situation is and how much care is needed in selecting the wood properties. This was emphasized in van Buijtenen's synthesis (1967b) on pulpwood properties and tree breeding.

2.3 Summary

Anything that is done to change the growth pattern or form of a tree may result in a change in wood quality and utility. Wood may also be changed directly by control of parentage. Although many wood characteristics respond to genetic manipulation, emphasis in applied programs has also been on indirectly influencing wood quality by improving tree bole straightness and limb form and by reducing degrade from pests. Genetic improvement of wood has focused on manipulating specific gravity (basic density) with limited emphasis on other traits such as tracheid length.

Results from manipulating wood by breeding have been good, producing trees with wood more suitable for the desired product, and, of great importance, more uniform wood. Changing wood through genetic manipulation is possible without any alteration in desired growth, since wood and growth characteristics in many species are usually poorly related or unrelated genetically so they can be developed independently. Thus one can produce fast growth, long tracheid, and high density trees or fast growth, long tracheid, and low density trees as desired. When characteristics such as wood density and growth rate are negatively correlated, inclusion of wood in a breeding program becomes more difficult.

This chapter has dealt mostly with basic concepts, sometimes illustrated with specific examples. Details of most factors covered are discussed in other chapters. Some items that are specific to this chapter are:

1. A short discussion on the physiological factors involved in wood production including the effect of growth termination and time of change from earlywood to latewood.
2. The kinds of genetic control were briefly described. Generally, genetic control of wood properties is strong enough that, along with the usual large amount of variation for most traits, moderate to good improvements in wood are pos-

Fig. 2.9. Although it is usually stated that there is a lot of wood variation among trees and among provenances, there are exceptions such as red pine (*Pinus resinosa*), shown here growing in Canada. Very little wood variation occurs in this species

sible by genetic manipulation. The different types of genetic controls, roughly separated as additive and nonadditive, and their uses are described. Only when additive variance is present can a traditional selection and breeding program be successful. Added to these was mention of the terms, and use of, general combining ability, narrow sense, and broad sense heritability. Gain is roughly indicated as selection differential × heritability ($G = SD \times h^2$).

3. The control of wood properties by the environment was briefly discussed and the use and importance of the genotype × environment interaction ($G \times E$) was covered. In general, $G \times E$ with wood properties is small except where grossly differing environments are involved.
4. Genetic control varies with age of the cambium producing the wood. The concept of juvenile and mature wood is important. Many results in the literature are not applicable to mature trees because they were calculated from data obtained with juvenile wood. There is sometimes a large difference in inheritance patterns between juvenile and mature wood from the same trees.
5. There was a brief discussion as to which wood properties should be included in a breeding program. Although this subject is covered in other chapters, especially in Chapter 12, it is clear that usually the most important wood property to breed for is wood density or specific gravity; but it is important to understand species differences. For example, variation in wood of red pine (*Pinus resinosa*) is small, and gains from breeding would not be useful (Fig. 2.9).
6. A general conclusion is that the most important decision to make in a genetics program involving wood relates to which properties should be included.

Chapter 3

Sampling and Analysis in Genetic Studies on Wood

3.1 Making Wood Studies – Sampling Methods

In some ways, working with wood is quite easy because the bole of a tree contains a permanent record of wood development. Thus one can see what wood was produced 5 or 10 years ago because it is still there and the actual cell structure has not changed. Cell chemistry can change with time but cell anatomy remains reasonably constant. Within-tree stress as the tree ages can alter cell anatomy to some degree.

A review of the material available indicates that those working on the genetics of wood must be very critical of the published literature and assess data carefully. It is an unfortunate fact that too many publications lack information on the number of trees assessed, how trees were chosen, how they were actually sampled (both within the forest and within the tree), and the laboratory methods used. The within-tree sample location is very important because of the huge radial and longitudinal variability in wood properties within some species (especially the conifers). Details are critical in interpreting results and must be carefully evaluated for each study. Although it may appear to be a negative approach, it is our policy that when a new wood study is reported the first question asked is: "What might be wrong with this study?" This approach has saved a good number of incorrect assessments when evaluating results in the literature or even for one's own studies. The warning of the authors is to "be careful and be critical" when assessing studies relative to the genetics of wood.

Most wood studies rely on small samples taken from standing trees. Many methods for obtaining such wood samples have been suggested, some of which are good, some not so good. A listing of numerous methods suggested for wood density is presented in Appendix Table 3.1, both to serve as a guide to the researcher and to show the variety of methods available. Despite all that has been done, there is still confusion as to the best way to sample wood for genetic studies. In order to help clarify the problem, TAPPI organized a subcommittee to summarize, on a worldwide basis, the methods used to collect small wood samples, measure anatomical, chemical, and physical properties, and prepare pulp and handsheets (Kellogg et al. 1982).

It is evident from the literature that many of the test designs used for the genetic analyses of tree characteristics have also been used for the genetic analyses of wood properties. Although statisticians often frown upon using a study designed to assess a given tree characteristic for the assessment of a second one, almost all wood studies have been done this way. For example, someone may have been interested in the genetics of growth rate, or of bole straightness, or pest tolerance, or provenance performance. After the specific study matured, the idea then occurred to use the same trees for a wood study. Although such a system has some faults, it will usually give a usable estimate of the genetics of wood properties since wood generally is not influenced by *small* environmental differences. The saving situation is that almost all wood properties are genetically independent of other tree characteristics. Those that are related, like bole straightness and fiber properties resulting from reaction wood, must be recognized and dealt with appropriately. However, if a spacing difference study has been established to determine effects on growth rate, the wood produced will be little affected by reasonable spacing differences and the wood relationships obtained will be useful (Zobel and van Buijtenen 1989).

The amount and complexity of sampling needed to integrate within-tree and tree-to-tree variation can be formidable. In order to aid tree improvement studies and provide a better understanding of the variability of wood, Kandell (1971) developed polynomial models to better present within-tree variation of wood properties. He stated that it provided a technique for statistical analyses and graphically presenting within tree variation for fibers, vessels, rays, and other traits.

3.2 Size of Sample

A frequently encountered problem is determining the number of trees that should be sampled. Actually, the number required depends on the uniformity of the wood. If nonuniform wood, such as a mixture of juvenile and mature wood, is sampled, the number needed would be larger than if more uniform, mature wood were sampled. The problem is one of determining sample size, given some measure of variability and some required accuracy of estimation. This difficulty was put into context by Freese (1967), who stated: "Samples cost money. So do errors." The question then arises as to how to determine the number of samples required to minimize costs and errors. There are several variations, but a commonly used formula for determining sample size was given by Freese (1967):

$$n = t^2 S^2 / E^2 \ ,$$

where: n = sample size,
t = student's t at the desired level of probability,
S^2 = an estimate of the population variance,
E = desired one-half width of the confidence interval around the mean.

Along with the question of the number of trees is the question relating to which trees in the stand should be measured. The usual method is to choose and measure only the trees in the dominant and codominant crown classes. If intermediate and/or suppressed trees are sampled, wood properties may vary considerably from the more vigorous trees. The justification for measuring the dominant and codominant trees is that they make up the bulk of the timber produced on an area, usually more than 80%, and inclusion of intermediate or suppressed trees would require a larger sample if unreliable results are to be avoided.

Experience from many studies by the authors working with the southern pines has shown that approximately 30 dominant or codominant trees are required to provide a reasonable estimate of a wood property representing a family, a provenance, or a treatment from unselected or half-sib populations. If clones, or closely related trees such as full sibs are used, good results can be obtained with 20 individuals. If rooted cuttings are used, where all trees are of the same genotype, then ten individuals are often found to be a sufficient sample. However, working with four eucalypt species in Tanzania, Hamza and Lewark (1992) report at the 95% confidence interval the number of samples from a stand could be reduced from 20 to 5 to give average values. If correlation and regression analyses are desired, a higher number of samples are needed. (They sampled basic density, fiber length, fiber wall thickness, vessel number, and vessel proportion.) Their results are greatly different (lower) than reported by nearly all other researchers, who find that five trees are too few to represent a population because of the large tree-to-tree variability.

The tendency of many, if not most, reported studies is to use (and report results on) fewer trees than the ideal; results are then suspect. That in itself is not so serious when recognized and reported as such, but numerous papers on the genetics of wood fail to record the number of trees used, which leaves great uncertainty in the value and interpretation of the reported results. Some researchers have even used one average-sized tree to represent wood properties of the whole sample population. Studies on the number of trees required to obtain a precision of ±5% of the mean, 95% of the time were made by Krahmer and Snodgrass (1967). They found that a sampling plan would require 39 trees for most physical wood characteristics. For crushing strength tests, 49 trees were needed while as few as 15 trees were sufficient for specific gravity determinations. However, most studies have shown that approximately 30 trees are needed for an accurate assessment of specific gravity of nonrelated trees (Fig. 3.1).

An example of the dispute on sample size was given by Goggans (1962) for a large number of *Pinus taeda* (loblolly pine) trees. Rather than the usual small and limited number of samples frequently used, he tried a series of sample sizes. He summarized: "... for springwood and summerwood tracheid length it is necessary to have 50 to 100 female parents, 2 to 4 replications, 10 or more progenies per plot, 2 cores per progeny and 20 tracheids per core to estimate the variance components precisely enough to allow for their effective use in tree breeding programs." Goggans recommended samples of the same size for double-wall thickness and for lumen diameter. Only rarely is such intensive sampling done.

For cell length, there are differences of opinion as to the number of cells that must be measured. Usually, 50 cells are sampled per specimen. However,

Fig. 3.1. Obtaining large numbers of high quality wood samples quickly and cheaply is a major problem in studies on wood genetics. Shown is an electrically powered drill, using a portable generator, which enables procuring large numbers of increment cores of good quality from a pine tree with a minimum effort of time and money

Phillips (1963) showed that a sample of 25 fibers is appropriate for determining mean cell length. The precaution needed in sampling cell length is to ensure that whole cells are measured, not those that have been cut.

3.3 Location and Age of Sample

The exact analytical design used to assess wood values is probably relatively unimportant as long as it is sound statistically; but it is of key importance just how and where the samples are obtained in the tree. A comparison of woods from different trees obtained from differing ring numbers from the tree center is not legitimate, especially when the wood involved is somewhere near the transition zone between juvenile and mature wood (Zobel and van Buijtenen 1989). The essential consideration then is to use trees, or more correctly wood, of a similar age. Unfortunately, papers have been published where this rule has not been followed, and the supposedly genetically sound results that were reported in actuality only reflect the differences between the types of wood used and not genetic differences. To avoid such a problem, Hancock (1962) used the portion of increment cores from 10 to 35 rings on some Canadian trees as a source of comparison of wood density. Similarly, the NC State Tree Improvement Cooperative used the wood outside the juvenile zone to compare older trees because density does not change rapidly with the number of rings from the pith in the mature wood. The importance of partitioning the juvenile and mature characteristics of wood cannot be overemphasized.

Although details of sampling location are thoroughly covered in Chapter 3 of Zobel and van Buijtenen (1989), it is of value to briefly repeat and summarize as follows: (1) One can never compare wood properties directly from locations at different numbers of rings from the pith, unless it is in the form of something like correlations of juvenile to mature wood. (2) One can never compare woods from different heights in the tree, even if care is taken to use the same relative group of rings. In conifers, as well as in certain hardwoods, wood properties vary with height. Much of this is due to differences in the proportion of juvenile wood. However, in certain diffuse-porous hardwoods, juvenile wood has only a minimal effect and a breast height sample alone can give a good estimate of the whole tree value. In contrast, even when restricted to the same ring number, some species, like *Pinus taeda*, have longer tracheids in the juvenile wood near the crown than in the juvenile wood at the base of the tree, although the wood densities may be the same. (3) In older trees, resin or other chemical deposits such as phenolics occur in the heartwood. These deposits, along with a greatly differing moisture content, create real analytical problems, especially with wood density. Usually, extraction of resins in conifers should be standard before the specific gravity is measured, and for some hardwoods, a hot water extraction should be added. (4) The area sampled must be knot-free or not from the area closely adjacent to knots, and reaction wood should be absent. Cell morphology of reaction wood and the zone near knots differs greatly from that in normal wood. (5) Never sample wood at stump height (unless of course this area is part of the study). Experience has shown that wood near the root collar is very erratic and nonuniform and normally should be avoided by sampling no closer than the 0.5 to 1.0 m (1.5 to 3.3 ft) above the stump.

How well wood from mature trees can be estimated from very young trees, such as seedlings, has been much debated. In a comparison of the wood of young (2 year-old) seedlings with wood of 15 year-old Douglas-fir trees, a genetic correlation of $r = 0.65$ was found (Anonymous 1990). The authors summarized that: "... selection for wood density at the seedling stage is about 50% as effective for improving juvenile wood density as assessments at age 15. But this ratio is low and will probably continue to decrease as the trees get older." Other references are given in Chapter 8.3.

3.3.1 Estimating Whole Tree Specific Gravity Values from a Single Sampling Point

One of the problems of working with wood is to know where to sample the tree. It is best, of course, to take multiple samples at different heights of the stem but this is frequently not possible. Thus, it is necessary to have some suitable sampling point that gives a representative relationship for whole tree values. The normal for all wood properties is to sample at breast height (4.5 ft or 1.3 m aboveground) because it is a handy and a widely recognized height; this is especially true for wood density (Figs. 3.2 and 3.3).

Thirty-three studies on sample location in relation to height were reported by Zobel and van Buijtenen (1989), most of which referred to wood density. The literature on single point sampling for wood density can be summarized as follows:

1. A breast height or similar sample is satisfactory to estimate whole tree values if a reliable regression equation, based on 50 or more trees, is constructed. Correlations between breast height and total tree values are usually high, with an *r* value about 0.9 (Zobel et al. 1972). Working with sugi (*Cryptomeria japonica*) and Todo fir (*Abies sachalinensis*), Kanô (1961) found that the bulk density of whole stems could be correctly estimated from increment cores taken near the base of the tree. This method has even been used to estimate whole tree pulp production from samples taken at 5 and 15 feet from the stump (Wangaard et al. 1967).

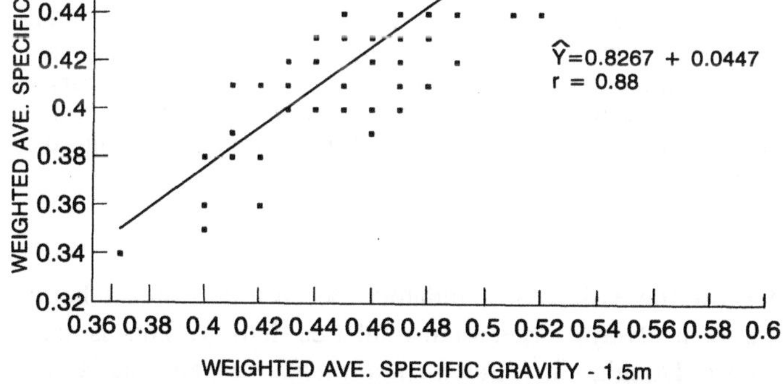

Fig. 3.2. There are always questions as to the simplest way to measure wood properties to ensure that they represent the entire tree. For most characteristics, such as the wood density in the graph, a sample at or near breast height (4.5 ft, 1.3 m) is satisfactory. A graph, as shown, enables calculation of tons of wood/ha from breast height samples. Usually, about 50 trees need to be sampled to develop reliable breast height to total tree regression lines

Fig. 3.3. To convert individual samples to total tree values, it is essential to take samples representing the whole tree, especially in the conifers, where juvenile wood has a large influence. Shown are samples taken every 1.3 m along the stem which is necessary to develop a breast height to total tree regression equation. No wood assessments will be made on the section with the large knots. Approximately 50 trees need to be sampled to obtain a meaningful regression

2. For most conifers, a regression is mandatory because the breast height wood density value is almost always larger than total tree density. This is especially the case for the pines, Douglas-fir (*Pseudotsuga menziesii*), and the larches (*Larix* spp.) because of the higher proportion of mature wood at breast height. A value that would represent the average of the whole tree must be obtained above breast height. However, the breast height value is most useful and standard since, as Paterson (1967b) found for *Pinus patula* (patula pine) in East Africa, the correlation of breast height to total tree values were: basic density $r = 0.94$,

fiber length $r = 0.85$, grain angle, $r = 0.80$. A sample at 10% tree height in Norway spruce (*Picea abies*) gave about 102% of the average dry density of the whole tree, according to Nylinder (1965). For birch (*Betula pubescens* and *B. verrucosa*) it was 100%. We have found for young loblolly pine that wood sampled at the top of the first log 5.1 m (17 feet) gives a good estimate of the weighted wood density of the whole merchantable tree bole.

3. Especially in the diffuse-porous hardwoods, the breast height to total wood density correlation is usually very high. Therefore, a breast height sample is accurate in estimating the wood density of the whole tree. This is possible because of only minor effects of juvenile wood. This relationship was shown for *Eucalyptus urophylla* by Harding et al. (1989), who obtained $r^2 = 0.93$ between disks or cores at breast height and whole tree values. Studying wood properties variation in 11-year-old *Eucalyptus nitens* in South Africa, Purnell (1988) concluded that the best position within the tree to measure wood properties, such as percentage heartwood, density, and moisture content, was between 10 and 30% of the total tree height. In their studies on aspen, Einspahr et al. (1963) found for 30-year-old triploid aspen that: "... specific gravity measurements on breast height 10 mm increment cores provide useful estimates of whole tree gravities. The increment core average fiber length and whole-tree fiber length correlation was significant at the 5% level." Similar findings were reported for hybrid poplar (*Populus* x *euroamericana*) by Beaudoin et al. (1992), where a highly significant correlation was obtained between the weighted wood density of the merchantable stem and the wood density of the increment core taken at breast height.

There are, of course, some exceptions to this generalization, such as that by Rudman et al. (1969), who found that breast height sampling was inadequate in three species of eucalypts (*E. regnans*, *E. calophylla*, *E. deglupta*); samples at 4.5 m (15 ft), were more representative. Yet, most reports on the eucalypts indicate that breast height sampling is adequate.

4. Breast height to total tree values have rarely been applied in ring-porous hardwoods, but those reported show it to be satisfactory.

Interestingly, the potential to estimate tree values from breast height values seems to apply to most forest trees, from the tropics to the boreal forests. As one example, Singh (1986) found that breast-height sampling was useful for wood density for six species growing in the Northwest Territories of Canada.

The generalization that there is a usable breast height to total tree relationship is of great value when making genetic studies of productivity. The key point to remember is that the regression equations are different for each species and often for different provenances within a species. In *Pinus taeda* in South Africa, it was found that 5 mm increment cores taken at 1.2 m compared favorably to destructive sampling with disks at breast height or at 2.4 m intervals along the stem length and resulted in a similar ranking of trees for wood density (MacLennan 1990) (Fig. 3.4). Likewise, Syzmanski (1991) found for *Pinus taeda* that an increment core taken at breast height is a reasonably good estimator of whole-tree specific gravity. Even though the most common method is to take measurements at breast height and relate them to total tree values, other methods

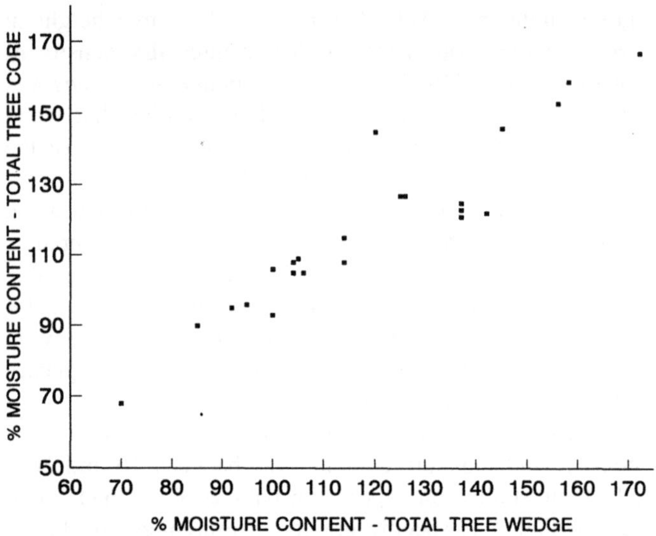

Fig. 3.4. There is controversy as to whether increment cores or larger and more representative wedge samples should be used. Individual observations are more variable for cores compared to wedges. Shown is a comparison of core and wedge moisture content values to indicate the close relationship between the two

are available, such as that used by Nylinder (1965), who used a measurement at 10% of the height to total tree values in Norway spruce. He reported: "... very strong relationships between the average dry density of the whole stems and dry density at 10% of the tree height." In an attempt to find the best sampling height for young Douglas-fir trees, Loo-Dinkins and Gonzales (1991) sampled at 0.4, 0.7, and 1.3 m. They found that the radial patterns varied with height up to six or seven rings from the pith; these differences subsequently decreased with age. The result was that the four or five outermost growth rings at the 0.4 m height gave similar inheritance patterns as the same number of rings at breast height although the heritabilities were lower. The authors suggest that the best sampling height to determine heritability for young trees is 0.7 m, giving reliable results 1 to 2 years earlier than if one has to wait for breast height sampling.

Characteristics other than wood density can be estimated from breast height values although almost all those reported are for wood density (Fig. 3.5). An example is Paterson (1967b), who found the following average correlations between breast height and whole bole values for several wood properties of some East African softwood species:

	Breast height to whole tree
Wood property	Correlation
Basic density	0.94
Latewood percentage	0.99
Fiber length	0.85
Grain angle	0.80

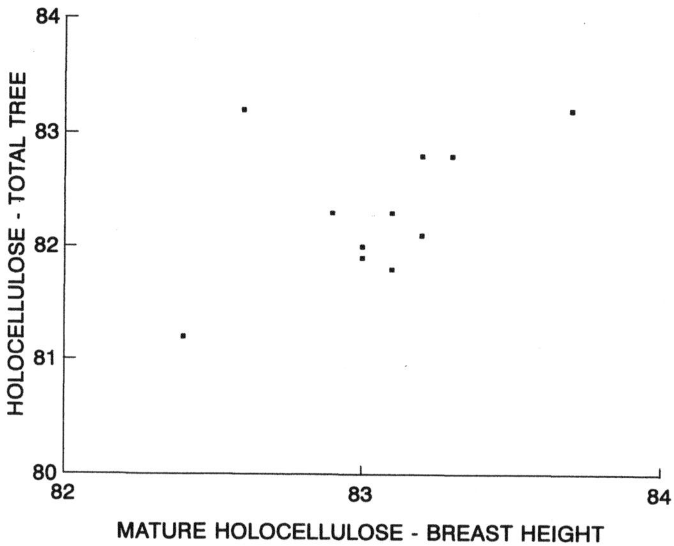

Fig. 3.5. Total tree to breast height value assessments are good for wood properties other than specific gravity. Shown is a graph for holocellulose content

Assessment of moisture content can be very accurately done if care is exercised (Zobel et al. 1968a). Results with three species are shown in Fig. 3.6.

A WARNING must be made here. If the trees are older and heartwood has formed, breast-height sampling for wood density *will not be satisfactory* unless the deposited chemicals are extracted. A very common error is to attempt breast height to total tree correlations of wood density on old trees without extraction. When this is done, correlations disappear. The need for chemical extraction is obvious for all assessments of wood density of old trees for selection or genetic manipulation as well as for estimating whole tree values.

The pattern of wood density from the base to the top of a tree is fairly well set genetically making it possible to assess the breast height to total tree relationship. These patterns, however, can vary by species, and this must be taken into account on studies of whole-tree dry weight production. For example, Okkonen et al. (1972) evaluated wood density changes with tree height in 28 important temperate timber species. In 17 of them, wood density decreased with tree height, in five species, it increased with height, in three species no change was observed, and in three species density at first decreased and then increased with tree height. The change of density with height in most tropical hardwood species is not well known. In some hardwood species, like *Populus* × *euroamericana*, the density is high at the base of the tree, decreases to a minimum at mid-height, and then increases again near the top of the merchantable stem (Beaudoin et al. 1992). This is similar to the pattern to hinoki (*Chamamaecyparis obtusa*) (Hirai 1958) where the wood is heaviest at the top and tree base.

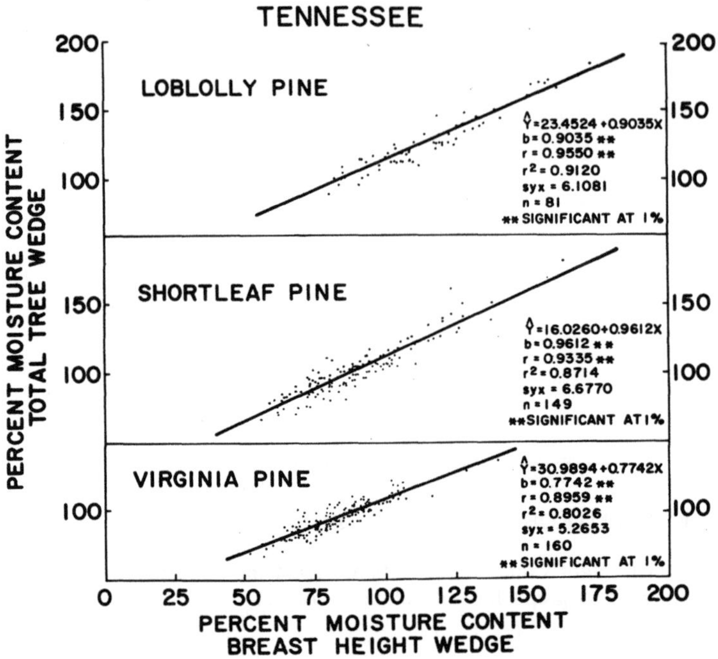

Fig. 3.6. A relationship that was questioned by some was breast height to total tree moisture content. It turned out to be excellent. Results for three species from Tennessee are shown. Note that the slopes of the regression lines are quite similar for all three species

Typically, wood properties in the juvenile zone of conifers are similar at the base and at the top of the tree. However, there can be differences. For example, Zobel et al. (1972) and Megraw (1985) both reported that tracheids of juvenile wood were longer at the top than those of juvenile wood in the lower bole of the same *Pinus taeda* trees. In radiata pine, Kibblewhite and Lloyd (1983) found that butt core-wood tracheids were generally shorter and thinner-walled, but more extensible than those in pulps from the top. They also noted that the low wood basic densities of both the top- and butt-wood samples resulted in their having low handsheet tear and high burst and tensile strength relative to corresponding slabwood (mature wood) pulps.

Not only can sample height in the tree influence wood properties, but the side of the tree where the sample is taken can also be of importance, especially for trees grown at high latitudes. The side of the tree that receives the most light or has prevailing winds blowing against it can be different from the other side. Working with Norway spruce, Nylinder (1953) showed a difference of 2% in basic density on the south side of the trees, while Göhre and Götze (1956) could find no differences in Douglas-fir. In hardwoods, Knigge and Koltzenburg (1965) found fiber length to be shorter on the sunny side of the tree, the same result as reported by Hildebrand (1960) for several species. At low latitudes, in tropical areas, the effect of the side of the bole is less pronounced (Bissett and Dadswell 1949). For trees in subtropical areas, Taras and Wahlgren (1963) state that sampling can be safely done on any side of the tree. As a general rule, it is safe to say that samples can be taken from any cardinal direction on the stem for genetic sampling except on trees growing at higher latitudes. One must always make sure that reaction wood is avoided in trees that are not round.

3.3.1.1 Age of Wood – Juvenile to Mature Wood Correlations

A problem, and danger, is in the definition of wood age. A tree has two major growth centers – the terminal meristem and the cambium. It is the age of the cambium, not of the tree itself, that is of importance in determining the properties of wood. Thus, the fifth ring from tree center in an old tree has fiber morphology similar to the last ring at the base of a 5-year-old tree. Also the wood of the fifth ring at the base of a 30-year-old tree is somewhat similar to that of the fifth ring near the top of the tree. These relationships are clear and there are hundreds of references to this relationship; many were summarized by Zobel and van Buijtenen (1989). The important point is to emphasize that *age of wood* is not the same as *age of the tree*. Age of wood refers to the number of annual rings (age of the cambium) from the tree center. This critical difference must be kept in mind when making studies relating to the genetics of wood.

There is a great need in forest genetics to estimate older wood properties (i.e., wood from older cambia) from the wood produced by young cambia (see also Chap. 2.1.1.2). For example, what is the relationship between wood density of the first ten rings to the wood adjacent to the bark? After a series of tests, it was found that selection for wood density in Douglas-fir at ages 10 to 11 years gave as good results as selection at age 15 years (Anonymous 1990). If selections were made for wood density at age 8, results were 88% as accurate as what would be obtained at age 15 years. It was suggested that such early selection potential could speed up a tree breeding program. However, when very young trees were used, the age 2 to age 15 years correlation was only 0.59. Although this correlation is fairly weak, such early evaluations would be of some value in eliminating those families with extremely undesirable wood density.

For specific gravity, the relationship between the wood at the stem center and outside is sometimes fairly strong. This was reported for *Pinus taeda* by van Buijtenen (1963a), Matziris and Zobel (1973), and Megraw (1985). Also for loblolly pine, Stonecypher and Zobel (1966) found that the specific gravity of older pine trees could be predicted well from the specific gravity of the trees at about 4 years of age. In clones of *Pinus radiata*, Burdon and Harris (1973) reported high correlations between wood density of the first and fifth rings from the pith. As an extreme example, in 23 year-old Norway spruce, Nepveu and Birot (1979) found that adult specific gravity could be predicted even from a single ring at the juvenile stage. A close relationship was found from the tree center to the outside for *Eucalyptus nitens* in South Africa (Purnell 1988). The small change that occurs in specific gravity with some hardwoods is indicated in Table 3.1.

There is also a question as to the relationship of cell length at the tree center with cell length further out, such as the 30th annual ring. Correlation of cell length between rings nearest the pith with the outer rings is often poor until several rings from the pith. We found that in *Pinus taeda* the predictability for the outer rings became good after about the 10th ring. In *Populus deltoides*, Boyce and Kaeiser (1961) reported that the fiber length of the 20th ring from the pith had the highest correlation with all other rings and was the best for comparing fiber lengths among trees. Despite this, the rate of change with ring number was the same for all trees, and they stated that fiber length of a tree can be estimated from the fifth annual ring.

Although only briefly discussed in this section, *understanding and avoiding problems* related to the juvenile and mature wood concept is of PARAMOUNT importance. Especially in the early years, it is conservative to estimate that over 50% of the reported inheritance estimates of wood properties were either wrong or highly skewed because the differing characteristics of juvenile and mature wood were ignored. Such errors cannot be tolerated; their possible magnitude can be seen from Table 3.2, which shows actual values for trees of various ages and indicates the error that would be made if age were not considered.

Table 3.2 shows the amount of variability in specific gravity and moisture content and the necessity of comparing material of only the same age in genetic studies and the effect of age and juvenile wood. The amount of juvenile wood varies considerably from tree to tree (Fig. 3.7).

3.3.1.2 Removing Extractives

One problem unique to wood density is related to content of resin and other extractives in the wood. Resin concentrations occur in the heartwood, in resin streaks, around ring shakes and cracks, and adjacent to large limbs. The presence of resin, as well as other deposits, has a major effect on wood density. Since

Table 3.1. Whole tree and breast height specific gravity for five hardwood species

Species	Breast height	Whole tree
Liquidambar styraciflua	0.47	0.46
Platanus occidentalis	0.47	0.46
Fraxinus pennsylvanica	0.54	0.55
Acer rubrum	0.47	0.46
Quercus nigra	0.59	0.59

Table 3.2. A summary of variation in three wood properties with strong inheritance patterns as affected by age in slash pine. (After Table 3, Howell et al. 1984)

Age	Bark/ton rough green wood		Wood specific gravity	Wood moisture content (percent)	OD Wood/ ton rough green wood	
(*yrs*)	(*lbs*)	(*kg*)	(*unextracted*)	(*green weight*)	(*lbs*)	(*kg*)
10	433	(196)	0.43	57.69	663	(301)
15	348	(158)	0.48	53.05	776	(352)
20	298	(135)	0.50	50.30	846	(384)
25	269	(122)	0.52	48.22	896	(406)
30	252	(114)	0.54	46.45	936	(424)
35	242	(110)	0.55	44.86	969	(439)
40	236	(107)	0.57	43.39	999	(453)
45	233	(106)	0.58	41.98	1,025	(465)

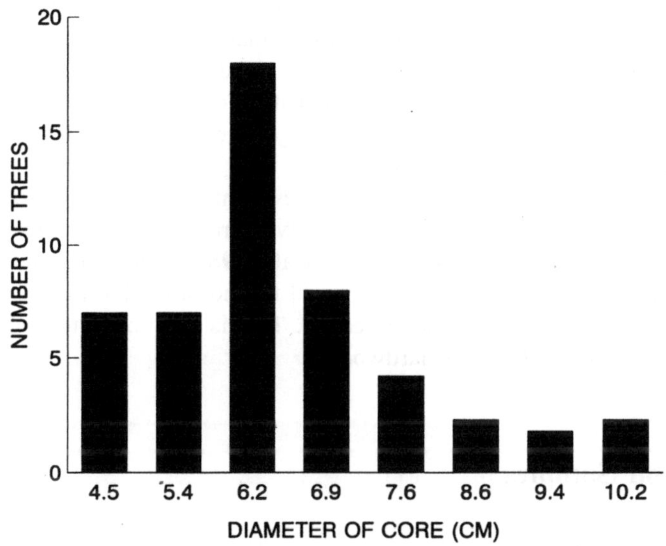

Fig. 3.7. Within a population, the size of the juvenile core varies greatly from tree to tree, indicating some genetic control of this characteristic. Shown are juvenile core sizes for a number of trees of loblolly pine of the same age growing on the same site

the resins and extractives are not an integral part of the cell, their presence is not directly related to true wood density. Sometimes the apparent wood density will be doubled because of the presence of resins. Therefore, to obtain an estimate of the true wood density for genetic studies, the resinous material must be extracted (see TAPPI Standard T204). Where resins are very heavy, such as in the heartwood of a high density hard pine, it is difficult to remove all the resins by any simple extraction process. Because of the effect upon accuracy, all studies of wood densities on pine must be labeled as "extracted" or "non extracted" and results assessed in that light.

It is also very important to note the extraction procedures employed in wood studies. Methods are changing because of concern for worker safety and hazardous material disposal, and direct comparison of extractives content using different extraction methods are difficult. Much of the existing literature suggests alcohol-benzene extractions (see TAPPI Standard T204), but today few laboratories use this extraction procedure because of the hazards posed by benzene. Direct comparison of percent extractions determined when different methods are used is not appropriate.

Trees that are young (up to about 15 years of age for southern pines) have low and similar resin contents from tree to tree, so comparisons of wood density can be obtained without extraction. Absolute values of specific gravity will be incorrect but comparative values will be satisfactory. It is *never* allowable to compare conifer trees of markedly differing ages, or those growing on grossly different sites, without resin extraction. It is most unfortunate that many published data do not indicate whether values refer to extracted or nonextracted wood. Sometimes comparisons result that reflect only the differences in resin content and are not related to the genetics of wood density. To add to the confusion, resin content of the wood of different trees varies genetically, sometimes to a significant degree (see Chap. 8.4.2 and 8.4.3).

Extractives in conifers are usually resins but those in hardwoods are also phenolics (polyphenols). This material cannot be easily extracted and requires methods such as hot water extraction or as Harding et al. (1989) reports for the eucalypts, the use of sodium hydroxide. It is not unusual to find in the hardwoods that some chemical deposits cannot be extracted easily. This is certainly true for silica deposits that occur in some tropical hardwoods.

3.4 Obtaining Wood Samples

Methods to obtain wood samples vary greatly and usually the emphasis is on nondestructive sampling because follow-up measurements are desired. The methods usually give different absolute values but useful comparative values. For example, a number of researchers have shown that increment cores give biased results because of the inclusion of a larger percentage of juvenile wood than actually exists in the tree, and this is particularly serious with younger trees. As an example, the specific gravity of tropical pines in Brazil is lower from increment cores than values obtained from disks (Amaral et al. 1977). This differs from other reports, in which the increment cores had higher densities than disks because of compression of the wood caused by the method of insertion and use of the increment borer (Fig. 3.8).

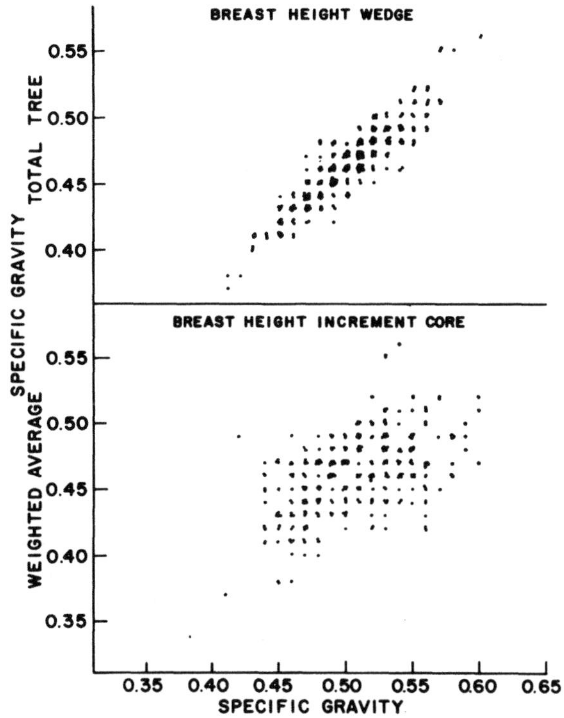

Fig. 3.8. There is considerable controversy as to whether increment cores give the same readings for wood density as are obtained from wedge samples. Actually, the results shown on the graph indicate that increment cores give a good estimate compared to the more accurate wedge when used on pine. Note the greater spread in values when increment cores are used

Surveys done during the 1960s of the existing methods of sampling wood and fiber summarize what was known at that time from a worldwide survey of the literature (Anonymous 1963, 1968a). The surveys indicated that better sampling techniques were required.

Increment cores are the standard method of obtaining wood samples (Kellogg et al. 1982). There are dozens of publications describing the use of increment cores, how to obtain them, where to obtain them and the best equipment to use (see Appendix Table 3.1). These can be summarized by saying that: (1) The increment borer must be sharp and clean at all times. For large studies, the borers can be "mechanized" by using a drill and generator. This results in rapid and easy collection of high quality cores, as illustrated in Fig. 3.1. (2) If tracheid lengths of conifers are desired, larger-sized borers (10 to 12 mm) are necessary. (3) Areas of reaction wood and the wood close to knots must be avoided; cores should be obtained at least 15 cm below a large limb. (4) In our studies with pines, it was determined that two cores, (bark to pith) taken at 90° to each other, are sufficient to categorize a single tree, such as for selection in a seed orchard. In their study on slash (*Pinus elliottii*) and longleaf pines (*Pinus palustris*), Taras and Wahlgren (1963) found that the increase in variation when using two increment cores was only 6%. They concluded that usually one core (bark to pith) was sufficient for a given tree, especially if only population averages are desired. However, enough trees must be sampled with two cores if an estimate of within-tree variation is needed. In studies in Italy, Ghisi (1969) found that the average values obtained were similar whether one to four cores were taken. Core-to-core differences were greatest with four cores because of radial density variation in the tree. For species in British Columbia, Gonzalez and Kellogg (1978) found that use of only the outer half of the increment core (the mature wood) taken at breast height was satisfactory. They recommended the use of two cores per tree.

Many alternative methods have been suggested for obtaining large samples where standard increment cores are not satisfactory (see Appendix Table 3.1). Large samples are usually needed for measurement of grain patterns, for cambial studies, or for chemical and micropulping tests.

Most of the methods developed for the determination of wood density are based upon small volumes of wood. Controversy has developed and a feeling has emerged that small samples will give an overestimate of the density. However, in his study of methods to determine basic density in small wood samples, Ericson (1966) found no differences between wood samples of various sizes.

Working on lodgepole pines (*P. contorta*), Gonzales (1987, 1989) tried different ways of taking samples for wood density determination. She found that two samples, taken 180° apart, gave the same or slightly lower values as when three or four samples were randomly selected at a given height, and that one core is not sufficient for estimating breast height density. Similar methods were developed especially for tropical pines (Amaral et al. 1977).

In an attempt to make early estimates of the wood of small trees in a non-destructive way, a number of geneticists have tried comparing the wood specific gravity of a limb with that of the bole. This method has also been used for tracheid length, (Jackson 1959) and for fibril angle measurements (Jackson 1964). The results from these studies are variable. In our own work with 2-year-old loblolly pine trees, only a weak relationship was found between branch wood properties and those of the bole, and Persson (1972) reported comparable results working with Scots pine (*P. sylvestris*). In contrast, Jackson and Warren (1962), working with 2-year-old slash and loblolly pines, obtained a highly significant correlation between branch and stem wood specific gravity. Similar results were found for maritime pine (*Pinus pinaster*) by Polge and Illy (1968), who reported that because of the relatively high values of heritability for branches (0.58) and the high genetic correlation between mean density of branches and of stem wood, that branches could be used to determine the relative wood quality value of the tree boles of families.

Although the branch-to-bole correlation has usually been found to be only marginally successful, Greene (1966) reported a correlation coefficient of $r = 0.89$ between the two. The tracheids of the first ring in the limb were also about the same as those from the first ring from the pith. Similarly, Hancock (1962) reported highly significant correlations for tracheid length between branchwood and stemwood of the same tree. Yet in *Pinus taeda*, Smith (1962), working with 2- and 3-year-old trees, found that: "The specific gravity of neither limb sections nor limb tips ... can be used to predict ... the specific gravity of increment cores of individual trees eight and nine years of age or boles of trees two or three years of age."

For hardwoods growing in the Southern USA, Taylor (1977a) found branchwood fibers to be shorter than fibers in the stem, with higher specific gravity in some species, and lower in others. It made no difference which part of the branch was sampled. No significant correlation between stemwood and branchwood densities was found for Canadian species (Hancock 1962). In sycamore, (*Platanus occidentalis*), Land et al. (1983) reported that limb-wood specific gravity was higher than that for the stem.

In summary, branch-to-bole assessments to estimate wood density of the bole are not generally used, and such usage should be approached cautiously. Branch wood always contains much reaction wood, and wood properties determined from branch wood can be atypical of the bole.

3.5 Methods of Determining Wood Density

Many methods are used to determine wood density; some of these have been listed in Appendix Table 3.1. Of the many that have been tried, some are useful and practical, while others are of marginal value. In 1965, Phillips (1965b) summarized methods and equipment for determining specific gravity of small wood specimens. He concluded that the most satisfactory measure of basic density is oven dry weight divided by the water-saturated or green volume. This definition was also used by Britt (1967). He stated that for a method to be satisfactory it should be simple, and use only readily available equipment capable of immediate adaptation by the majority of the workers in the field. Before the more sophisticated methods now being used were refined, Phillips (1965b) stated: "It is clear that various radiation and other specialized techniques ... are not ripe for standardization and ... a gravimetric method should be favored." This idea was also expressed by Lannan (1960), who emphasized that the pulp yield per unit volume is directly related to the density of the wood.

The most common concern about the use of increment cores is related to determining the volume of the increment core. Bruckman and Walters (1964) evaluated ten different methods of determining increment core volume and found real differences among them. Part of the problem resulted from compression by the increment borer followed by "springback", which increased the volume. They suggested that the best method is to determine volume by measuring length and average mid-point diameter of the core. Actually, in practice, the water-displacement method has been most used and is still popular today. The need for improvement of methods was emphasized in Anonymous (1968a), just one of many publications to do so.

A frequently used method of assessing wood density is by X-ray densitometry (Heger et al. 1974, Clauson and Wilson 1991). Good differentiation is possible,

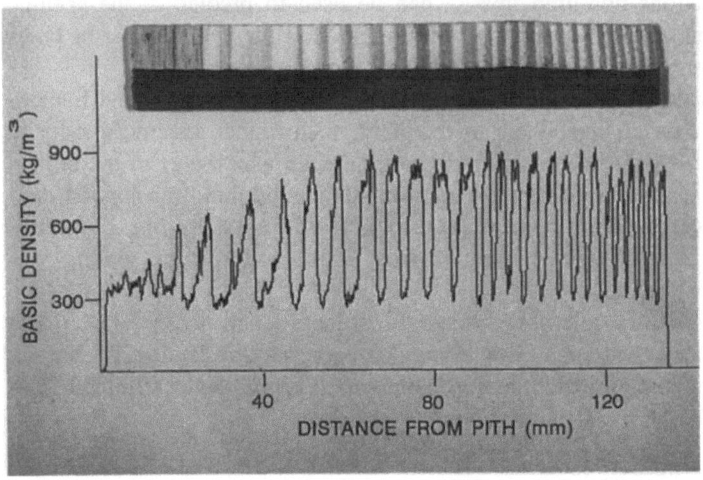

Fig. 3.9. The use of X-ray analysis is very accurate and widely employed. Illustrated is a section from tree center to bark showing wood density in loblolly pine. Juvenile wood is easily observed. (Photo Courtesy Kevin Harding, Graduate Student, NC State University)

as shown in Fig. 3.9. Additionally, the method allows ring-by-ring analyses and provides a clear pattern of radial variation in density and ring width.

3.6 Methods of Determining Other Wood Properties

3.6.1 Spiral Grain

One wood property of primary interest for solid wood products is spiral grain because of its influence on timber strength and stability (Harris 1984). Many workers have investigated spiral grain; there is a strong inheritance pattern (Noskowiak 1963). The study of this characteristic of wood is made very difficult by the complexity of the measurement procedures. This was emphasized by Nicholls (1967e), who stated: "... inclusion of spiral grain as a selection feature in tree improvement programs has brought a degree of urgency to the need for more information.... Between-tree comparisons should be done by comparing variation curves from pith to bark rather than by attempting to represent grain slope in the tree by a single value". Nicholls warned against the error of measuring spiral angle on the outside of large old trees. In contrast, Fielding (1967), also working on *Pinus radiata*, found that measurement of the grain angle on the wood surface provided a convenient and quick estimate of spiral grain.

There have been many suggestions as to how best to measure spiral grain. This book is not intended to describe such methods, but it is necessary to know them if genetic studies are to be made. Therefore we have cited in Appendix Table 3.2, with very brief descriptions, a number of methods that have been used for making spiral grain measurements.

3.6.2 Tracheids and Fibers

Measurement of the length of tracheids and fibers is rather routine and need not be described; there are numerous publications, such as that by Jackson and Greene (1957) that describe such methods. The most important precaution in genetic studies is to assure that only *uncut* cells are measured. Especially in the conifers with their long tracheids, use of a standard small increment borer (4.5–5.0 mm) to take wood samples will result in a large percentage of cut tracheids. Only 10- to 12-mm borers provide an adequate number of uncut tracheids. Although small increment cores (4.5 mm) have been uniformly rejected for tracheid length measurements of conifers, Polge (1967) lists a method of taking 5-mm cores at an angle of 30° to the grain direction and only keeping the central part of the cores. He reports that the percentage of uncut fibers is about two times higher than when 10-mm cores are used in the standard manner, perpendicular to the grain. When using a microscope, an experienced operator can avoid measuring cut cells, but modern electronically controlled cell measurement methods assess all cells including the cut ones, which in turn can create major inaccuracies that will negate accuracy and speed. As was pointed out by Britt (1967), the weighted average of cell length is more meaningful than the arithmetic average; it gives the average value rather than the number of cells in each length category.

Various systems have been developed to view cells and to determine whether they are intact and to measure wall thickness. One system was described by Clark (1961) for obtaining a 25 to 1 image. When measuring cell characteristics, it is often difficult to arrange them so they can be easily seen. A liquid dispersion method was developed by Echols (1961) to help solve this problem. Since longer fibers have a greater probability of being cut in an increment core sample, the measurement of only the uncut fibers from macerated increment cores results in both the means and variances of fiber length being underestimated. To overcome this problem, Hart and Hafley (1967) present a method of measuring wood fiber length that is unbiased. Beta rays have been tried but found to be of little value in assessing cell dimensions (Phillips 1965a).

A successful method for determining cell wall areas is the dot-grid integrating microscope eyepiece, as described by Quirk (1975) and Quirk and Smith (1975). It is used directly on the surface of the wood. In 1974, Scallan and Green described a technique for determining the transverse dimensions of wood fibers. Counts of fiber width are made, using light microscopy, by making counts of the number of cells per unit area on the face of a small block of wood. Frequently, a microscope is used to measure cell dimensions but this method might entail a number of possible errors according to Smith (1967), who lists a number of helpful hints for obtaining more accurate results. A modified Coulter particle counter gives values comparable to microscopic measurements for most studies of cell characteristics (Valley and Morse 1965).

Usually, individual cells are obtained by standard chemical maceration techniques when inheritance patterns are studied. To obtain representative fibers, Wooten and Barefoot (1965) used a mechanical maceration method; it consists of sectioning and thrashing or mechanical agitation. The cells obtained do not have the severe chemical decomposition often found, although some cells are sometimes cut or split by the thrashing. There are always problems in measuring wall thickness or cell width of macerated cells, especially when tangential or radial walls are desired separately. The tracheids of latewood, particularly, tend to lay on their widest face (tangential) so radial wall thicknesses are seen. Methods developed by Kallmes and Bernier (1963) for pulped fibers may be useful in overcoming this problem. Genetic studies have been made related to the distribution and shape of cell wall material. A method to do this, called point sampling, was reported by Ladell (1959). Using this system the diameters and cross sectional areas of cells can be obtained at the same time.

Another nonstandard method of assessing cell elements is with the dimension-calculator by facsimile process (Yazawa and Kurata 1963). The method involves two photomicrographs. The results were similar to those obtained by the dot-counting method. In light of the difficulties involved in these measurements some of the differing results reported in genetic studies of wood may be explained by the vagaries of the techniques that were used.

3.6.3 Moisture Content

Measurements of moisture contents and their relation to dry wood weight are frequently made and the genetic patterns determined (Zobel et al. 1968a). Moisture content is an important wood property; it usually is calculated based upon dry weight of wood by foresters, but sometimes as green weight by paper mills. Exception is taken to this method by Franklin and Squillace (1974), who state that accurate, meaningful calculations of moisture content should always be based upon volume, not weight.

In 1977, Kininmonth and Williams published a paper on how to measure the moisture content of wood. Although this would seem to be simple and straightforward, there are many pitfalls, such as basing moisture percent on weight rather than volume, as Franklin and Squillace (1974) pointed out. Despite the indicated inaccuracy, essentially all published data are based upon dry weight. There is concern about the frequency of sampling needed to determine the moisture content along a tree stem. The value of breast height measurements is doubted, and as pointed out by Comerford and Leaf (1981), measurements of wood moisture (of *Pinus resinosa*) should be taken at several locations along the bole;

those authors recommend the use of five disks to represent the moisture content of a tree.

There are several methods to measure the moisture content. The standard is oven drying, and it usually works well; but for species like eastern red cedar (*Juniperus virginiana*), Smith (1992) reports that oven drying is not satisfactory because of the presence of volatile organic compounds. The toluene extraction distillation method was more suitable, especially for the heartwood.

3.6.4 Pulp Yield

There has been considerable interest whether there is strong enough genetic control of wood properties to increase pulp yield per se. Several methods have been tried, especially to establish if this could be achieved by obtaining nondestructive wood samples from the tree. Care is necessary where the sample are taken because cellulose yields often vary with tree height (Fig. 3.10). As was shown in Fig. 3.5, holocellulose at breast height is weakly related to holocellulose of the whole tree. One suggestion was made by Echols (1959) to use only a pair of 10-mm increment cores. In 1961, van Buijtenen et al. published a method

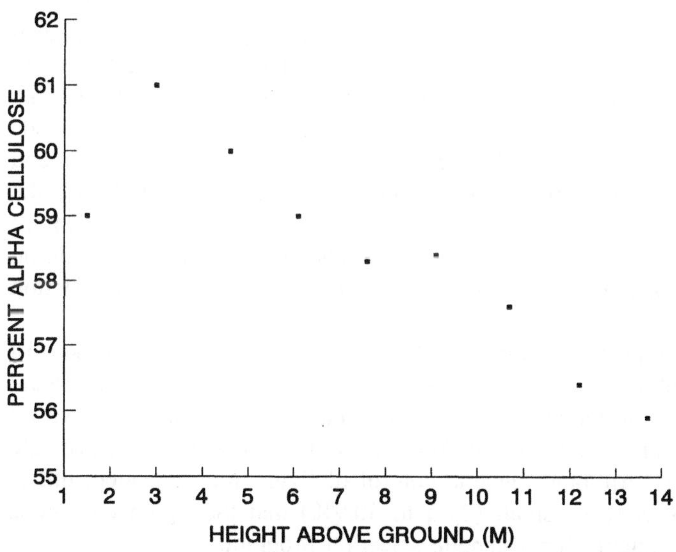

Fig. 3.10. Extreme care is needed when comparing cellulose samples from the same-aged trees, even at the same location. Note the difference in alpha cellulose yields at differing heights in the same tree; this is related to the percent of juvenile wood. Comparisons must always be made with wood from the same height in the tree

using a small-scale kraft pulping procedure with increment cores of loblolly pine. Considerable variation in pulping yields was found among trees, but selection for growth rate or form had no effect on pulp yields.

There has been a great deal of controversy as to the diagnostic worth of micropulping methods to predict operational pulping in the mill. Some mill managers feel that micropulping, which must be used to make genetic studies, is essentially worthless. However, Wright et al. (1989b) and Wright (1991) reported differently, based upon micropulping increment cores, disks at breast height (1.4 m), and disks from the whole tree. They pulped 400 pine and 900 eucalypts and found excellent correlations among them and concluded that micropulping can be used to identify superior phenotypes for cellulose production in a tree breeding program. They found that tests of paper made from pulp from increment cores gave higher values for breaking lengths, burst index, and tensile energy absorption index, but lower values for tear index than paper made from whole tree pulp. Despite this, there is still a strong bias by paper makers against accepting micropulping as representative of operational pulping.

There have been a number of "micro" methods for chemical analyses and pulping used for genetic studies such as the micropulping method used by Zobel et al. (1966). This assessed the water-soluble carbohydrates in an attempt to determine if there are genetic differences in sulfate pulping yields. This early and crude method was refined by a number of researchers, such as Byrd et al. (1965). A method developed by van Buijtenen et al. (1968) was used to test grafts that had been selected for extremes in wood specific gravity. They found that the main effect of specific gravity was on hand sheet apparent density. The cross-sectional dimensions of the pine tracheids also had a strong effect. In a detailed study to determine the efficiency of micropulping compared to macropulping, Clarke (1990) found that the linear correlations between macropulping (1 kg dry wood) and micropulping (100 g of dry wood) were highly significant for all the pulp and papermaking properties except for breaking length in *Eucalyptus grandis*. Clarke (1990) concluded that micropulping is satisfactory for a genetic assessment of pulp and paper characteristics. There are several other methods of micropulping that have been used, some quite good, such as the one by Keays and Hatton (1971).

Some researchers report that lignin analysis provides a better assessment of pulp yield than do cellulose determinations. For example, Marton (1967) found that the lignin content correlated well with kappa number determinations, with a precision of ± 2%. Lignin analysis was found to be relatively easy and accurate (van Buijtenen 1967b). Although genetic control of lignin was indicated, differences between clones were so small (29.5 to 30.9%) that the lignin content is not considered to be suitable for a genetic selection program.

There is great interest in the pulp and paper mills in fiber size and the number of fibers per gram of pulp. Little has been done on this genetically, but it has been suggested as an excellent approach that should be explored. For example, average cell diameter has an important effect on sheet bulk and Canadian Standard Freeness (Ifju and McLain 1982). There are considerable differences among trees of the same species in cell number per unit weight of wood and a method to assess this has been developed by Horn and Coens (1970).

3.7 Indirect Selection for Wood and Pulp Properties

Tree breeders are always hoping to find methods of indirect selection to estimate wood properties. A method that works well is to use mid-parent values to estimate the specific gravity of the progeny of loblolly pine. It is approximately 70% accurate and saves the time-consuming job of measuring the progeny trees. The method is successful because of the high additive variance of wood density (Jett and Talbert 1982). Working with Austrian pine (*Pinus nigra*), Nanson et al. (1975) made a similar observation and stated: "... paradoxically the phenotypic selection of mother trees provides the highest gain and not the selection of the progenies". The potential importance of indirect selection for pulp quality has been promoted by many, but applied by few.

There have been developments to try to estimate biomass produced by assessment of other characteristics. One was suggested by Fries (1986), who related biomass directly to height growth and found it to be more effective than using wood density.

A recent paper (Williams and Neale 1992) describe a case study on loblolly pine in the Atlantic Coastal Plain using maker-aided selection. The method still requires a lot of development and is quite complex, but will be tried in small specialty populations. Although no details will be described here, the most useful way of working with correlated characteristics is the coefficient of genetic prediction, described on p. 276 and Tables 8.5 and 8.6 of Zobel and van Buijtenen (1989). The method was developed by Baradat (1976).

The method of using indirect selection of anatomical characteristics to assess pulping yields and quality was suggested by Burdon (1992). Using both chemical properties and anatomical properties of the wood, Bertolucci et al. (1992) found that wood density could be a good means of indirect selection for pulping qualities because of its high heritabilities and correlations with them. However, Wright and Sluis-Cremer (1992) found that direct testing of the wood by micropulping would be better than indirect selection with tracheid characteristics in pine.

Marker-aided selection currently is a popular method of indirect selection. It is complicated using randomly amplified polymorphic DNAs (called RAPDs). The use of markers to link quantitative loci (QTL) and their application in marker-aided selection is described by Strauss et al. (1992); it will not be described further here. When working with *Eucalyptus grandis*, Grattapaglia (1994) determined that 35% of its wood specific gravity variation could be accounted for by using these new methods. There is great hope that such biotechnological methodologies will contribute to genetic assessment of wood in the future.

Appendix Table 3.1. Some methods used to determine wood density in trees

Author	Date	Comments
Anonymous	1990	The pilodyn has particular promise for indirect selection of mature wood density in Douglas-fir
Bauer	1975	An autopsy saw is good for a study of cambial activity and wood formation
Brazier	1969	X-rays and beta particles were used to produce a continuous record of density change within the ring. The choice of technique depends on a series of considerations listed in the article
Bruckman and Walters	1964	The use of increment cores, especially the measurement of core volume, was described
Cameron et al.	1959	Beta-ray technology for determining wood density was described and explained
Clauson and Wilson	1991	Video scanning and X-ray densitometry were compared for use in commercial grading and inspection of wood density by scanning cross sections of Douglas-fir by both methods. The two produced comparable results but the X-ray method had several advantages
Echols	1969	A power drive for obtaining large increment cores was described
Ellis	1971	Use of X-rays to measure wood density on increment cores was illustrated
Ericson	1959	A description of techniques for determination of increment core density by using mercury immersion was given
Ghisi	1969	All known methods of determining wood density were tried and found satisfactory, with only minor variation in results obtained among them
Green	1965	Studies of wood using photometric techniques with a microphotometer, digitizer, and computer were described
Hall	1988	The pilodyn tester was compared with increment cores for determining specific gravity and found to give different results. Recommendation is to use four penetrations of the pilodyn
Hamilton et al.	1984	A commercial plug cutter was used to obtain cores 2.5 cm in diameter and 15 cm in length from oak trees
Harris	1969	A practical, operational densitometer using beta rays with radiata pine samples in New Zealand was discussed
Harris and Polge	1967	A comparison of X-ray vs. beta ray techniques for measuring wood density was made
Hart and Harding	1990	A system for measuring basic density, spiral grain and fiber dimensions for use with a personal computer was studied

Appendix Table 3.1. (*contd.*)

Author	Date	Comments
Heger et al.	1974	Describes a computerized X-ray densitometer system for measuring wood density with an example of its application
Hodge et al.	1992	The depth of needle penetration of the pilodyn is genetically controlled ($h^2 = 0.46$) and correlated well with mature wood density
Julien et al.	1971	Methods to separate bark from wood and obtaining density are described
Kanô	1961	Increment cores gave values within 5% of the values obtained from disks
Klein	1966	It was found that values from small increment cores (5 mm) gave the same results as stem sectors 2.5 cm thick
Lee and Wahlgren	1979	Measuring only the five outer growth rings was acceptable in an assessment of specific gravity of *Pinus resinosa* compared to a full radial section
Marian and Stumbo	1960	A method based on graphical recording obtained with a stylus tracer-type electronic instrument, on the surface of the wood, was described
Marts	1955	Flourescent microscopy and photomicrography showed wood and fiber structure differences very clearly
Muller-Stoll	1947	The photometric method gives an objective measurement of latewood avoiding the usual problems when visual delineation is used
Nepveu and Velling	1983	In the forest, it is possible to select tree groups for wood density by means of a pilodyn without disturbing the bark
Nicholls	1977	A window is cut in the bark and a special nail is hammered into the tree. The maximum withdrawal force is measured. It is quick, simple, and does not harm the tree
Olesen	1971	Water displacement is a rapid and accurate method of determining volume of green wood specimens, especially irregularly shaped ones
Parker and Jozsa	1973	X-ray scanning produces radiographs for densitrometric analysis for both in motion and stationary techniques
Parker and Kennedy	1973	Both X-rays and beta rays were used to measure intra-increment wood density. The X-ray method was favored for both accuracy and speed
Pawsey	1965	A plug-cutter, using a chain saw hung from a bracket on the tree, produced cores that were very usable
Phillips	1965a	Beta ray techniques were used for wood density but they are not good for cell dimensions; the recommendation was not to use them

Appendix Table 3.1. (*contd.*)

Author	Date	Comments
Polge	1963a	Radiographs were scanned by a microdensitometer giving a record of the variations of optical density along the analyzed sample
Polge	1980	Most quality measurements can be made on 5-mm increment cores. Use of a torsion torque at the time of boring gives the first good information on wood hardness. X-ray densitometry gives many useful density parameters
Polge and Keller	1970	Standing trees were evaluated by a torsionometer which measures the torque required to drive an increment borer in the tree to a given depth. The method was described as quick and easy. The torsionometer is a powerful tool for tree improvement, especially for specific gravity
Prestemon	1965	The mechanized increment borer for hardwoods utilizes a modified plug cutter at the cutterhead and a battery-powered impact wrench as the power unit. The author stated that his tool was superior to others evaluated
Scallan and Green	1974	A count was made of the number of cells/unit area in softwoods; the bulk density was calculated by a formula developed by the authors
Scaramuzzi	1966	No differences were found in basic density when the water displacement and maximum-moisture methods were used. The maximum-moisture content method is recommended for its greater simplicity and speed. Measuring increment core diameters gave differences in volume
Smith	1955	Described methods to determine wood density of small wood sections, including the maximum moisture content method, which is widely used
Smith	1961	A modification and simplification of the TAPPI standard method for wood chips was outlined
Sprague et al.	1983	The utility of the pilodyn in estimating specific gravity for selected loblolly pines was tested. Three types of pilodyns were used. The 18-joule pilodyn was found to be 84% as efficient as direct testing. The instrument was recommended for genetic selection
Stonecypher and Cech	1960	Use of a power drill enabled taking three cores from each of 40 trees in 6 h. It saved half the time required to take the cores by hand
Taras and Wahlgren	1963	One core gave nearly as good a specific gravity determination on slash and longleaf pine as two cores. Using only the outer two-thirds of the core was of little help in estimating wood density
Taylor	1981	The pilodyn wood tester gave values that correlated well with specific gravity

Appendix Table 3.1. (*contd.*)

Author	Date	Comments
von Wedel	1962	Hydrostatic methods proved to be reliable and precise for measuring wood density. Volume of any shaped specimen could be determined by immersion of saturated wood samples in water
Yanchuk and Kiss	1992	The correlation between wood density determined by the maximum moisture content and use of the pilodyn was 0.80. The pilodyn was 67% as efficient as the maximum moisture content method
Zoerb and Remeta	1968	A ratchet drive assembly was described which is light, inexpensive, and can be made from easily available materials. It can be adapted for any size increment borer

Appendix Table 3.2. Some methods that have been used to determine spiral grain

Author	Date	Brief description of method
Birot et al.	1979	Radio-isotopes were injected and auto-radiography was employed to trace the angle of transport
Burley et al.	1967	Successive annual ring increments are gripped in a vise permitting the removal of successive annual ring increments. These are read on the tangential face with a protractor attached to a vise
Foulger	1969	Sonic probes are inserted at specified distances and angles to the grain direction
Harris	1984	A nondestructive method for use in standing trees was outlined
Keller et al.	1974	The tracer method employs a dye injected into the tree which is carried in the direction of the grain and marks the bark, which is then easy to evaluate
Krempl	1970	The methods to assess spiral grain in spruce (*Picea abies*) were described
Northcott	1958	On dry wood, a cross section is split through the pith by a straight-edged tool measuring the angle between the split and pith being assessed
Noskowiak	1968	A method using 12-mm increment cores was described. The standard 4.5-mm core does not work
Pelle	1967	Described a special style to measure spiral grain in *Pinus radiata*
Wheeler	1964	A mirror was aligned with the image of the tree's axis. A chisel edge is pushed into the wood and grain direction is measured relative to the cut line

Chapter 4

The Importance of Wood Density (Specific Gravity) and Its Component Parts

4.1 General Concepts and the Importance of Wood Density

A general definition of *specific gravity* and *wood density* was given in Chapter 1, but it will be of value to repeat a summary of it here. The two terms will be used interchangeably, since they measure the same thing although results are expressed using different units. Wood density and wood specific gravity both indicate the amount of actual wood substance present in a unit volume of wood (see Chap. 1.3.1). Although there are several different methods used to measure wood density, the standard way is to calculate the ratio between the dry weight of wood divided by the green volume of the same wood. This is often referred to as basic density (Tsoumis 1991). Thus, grams/cubic centimeter (specific gravity), pounds/cubic foot (wood density, English units) or kilograms/cubic meter (wood density, metric system) are all related. Because one ml of water equals 1 g, specific gravity is a unitless measurement and is expressed only, for example, as 0.45, not as 0.45 g/ml. The above classifications will be used throughout this book but the reader must observe carefully all data on wood density referred to in publications. Occasionally, one will find specific gravity or wood density expressed as the dry weight of wood divided by dry volume, or green weight or air-dry weight of wood over green volume or, very rarely, green weight of wood over dry volumc. Thc values for wood density or for specific gravity will be quite different when such different methods of measurement are used.

Wood density is an important wood property for both solid wood and fiber products in both conifers and hardwoods (de Guth 1980). It is without doubt the single most important wood property because of its strong relationship to both yield and quality as well as its large variance and high heritability (Zobel 1963, Einspahr et al. 1969, Barefoot et al. 1970, Kanô and Saito 1970, van Buijtenen 1982). It has an effect on yield and quality of both fibrous and solid wood products (Davis 1961, Barefoot et al. 1970). An example was given for North American conifers in general by Mitchell (1963) who reported that a 0.02 change in specific gravity represents a change of about 1000 pounds/in^2 (70 kg/cm^2) in the modulus of rupture of clear wood. Similarly a change of 0.02 in specific gravity results in a change of 100 lbs (45.4 kg) of dry weight per cord or 50 lbs (22.7 kg) of dry processed kraft pulp (1 cord = approximately 2 m^3).

There has been special emphasis on the effect of specific gravity on pulp and paper; three publications summarizing the findings are by Kirk et al. (1972), Bendtsen (1978), and Zobel (1981). Paper strength properties such as tear or burst (mullen) are directly related to wood density. The importance of specific gravity was reported for *Pinus radiata* (radiata pine) as: "Variation in chip basic density accounted for 70 to 80% of the variation in handsheet tear, burst and apparent density" (Kibblewhite and Lloyd 1983). In 1973, Zobel emphasized that specific gravity differences affect yield, quality, and cost from harvesting to the end product in loblolly pine (*P. taeda*). For black spruce (*Picea mariana*), Boyle et al. (1987) report that: "There is a high heritability for wood density, giving it great importance." The importance of wood density was stated in general terms by Burdon and Thulin (1965) for *Pinus radiata* as: "There was general acceptance that increased basic density was the key requirement, especially of the juvenile core." The importance of wood density was strongly emphasized by Gonzalez and Kellogg (1978), who suggested that if one were to select a single characteristic for a future softwood raw-material source for pulp and lumber, it unquestionably would be wood specific gravity. They explain its relationship to pulp yields, paper-making properties, and clear-wood and lumber strength, and summarized that no single wood property has been more thoroughly investigated and found to have a wider application in the evaluation of wood quality than does wood density. In the tropics, knots, wood density, and spiral grain are three of the most important factors that affect wood quality in fast-grown conifers according to Wilcox (1977). He concluded that little could be done by tree breeding to counter the bad affects of knots, but breeding for improved wood density would be effective.

The importance of a knowledge of wood density was also stressed by Nylinder (1965) as: "... the common way of reporting forest resources, growth, and size of trees in terms of volume is not sufficient. It is necessary to convert data from volumetric to weight units." This can only be done by knowing wood density and assessment by weight has now become standard in forestry in some parts of the world.

Because wood density can generally be quite easily altered by genetic manipulation, it is important to have information on the magnitude of product alteration produced by changes in wood density. The subject of the relationship of wood properties to the quality of the final product cannot be covered in detail here; it is extensive enough for a whole book; but enough examples will be discussed in Chapters 4.2.1 and 4.2.2 to give the reader an idea of the extent and value of product changes that result from wood density alterations, whether from genetic or silvicultural manipulation. Two good examples were already outlined relative to latewood and earlywood in Chapter 1.2 (Watson and Dadswell 1962, Gladstone et al. 1970).

As a result of studies of wood from both hardwoods and conifers, it is clear that wood density has by far the most important effect of any wood property on product quality; some of these results are listed in Table 4.1 (Chap. 4.2). In hardwoods, the arrangement of the cells within the wood has a major influence on wood properties, and frequently the wood density per se is not as important in determining product quality as is the *arrangement* of the cells. The distribution and proportion of vessels and rays within the growth ring in hardwoods may be dominant in determining the kind and value of the product produced.

Wood density can vary among provenances and is very variable among trees and within individual trees of a species within provenances (Zobel and van Buijtenen 1989) (Fig. 4.1). This variability greatly complicates studies of the genetics of wood and makes it difficult to draw clearcut conclusions about inheritance patterns. For an example other than pine, Kanô (1957) reports that in Todo fir (*Abies sachalinensis*) the variation in the bulk density of wood within a specimen amounts to over 50% of the total variation.

Breeding for desired wood density is relatively easy when the final product is well defined and the effects of wood density are known. However, for governmental organizations and those producing trees for clients with very different wood needs, breeding for wood density is rarely done. Instead, screening is carried out to eliminate only the poorest end of the range for any given property. This was clearly stated by Harding (1990), who discussed the goals of the breeding program in Queensland, Australia. He stated: "Tree breeding studies have produced heritability estimates for wood properties in *Araucaria cunninghamii* (hoop pine) and *Pinus caribaea* var. *hondurensis* (Caribbean pine), as well as parent trees and hybrids ... Although basic density, spiral grain and percent latewood are capable of genetic manipulation, active breeding to modify species wood properties of different species is not currently adopted due to the diversity of end user needs."

Although the common emphasis is breeding for high density because it is beneficial for strong lumber and plywood, high yields in pulp, and better tear in paper, the fact is that some products, such as tissues, newsprint, and quality writing papers, are best made from low density wood. Such wood yields increased bursting strength, tensile strength, and apparent density of the paper produced. Trees from shorter rotations have lower densities because there is more low density juvenile wood. As a result of shorter rotations along with greater use

Fig. 4.1. Wood density is strongly inherited. Shown are increment cores from high density trees (*above*) and low density trees (*below*). Although these are the same species, the products from them will have greatly differing qualities. (Courtesy Kevin Harding, Graduate Student, NC State University)

of thinnings, the trend in industry is toward selection for higher density wood (Harris et al. 1976). There is a special need to develop high density juvenile wood to increase yields and strength of linerboard and for improved strength of solid wood products (Zobel et al. 1978).

4.1.1 Earlywood and Latewood

It was emphasized earlier that wood density is a complex characteristic, being a combination of proportion of latewood, wall thickness, and cell size (van Buijtenen 1964, van Buijtenen et al. 1968, Zobel and van Buijtenen 1989). Each of these components has an inheritance pattern of its own but in this section we will deal primarily with the ratio of latewood to earlywood, usually referred to as latewood percent. Cell wall thickness is covered in Chapter 4.1.1.3 while cell size is dealt with in Chapter 6. It is necessary to point out again the interchangeable use in the literature of the terms latewood and summerwood and earlywood and springwood. Although latewood and earlywood are the preferred terms, it is common to find summerwood and springwood still used in the literature, especially in the United States.

A generalized statement by Pillow (1954) stresses the close relationship between specific gravity and latewood percent in many conifers; r^2 is usually near or above 0.50 (r = 0.70) for loblolly pine. Sometimes the relationship is quite high, as in Norway spruce (*Picea abies*) clones, where Kennedy (1966) found latewood percent highly correlated to specific gravity (r = 0.90). A similar result was reported for European (*Larix decidua*) and Japanese larch (*L. leptolepis*) by Pearson and Fielding (1961), where latewood was correlated with specific gravity with r = 0.70. Latewood was better correlated with specific gravity than was earlywood. This relationship was also found for Douglas-fir (*Pseudotsuga menziesii*) by Kennedy and Warren (1969) and for *Pinus taeda* by Pillow (1954). The closeness of the relationship varies considerably, but for *P. taeda* the authors of this book find a correlation coefficient of about 0.66 between latewood percent and specific gravity based on thousands of samples. In his analyses, van Buijtenen (1965) reports that specific gravity for most North American conifers is highly correlated with latewood percent and that the latter might be even more strongly inherited than wood specific gravity. In *P. taeda*, Syzmanski (1991) found a good correlation between latewood and specific gravity, while Nicholls et al. (1980) reported only moderate correlations with r equal to 0.44 to 0.67 in *P. radiata*. In *P. elliottii* (slash pine), Squillace et al. (1962) obtained relatively low correlations of 0.30 and 0.37 between specific gravity and latewood percent.

Sometimes the relationship between specific gravity and latewood percent in conifers is not as high as would be expected. The reason is that because of its complexity, specific gravity has several other factors determining its value in addition to latewood percent. This is particularly true in the juvenile wood zone, where latewood is recognized but technically sometimes does not exist based upon Mork's (1928) definition. Also, the methods of measuring latewood percent are often not precise (see Chap. 3.5).

The earlywood and latewood themselves are usually very different, with the latewood having a high density. Frequently, the relative densities of the earlywood and latewood within a tree are strongly correlated (Gladstone et al. 1970); this was also true for lignin, holocellulose, and alpha cellulose fractions. Thus, usually a tree with high density earlywood will also have high density latewood, although occasional *Pinus taeda* trees have relatively high density earlywood and low density latewood, giving the tree more overall wood uniformity.

Because of differences in cell wall thickness, and to some extent shape and cell size, earlywood pulps are very different from those made from latewood. A special study which assessed these two wood types in loblolly pine showed that latewood yielded 2 to 7% more pulp per unit dry weight of wood than the earlywood, on an oven-dry basis (Gladstone et al. 1970). Most of this difference resulted from the greater resistance of latewood cellulose to degradative pulping reactions. In 1962, Watson and Dadswell made somewhat similar studies on *Pinus radiata* and *P. taeda.* They found that paper made from over 50% of latewood (or under 20%) would not produce satisfactory burst and tensile strength. For radiata pine, the maximum amount of latewood normally found in this species is less than 50%, and Watson therefore concluded that latewood percent generally does not have a marked influence on paper properties of this species. In 1966, Harris stated that latewood percent of radiata pine usually is about 20%, even though the density of the 100 trees he studied varied in specific gravity from 0.33 to 0.50. However, in other hard pines and Douglas-fir, it is not unusual to have 50% latewood.

The strength characteristics of earlywood and latewood were described as follows by Kennedy and Warren (1969): "Parallel-to-grain tensile strength and elasticity increase markedly across the earlywood toward the latewood boundary without a significant corresponding increase in specific gravity." In Brazil, Barrichelo and Brito (1979) reported that latewood specific gravity of *Pinus caribaea* var. *hondurensis* was 30 to 100% higher than that of earlywood in the same annual ring; for two larch species, Pearson and Fielding (1961) found latewood to have a specific gravity of 0.70, compared to only 0.30 for earlywood. These differences are illustrated by a photograph of the juncture between latewood and earlywood in pine (Fig. 4.2).

4.1.1.1 The Ratio of Latewood to Earlywood and Its Value

The question is frequently asked: "What is latewood?" There are numerous definitions but the one by Mork (1928) is the most universally accepted. This says that cells are classified as latewood when the double wall thickness is greater than the lumen size. Although this definition is frequently cited, it is often overlooked. For example, the dark brown ring of cells in the outer part of the growth

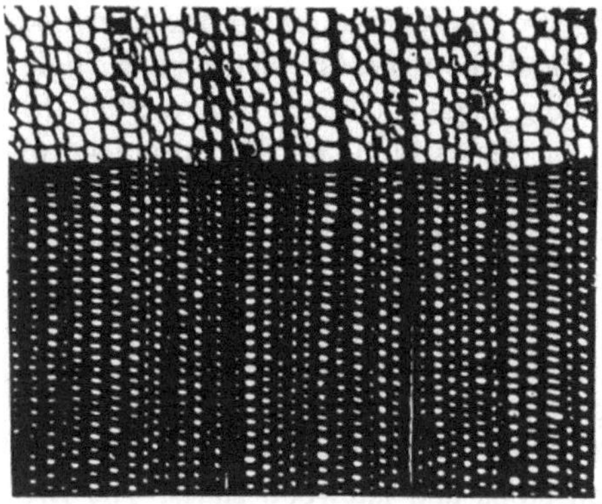

Fig. 4.2. Differences between earlywood and latewood are dramatically illustrated. Shown is the end of one year's growth followed by the start of the next year's growth. These differing woods produce grossly differing types of products. Wood density of both earlywood and latewood is strongly inherited. (After Zobel and van Buijtenen 1989)

ring in juvenile wood of pine is referred to as being latewood even though the cells often do not fit Mork's definition. Also confusing is the fact that some latewood cells barely meet Mork's definition while others have very thick walls, appearing nearly as rods of cell wall substance with a very small lumen. The thicker the cell walls, the darker the appearance of the latewood band. High latewood densities are associated more with thick tracheid walls than with reduced radial diameters, although near the end of a year's growth cell diameter often is quite small.

Numerous studies indicate the importance of the latewood to earlywood ratio. According to van Buijtenen (1964), the percent latewood has by far the largest influence on wood specific gravity. This confirms Hildebrandt's (1960) observation that the specific gravity of the latewood itself plays a role in determining average specific gravity. For *Pinus elliottii*, Foelkel et al. (1975) state that latewood must be considered the most important factor in determining wood quality. In loblolly pine, classification of wood by the latewood to earlywood ratio alone is not sufficient to characterize the wood for pulping (Pew and Knechtges 1989). It is necessary to relate the ratio to specific gravity to obtain reasonable relationships with pulping.

In a study on loblolly pine, van Buijtenen et al. (1968) found that wall thickness in the latewood is especially important. This observation is in agreement with statements by many scientists who point to the huge variability between latewood cells from different trees, indicating that the wall thickness of the latewood is the key to the differences between woods. Others believe, however, that major differences between woods are frequently caused by the variations in earlywood, which also has a strong genetic component. Occasionally, one finds the earlywood of one tree with a density equal to or greater than the latewood of another tree of the same species and age growing under similar conditions.

An experienced wood anatomist can observe the relative size of the latewood band of cells (compared to the earlywood) and the darkness of the band (light brown to nearly black) and, with an amazing degree of accuracy, predict the specific gravity for species with which he or she is familiar. The apearance of the latewood zone has routinely been used as a visual method of grading timber quality or the quality of boards. Certainly, it is essential to grading wood for esthetic uses, and helps to determine the wood grain or pattern in a manner similar to the rays and vessels in the hardwoods.

Latewood and earlywood are difficult to measure accurately since there is a transition zone between them. Different persons can obtain quite different percentages when measuring the same cross section of wood, depending on the individual's opinion as to where earlywood stops and latewood starts. The two types of woods are especially difficult to assess in the low density conifers, the soft pines, and the diffuse-porous hardwoods. To put it simply, latewood percent is not the best measure of wood density to use for these groups because its accurate assessment is difficult. Despite this, it still is commonly employed, and extensive literature has been published on it. Although it cannot be precisely measured, latewood percent does categorize wood into broad groups which can be quite useful.

4.1.1.2 Inheritance in Earlywood and Latewood

The genetics of these two types of wood has been determined for several species. Even though the latewood percent is influenced by many environmental factors, it appears to be under fairly strong genetic control.

The inheritance of latewood percent in slash pine is strong to moderate with $h^2 = 0.26$ (and $H^2 = 0.53$), which is about half of the narrow sense heritability of wood density (Squillace et al. 1962). Also for slash pine, Echols and Dorman (1962) found that the heritabilities of specific gravity were roughly twice those of latewood percent. The broad sense heritability of latewood percent in slash pine was reported by Einspahr et al. (1964) to be $H^2 = 0.53$, the same as found by Squillace et al. (1962). In his 1962 paper, Goggans reported heritabilities for latewood percent in loblolly pine ranging from $h^2 = 0.25$ to 0.92. Reports on inheritance of latewood percent are also variable for *P. radiata*, where Dadswell et al. (1961) and Nicholls et al. (1964, 1980) found h^2 from 0.15 to 0.72. In *Picea abies*, Kennedy (1966) reported a broad sense heritability of $H^2 = 0.85$ for latewood percent. Even though latewood percent responds to differing environments (see Chap. 9), it appears to be under fairly strong genetic control, as indicated by the heritabilities cited above.

Although most work on inheritance has been with latewood, a few studies have also been carried out with earlywood. In his work on Norway spruce, Kennedy (1966) found that clones that had dense latewood also had dense earlywood. As a result, the density of the clones he studied was determined by relatively high densities of both the earlywood and the latewood.

Summarizing for North American conifers van Buijtenen (1962) stated: "It appears that latewood percent is under very strong genetic control and is one of the primary factors controlling the average cross-section tracheid dimensions... ."

The genetic control of the wood density components, which consist of earlywood density, latewood density, and latewood percent, all have relatively strong genetic controls. There is discussion as to whether the overall density would be more accurately determined by inclusion of its components. A study of this was done by Vargas-Hernandez and Adams (1991) for 60 open-pollinated families of Douglas-fir, 15 years of age. Their summary was: "Although wood density components varied significantly among families and were under moderate genetic control ... ($h^2 = 0.24$), none had a higher heritability than overall density ($h^2 = 0.59$). ... Density components have limited value in improving the efficiency of selection for overall density."

4.1.1.3 Wall Thickness

In addition to latewood percent, the actual thickness of the cell walls has a major effect on wood density, which in turn relates to the quality of wood products (Fig. 4.3). This was stated clearly by Dadswell and Watson (1961) as: "... the thickness of the cell wall has an important bearing on most paper properties. ... Basic density ... is indicative of cell wall thickness ... it may be used for

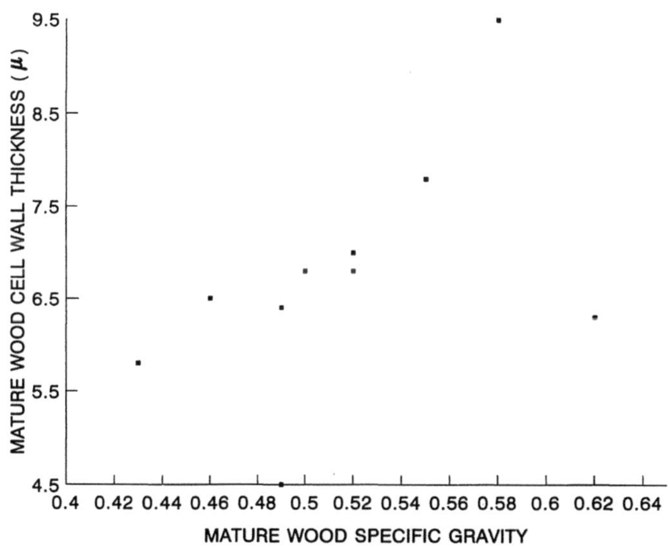

Fig. 4.3. As expected, cell wall thickness of mature wood is directly related to the specific gravity of the mature wood, as shown; yet the correlation is not high because the thickness of the walls of earlywood, cell size, and latewood percent all also have a major effect on specific gravity. Cell wall thickness is only one determinator of specific gravity

assessing the value for papermaking." They emphasized in a number of reports that the main difference in pulpwood quality, especially in the conifers, is caused by the thickness of cell walls that make up the latewood component of the pulp furnish. Within a species, despite the exceptions listed above, the earlywood at a certain tree age usually tends to be quite similar among trees in a stand, with the major differences among them being the thickness of the cell walls of the latewood.

Differences in cell wall thickness are genetically controlled but also vary with the environment. Wall thickness is usually well correlated with earlywood and latewood as well as with juvenile and mature wood (Fig. 4.4). Wall thickness varies greatly by species, provenance, and individual tree (see Fig. 4.5).

4.2 The Effect of Genetic Manipulation of Wood Density on the Final Product – General

It would be very misleading to give the impression that altering wood density through genetic manipulation is simple or easy. First, wood density itself is not a simple characteristic, as has been emphasized previously, since it is composed of a combination of several characteristics, each of which has its own inheritance pattern. However, as briefly explained in Chapter 1.3.1, those working with

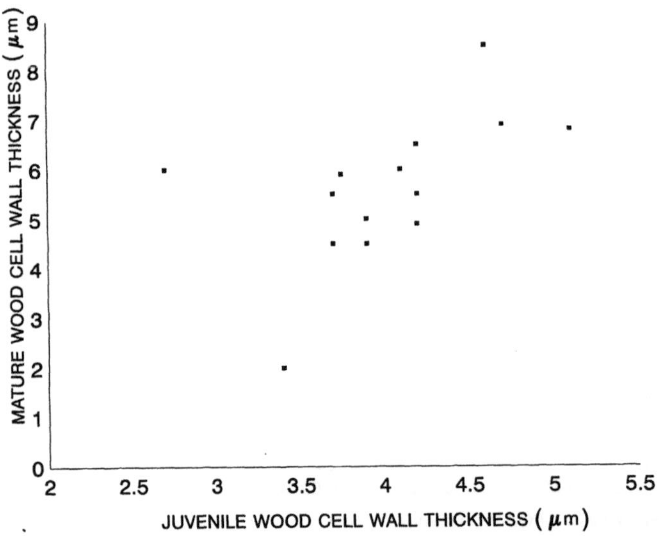

Fig. 4.4. Wall thickness of juvenile wood is somewhat related to that of the mature wood, as shown in the graph. However, the correlation ($r = 0.61$) is lower than would be expected, indicating differences in genetic control for wall thickness at different ages of the cambium

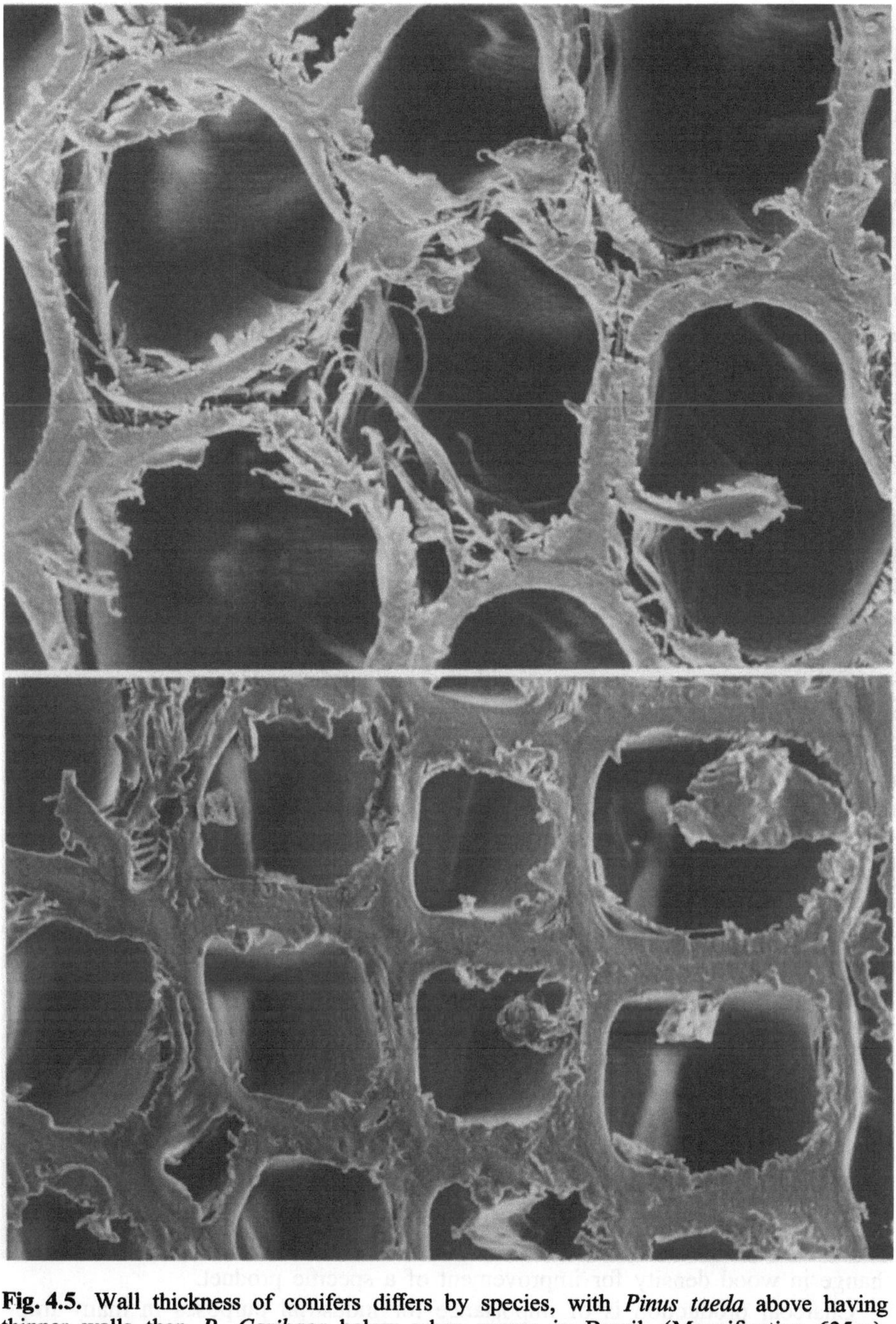

Fig. 4.5. Wall thickness of conifers differs by species, with *Pinus taeda* above having thinner walls than *P. Caribaea* below when grown in Brazil. (Magnification 635×). Similar differences occur among trees within species and have a strong genetic component. (Courtesy of Companhia Florestal, Monte Dourado, Brazil)

the genetics of wood density are exceedingly fortunate that the combination of characteristics of wood density react as a unit genetically, despite their complexity. This enables the efficient manipulation of wood density through genetics to produce more desirable wood.

It is essential to keep in mind the potential genotype × environmental interaction, as well as the direct effect of the environment on wood properties. Wood density has a rather strong additive genetic component and thus is not easily affected by relatively minor environmental differences. It can, however, be greatly changed by large differences in environment (see Chap. 9). Such a reaction frequently occurs when exotic species are moved into areas grossly different from their indigenous environment. One can never assume that a species, or a provenance, will have the same wood in the new environment that it had in the area from which it came. Only tests in the new environment will give the answer. The low density of the wood of *Pinus caribaea* grown in Zululand, South Africa, compared with the high density of *P. elliottii*, also growing in the same environment (described in Chap. 2) is a classical example of this phenomenon. In Zululand the environment does not induce the *P. caribaea* trees to produce thick-walled latewood cells, while *P. elliottii* has a completely different physiological response to the same environment and initiates production of thick cell walls early in the annual growth period. It is important to geneticists to note that even in the environmentally altered wood from such stands, there still is a lot of variability among individual trees within the plantations. Most of this variability results from genetic control of wood density.

There are extensive data, especially with conifers, about the effects of specific gravity on the final product (Fig. 4.6). Wood density also has a major effect upon the costs of transport of wood and the manufacture of a product. The effect of wood density on the final product is not the subject of this book and will therefore only be dealt with briefly. Table 4.1 contains only a few of the hundreds of references available, and more are included in Chapter 13. Those listed were chosen to illustrate the importance of wood density to the final product for various species or for various products or technologies of manufacture. There is a need for a book covering the effects of wood properties on final product quality; it would be a long one because this subject has been well researched, especially for the conifers. The interest here is how wood properties can be altered by genetic manipulation and thus be helpful in altering and improving product qualities.

It becomes obvious from a glance at Table 4.1 that wood density has an effect on pulp and paper production, as well as on lumber quality. Although not listed, density also affects the costs of production and the quality of particleboard and other fiberboard, on the strength of lumber, plywood, and poles, and on the quality of charcoal. The latter will be used to illustrate the value of a genetic change in wood density for improvement of a specific product.

Charcoal production is of importance for industrial purposes in many tropical countries, but especially so in Brazil because of its use in the iron and steel industries. To produce trees efficiently for charcoal, fast growth is essential. However, these trees must have high wood density if high yield and charcoal

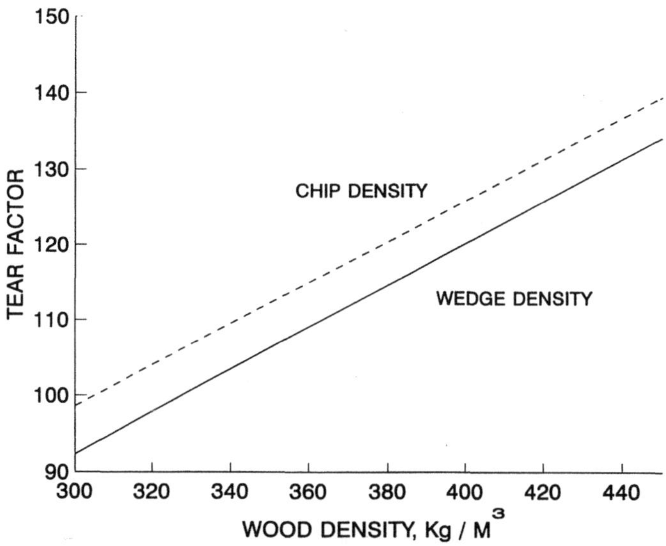

Fig. 4.6. Breeding for different wood properties can have a major effect on the ultimate quality of the final product. The most important is wood density. Shown schematically is the effect of density differences on the tear factor in paper. Close relationships of wood density with product quality are evident with most paper products

Table 4.1. A small sample of publications relating wood density to product quality and yield[a]

Species	Author	Comments
Eucalyptus regnans *E. gigantea*	Nicholls 1967b	No increase in basic density is desirable for these eucalypt species and basic density should be reduced in many other species
General	Bunn 1981	Basic density affects yield, wood strength, tear of paper, and other papermaking properties
General	Wilson and Ifju 1965	Specific gravity is the most important factor for tensile strength
Picea abies	Kennedy 1966	Tensile strength and stiffness varied significantly among clones *after* adjusting for differences in specific gravity. These increased 50 to 60% across the radius
Picea abies	Lewark 1981	Specific gravity is the determining factor in strength properties
Picea abies	Nylinder 1965	Wood density is considered to be the most important wood property because both yield and quality of pulps are closely related to it
Pinus ponderosa	McKimmy and King 1980	Trees with denser wood will produce stronger lumber

Table 4.1. (*contd.*)

Species	Author	Comments
Pinus radiata	Dadswell and Watson 1961	Thick cell walls give bulky, coarse-surfaced sheets; they adversely affect burst, tensile, and folding endurance but improve tear. The lower the basic density, the better are the general papermaking properties
Pinus radiata	Nicholls 1967c	Basic density provides the most important general guide to paper properties
Pinus radiata	Barefoot et al. 1970	Wall thickness (wood density) accounts for most of the variability in paper properties. In general, fiber characteristics associated with wood density were predominant in determining paper properties of the pulping fractions examined
Pinus taeda	Ifju and McLain 1982	Wall thickness greatly influences sheet bulk and Canadian Standard Freeness
Pinus taeda	Lannan 1960	Pulp yield and pulp quality are influenced by wood density. Yield per unit volume is directly related to density
Pinus taeda	Pillow et al. 1941	Bursting strength decreases and resistance to tear increases with a greater amount of latewood. Early-wood tracheids collapse upon pulping, because of their thin walls, and form a dense sheet with a smooth surface while latewood increases bulk, uneven formation, and roughness in the sheet
Pinus taeda	Zobel et al. 1978	For 10-year-old trees a gain of 2.0 lbs/ft^3 (32.0 kg/m^3) was obtained in juvenile wood from selected parents with high density juvenile wood
Southern pine	Mitchell 1963	An increase or decrease in wood specific gravity of 0.02 reflects a change of about 1000 lbs/in^2 (70 kg/cm^2) in the modulus of rupture

[a] The reference shown were chosen out of the hundreds available to illustrate different product qualities to species, location, wood density or to show the extent of information available on the relationships of wood density to product quality. (More are discussed in Chap. 13).

quality are to be obtained. In areas such as Minas Gerais, Brazil, the species that grow rapidly like *Eucalyptus grandis* or *E. urophylla* have, on the average, medium density wood (average specific gravity about 0.48) but with huge variation among individual trees, on the order of 0.35 to 0.75. Certain other eucalypt species, such as *E. cloeziana* or *E. citriodora*, have denser wood with specific gravity over 0.60, which is more suitable for charcoal production, but they grow slowly. Since the heritability for wood density is high, the best alternative is to select for a high-density strain of the fast-growing *E. grandis* or *E. urophylla* and reproduce it vegetatively. This will result in a group of trees that has reasonably fast growth and reasonably high density and is therefore usable for charcoal production. Growth rate has a low heritability, so it is much more difficult to develop fast growth in a high-density and slow-growing species than to develop higher-density wood in an inherently fast-growing, lower-density species. These basic concepts of inheritance of growth and wood density are either not known, or have frequently been ignored, resulting in much wasted effort trying to produce a faster-growing strain of high wood density eucalypt species rather than a higher density fast growing strain from a lower density species.

4.2.1 The Effect of Wood Density on the Final Product for Conifers

Specific gravity emerges as the most important wood property on which to concentrate genetic efforts for improving the final product in conifers because of its high heritability, relative ease of measurement, and strong influence on product quality. In fact, selection to improve wood products has been very successful; a few examples were listed in Table 4.1.

Most wood density studies on the final product of conifers have shown that there is an effect on both fiber yield and paper strength, and on strength and utility of solid wood products. The importance of wood density for conifers has often been considered primarily in relation to its effects on the yield of fibers, while its importance to paper and board quality has sometimes been overlooked. The predominant contribution of wall thickness (density) to paper quality for loblolly pine was stressed by Barefoot et al. (1970), who found that most of the quality of the paper could be related to specific gravity. Similar results were reported for slash pine by Schwartz and Bray as early as 1941 when they said: "The kraft pulps from this high-density wood were comparatively low in bursting, folding, and tensile strengths, but their coarse, bulky fibers possessed high resistance to tear and, in addition, required only short beating schedules to develop their strength."

According to van Buijtenen (1967b), an increased specific gravity and latewood percent results in paper with greater tear strength but in decreased tensile and bursting strengths and decreased apparent density. For young radiata pine, Kibblewhite and Lloyd (1983) found that chip basic density was closely related to pulp handsheet qualities, and its variation accounted for most of the differences in handsheet tear, burst, and apparent density. They also reaffirmed the effect of juvenile wood whose low density results in low handsheet tear and high burst and tensile relative to mature wood from slabs. One main effect of specific gravity is on the apparent density of the paper produced.

A difference in specific gravity of 0.02 on a forest growing 2 cords/ac/year (about 10 m^3/ha/year for young trees) will mean a weight difference of 0.1

ton dry wood or 0.25 ton of green wood produced (Mitchell 1963). Through genetics, it is possible to increase specific gravity by 0.05, which will be of considerable economic value, both from increased yields and more desirable/ fiber characteristics (Zobel et al. 1969). Although the differences cited are based upon young trees, results from breeding for wood density can be illustrated by those shown below by Stonecypher et al. (1964) for loblolly pine.

Gain in	Increased green weight	
Specific gravity	Pounds/cord	kg/m^3
0.008	80	18.2
0.016	160	36.3
0.030	300	68.1

The gain obtained is dependent upon the selection intensity. An additional value relates to the effects of specific gravity on the quality of the products produced; as stated by Lewark (1981) for spruce: "... specific gravity and knottiness are the determining factors for strength". Similarly, Wilson and Ifju (1965) found specific gravity to be the most important factor for tensile strength and its effect on tear of paper.

In Australia, age effects of annual rings of loblolly pine on kraft pulp and paper properties were well correlated with the basic density associated with tracheid wall thickness of the annual ring involved according to Watson and Hodder (1954), who concluded that genetic manipulation of wall thickness is of major economic importance. In another study in Australia, Watson et al. (1952) obtained good correlations between basic density of the wood and most sheet properties of the unbeaten pulp; tear and bulk exhibited positive relationships while burst, fold, and air resistance showed negative relationships to wall thickness. Basic density had a greater influence on paper properties than did tracheid length or microfibrillar angle.

As mentioned earlier, a few of the hard pines have wood that is somewhat similar to that of the soft pines. The most important exotic pine species, *P. radiata*, is a hard pine and its intermediate density wood is considered by many to be ideal. It is more uniform across the annual rings with a less abrupt latewood zone than most hard pines. Some foresters recommend leaving its wood alone, while others prefer a higher wood density in the species. Despite its greater uniformity within the tree, the average tree-to-tree variation in wood density in radiata pine is huge. According to Harris (1966), the latewood specific gravity varies from 0.33 to 0.50 in the dominant trees when grown on a uniform site. A similar variation was not found in the latewood percent, however, because this is approximately 20%. The differences in wood density is a reflection of cell wall thickness of the latewood. In radiata pine, more than most hard pines, differing wood densities are related to relatively small environmental changes. This is clearly illustrated in Chapter 9 on provenance variation.

In his 1981 paper, Bunn summarized the value of wood density with the statement that: "... basic density is probably the single most important intrinsic wood property for most wood products ...". An almost exactly similar comment was made by Nicholls (1967b) in that: "... basic density provides the most important general guide to paper properties ...". Probably the most graphic explanation of the importance of wood density in conifers was given by van Buijtenen (1982) when he stated: "... the three most important wood properties of southern pine are wood specific gravity, wood specific gravity, and wood specific gravity".

4.2.2 The Effect of Wood Density on the Final Product for Hardwoods

The general statement can be made that inheritance of wood density in hardwoods is moderate to strong, but overall not as strong as in the conifers. It is not easy to induce major changes in the wood properties of hardwoods by varying an individual cell characteristic or group of characteristics (Zobel 1965a). It was also pointed out by Lange (1959) that the great complexity of hardwood morphology makes relating cell characteristics to product characteristics a difficult problem. This was illustrated for specific gravity by Dadswell and Wardrop (1960), who stated that the critical knowledge of the arrangement of thick-walled cells in the annual ring of hardwoods was more important than the wood density per se. The importance of the great variability in hardwoods was also pointed out by Schreiner (1935), who indicated that in hardwoods, vessel volume, ray volume, fiber diameter, wall thickness, fiber length, and parenchyma cells all affect specific gravity. Examples are given by Taylor (1965, 1969), who found a direct relationship between specific gravity and vessel volume, as well as between ray characteristics and specific gravity.

Wood density is very important for energy uses of wood, as described by Boulding (1977) and Goldstein (1980). As the production of charcoal from plantations becomes more important, breeding programs have been started to develop high density strains from the faster-growing species.

Genetic variation in specific gravity occurs in some tropical species with relatively unknown woods. A good example is *Gmelina arborea*, for which Akachuku (1984) reported enough variability to justify a selection program in wood density. Nevertheless, this does not always hold. For example, Chudnoff and Geary (1973) found the rather unusual situation where there were huge differences in density among mother trees from natural stands of mahogany (*Swietenia macrophylla*) but very minor differences among their progenies when grown on four different sites in Puerto Rico. The authors concluded that they had found no evidence that wood density is a strongly inherited trait in mahogany.

One frequently hears that a breeding program should aim for higher basic density. While this is frequently true for conifers, it is less true for hardwoods. For a number of final product qualities in hardwoods, a medium or lower density wood is desirable. This difference in objectives is most clear for the eucalypts. For charcoal production, and for some papers, higher density wood is more appropriate. However, a major use of *Eucalyptus* is for tissue paper, where a medium to low density is preferred (Zobel et al. 1983). It was plainly stated by Nicholls (1967b), who said: "... any tree breeding investigation aimed at the production of eucalypt pulpwood should ensure that there is no increase in basic density in species such as *Eucalyptus regnans* and *E. gigantea* and endeavor to reduce basic density in (some) other species". In the eucalypts, it was concluded that the quality of groundwood pulp was correlated with thick-walled cells, in addition to long fibers (Dadswell 1951). For oak (*Quercus* spp.) in Europe, wood density was correlated with mechanical strength and shrinkage of lumber (Polge and Keller 1970). The wood of framire (*Terminalia ivorensis*) has considerable variation in technological qualities (Durand 1983).

Much of the information about the relation of wood to the final product in hardwood relates to specialty, high quality products that are affected by wood characteristics for which it would be difficult to economically justify the use of a genetics program to make changes. There may be two reasons for this. First, the wood characteristic has a genetic foundation but we do not understand the genetics well enough to justify spending a lot of money for improvement. Note, however, that this is not of concern if rooted cuttings or other vegetative propagation methods can be used. A good example in this regard is interlocked grain, a common feature in many hardwoods which may negate their use for some high quality finished products though highly prized in others such as the "stripe" figure in sapele mahogany (*Entandrophragma cylindrion*). This characteristic is under a rather strong genetic control (Webb 1969) but it is rarely used operationally. Second, to apply extensive genetic studies on wood, the market for the improved product must be large enough to justify the effort. Often the speciality products markets are restricted in size, and even more seriously, most manufacturers of specialty products do not grow the wood; they buy it on the open market. A small market not under control of the manufacturer nearly always precludes the use of a genetic breeding program to produce special woods.

We know essentially nothing about the genetics of serious defects in hardwoods such as end splitting or internal growth stresses that deform the wood when the trees are logged or manufactured; these defects may be related to density. The end-splitting problems is being approached genetically (van Wyk 1977, van Wyk and Hodgson 1983) using seedling progeny from special eucalypt seed orchards. Work is also underway using clones of "nonsplitters" from which great gains in lumber quality are possible.

4.2.2.1 Wood Density in the Diffuse-Porous Hardwoods

The specific gravity pattern for the diffuse-porous hardwoods is in many ways quite similar to that of the conifers. The key is the wall thickness of the fibers, and density is less dependent on the rays and vessels than in the ring-porous hardwoods. This was shown in a study of sycamore (*Platanus occidentalis*) by Huber and Bongarten (1981), who stated: "Substantial gain in specific gravity could be made by mass selection in the progeny test accompanied by increases in ray and fiber content and decreases in vessel content. Changes in tissue proportions are probably too small to seriously affect pulping characteristics of the wood". Fiber wall thickness can vary greatly, just as in the tracheids of conifers. Generally, the variation among trees within species is great, as shown by Bhat et al. (1990) for *Eucalyptus grandis*.

The bulk of the studies done on the genetics of specific gravity of diffuse-porous hardwoods has been in the genera *Populus* and *Eucalyptus*, although in recent years some has been done on other species. In the temperate climates, most diffuse-porous hardwoods have a wood of relatively low density and foresters often consider this to be general for this class of hardwoods, but it certainly is not the case in the tropical hardwoods, many of which are diffuse-porous and exceedingly hard. In this latter group, very few genetic studies on wood have been undertaken, and only suppositions can be made (Watson 1965). However, according to Mottet (1965), of all the physical properties in tropical woods, density is by far the most important. It is of interest that often tropical hardwoods are stronger than temperate zone hardwoods of the same basic density; in addition, they show less shrinkage. The exact reason for this is not known, but must be related to cell orientation. Basic density is a much better guide to strength properties of the wood than other cell properties. However, wood uniformity is most valuable and can be obtained when vegetative propagation is used (Fig. 4.7).

4.2.2.2 Wood Density in the Ring-Porous Hardwoods

Relatively few genetic studies have been made of the wood of the ring-porous hardwoods, which occur mostly on drier sites in the more temperate climates. They are relatively rare in the tropics. The wood density patterns are unique and interesting and are strongly influenced by rays and vessels. However, the wood strength patterns are sometimes not correlated well with the density; the arrangement and relative size of the vessels, relative to fiber characteristics, bring about this unusual situation.

For many years the ring-porous hardwoods were preferentially used for certain specialty, solid wood products, but were not particularly in demand for fibers and pulp. This has changed now, both because of the need to use more hardwoods but especially because of advanced technology in pulping and in papermaking. A good example is "picking", which sometimes occurred during printing of paper containing many large vessels. Picking creates a defective paper surface which results in poor ink distribution (Brown and Panshin 1940, Colley 1973). Now there are printing methods that avoid picking altogether so the presence of vessels has less importance. Although there is generally no major effort to grow ring-porous hardwoods in plantations, there are activities in that direction, particularly for certain types of specialty woods. If such wood is to be grown, it must have the best qualities, and genetic improvement programs will be required to achieve this. Hopefully, this will be an economically justifiable and therefore attractive emphasis for the future.

Fig. 4.7. Uniformity of wood density is desired for all kinds of wood. It is relatively easy to obtain with species that can be vegetatively propagated, as illustrated for *Eucalyptus* at Aracruz, Brazil. Every tree within a clone (as shown) has essentially the same wood density. (After Zobel and van Buijtenen 1989)

4.2.2.3 The Effects of Rays and Vessels

The size and distribution of rays and vessels have a major effect on wood properties, as discussed for ring-porous vs. diffuse-porous woods. They have a particularly large influence on specific gravity; as a result, some genetic studies have been made on their presence and distribution patterns, but little has been done concerning their effect on the final product.

4.3 Summary

Wood density is the most important wood property for both yield and quality of fiber products, solid wood products, and energy production. Typically, it is strongly inherited and variable, thus enabling good gains from genetic manipulation.

Most publications emphasize breeding for high density wood. However, this is a need for only some products; others are best made from low density wood.

Often it is necessary to breed for high density to offset the high proportion of low density juvenile wood harvested from short rotations (young plantations) and top wood.

Wood density is a complex characteristic, with one of its main components being thickness of the cell wall, which has a modest inheritance pattern for some species and strong inheritance for others. Wall thickness of the latewood is commonly credited as having the greatest influence on wood properties, but occasionally the thickness of the earlywood cell walls is found to be important. The percentage of earlywood and latewood produced is under rather strong genetic control, although extreme environments can have a major effect. Some reports show a very high correlation between latewood percent and specific gravity in species like Douglas-fir, while it is much lower for radiata pine. It is difficult to accurately assess the percentage of latewood, since measurement techniques are highly subjective.

The genetics of wood density has been widely studied, especially in the pines, the eucalypts, and the poplars because of the strong effect on the final product. Heritabilities are some of the highest and most consistent for any tree characteristic. Improvements in product quality by the use of genetics to alter density have often been documented and are large enough to have considerable economic value. The wood of both the diffuse- and ring-porous hardwood species respond well to genetic manipulation although the latter have been less well studied. The complex cellular structure of the hardwoods make specific gravity less predictive of product quality than in the conifers. In the hardwoods the arrangements of the vessels, fibers, and rays often have a greater effect on product quality than density itself. Especially in the tropics, hardwoods are frequently found that have the same specific gravity but produce greatly differing products. Breeding for or against certain cell types (like vessels) can be quite rewarding.

For all species, the value of wood uniformity has been emphasized. Much of the breeding for wood density is primarily aimed at producing more uniform wood, as desired by the processors and manufacturers of wood products.

Chapter 5

The Genetics of Wood Density

5.1 General

As will be evident in the following sections, the inheritance of wood density is uniformly strong. Most of the genetic variability is of the additive type, enabling good gains from a selection and breeding program as well as from a vegetative propagation program. Because of its major effect on paper, solid wood products, and energy programs, the genetics of wood density has been more studied than that for any other wood property.

Despite its complexity, wood density reacts like a single highly additive characteristic even though its component parts also have independent, and sometimes strong, inheritance patterns. These parts will be covered briefly, but the bulk of all studies and conclusions about the use of genetics will deal with wood density as a single characteristic.

Many papers report on the intensity of inheritance and heritabilities of wood density. These must be assessed carefully because some are merely estimates and many are made on very young trees so their broad applicability is suspect.

Although it was covered in detail in Chapter 2, two paragraphs have been inserted here as a short review as to what heritability means. The discussions in this section will not impart the information that is intended unless the exact interpretation of heritability is understood.

The strength of inheritance is normally expressed as *heritability*; i.e. the ratio of genetic variance to total variance (see Chap. 2.1.1.1). This ratio expresses the degree to which a characteristic is passed from parent to offspring. It may be based upon individual tree or family performance; the values from individual and family assessments can be quite different, with the latter always being the larger. Heritability can be expressed as *narrow sense heritability*, (h^2), where the additive variance is divided by the total variance, or *broad sense heritability* (H^2), where the additive plus the nonadditive variances are divided by the total variance. Broad sense heritability is used when vegetative propagation is measured or for certain control pollinated crossing programs. Thus, *when heritability values are presented for wood density, it is absolutely essential* for one to know *what type of heritability* is referred to for a proper understanding. In a detailed study on controlled crosses of Douglas-fir (*Pseudotsuga menziesii*), King et al. (1988) summarized as follows: "Analyses ... indicated additive genetic variance as the only important and significant genetic source of variation after 12 growing seasons." Most studies support this finding that wood density has a high proportion of additive variance.

Since heritabilities are only estimates and are influenced by environmental factors, small value differences in the heritability of wood density are not meaningful. Heritabilities change with the environment; the greater the effect of the environment, the smaller will be the heritability value. Because of the many environmental factors that affect heritability, there is always a question as to what extent heritability values determined for one area represent those from another area. Sometimes they are quite different, sometimes similar. The latter was reported for two different locations in Fiji and Queensland, Australia, where open-pollinated tests of the same material yielded approximately the same heritability values (Burley et al. 1983).

Age can greatly influence the heritability of wood density, and species response to age is very variable. The effect of age was clearly shown by Zobel (1964), who stated: "Inheritance patterns may change with age, and one may obtain erroneous results from estimates based on juvenile performance"; and by Vargas-Hernandez and Adams (1992), who summarized: "... the degree of genetic control for wood density in Douglas-fir (*Pseudotsuga menziesii*) changes with age." They emphasize the importance of the fact that there is a change in heritability with early selection compared to the usual assumption that heritability remains constant with age (Squillace and Gansel 1974, Lambeth 1980, Gonzales and Richards 1988).

Both the narrow (h^2) and broad (H^2) sense heritabilities of wood properties change with the age of the annual rings, but the pattern varies and often differs for different wood characteristics compared to wood density. For example, in radiata pine (*P. radiata*), broad sense heritability for tracheid length was low ($H^2 = 0.20$) at tree center and increased outward to $H^2 = 0.50$ at ring 9 and then decreased to $H^2 = 0.20$ at the bark. Spiral grain had $H^2 = 0.55$ at tree center and only $H^2 = 0.08$ at the 9th ring according to Nicholls (1967c). He reported that heritability for wood density was 0.60 near tree center, decreased to 0.24 at the 9th growth ring and then increased again to 0.60 at the bark. In contrast, the narrow sense heritabilities for specific gravity in the southern pines increased rapidly up to 15 years, from about $h^2 = 0.25$ to $h^2 = 0.50$ or above (Zobel 1964). The same trend was reported for Douglas-fir by Vargas-Hernandez and Adams (1992), who found a steady increase in heritability estimates, especially between the 7th and 12th rings. A pattern of increasing density with age was found for Douglas-fir by McKimmy (1966) for the first 15 rings from the pith.

One variable in wood density rarely considered genetically is packing density, i.e., the actual density of the cell wall. The ratio of genetic variance to total variance of packing density in loblolly pine (*P. taeda*) ranged from near zero for earlywood to 0.62 for latewood (Boyd 1968). Earlywood packing density differences are reported as dependent upon the environment whereas that of latewood has strong genetic control. Clonal differences were due to specific gravity but not to cell wall material per se. The packing density of the cell wall affects wood density, according to Jayme and Krause (1963). The apparent cell-wall packing densities of wood and pulp fibers were assessed by Yiannos (1964).

Data on inheritance of wood are so abundant that they have been summarized in the form of several tables and the few that have been discussed in the text are to illustrate some of the generalizations made. It is important to remember that selection for wood density is rarely the primary objective of a tree improvement program. As pointed out by Nanson et al. (1975), who worked with Austrian pine (*Pinus nigra*), genetic gains in wood density were smaller because of the reduced variability resulting from previously imposed selection for other properties of growth and form. Yet the heritabilities of wood density are high enough and the individual tree variabilities great enough, that useful gains can be obtained when wood density is included in a breeding program.

The heritabilities of wood density are usually high, some of the highest for all properties of forest trees; they are generally higher than those for form or growth traits (Harding 1990). This indicates that genetic manipulation of wood density can result in good gains (Burley et al. 1983). Best of all, it enables the manipulation of wood *after* intensive selection for tree growth and tree form because density has a large variance along with its high heritability. It also means that gains for wood density from a selection program can be good and that gains from use of vegetative propagation methodologies will be excellent.

As mentioned earlier, despite the complexity of wood density, it responds genetically as a simple characteristic; it is, in fact, one of the easiest of wood properties to work with genetically. The strength of inheritance of wood density in loblolly pine was obtained for a 6-year-old progeny test by van Buijtenen (1963a), where he concluded: "The data showed a strong inheritance of wood specific gravity... ." However, Nanson et al. (1975) take a different and more pessimistic approach when they state that it seems unprofitable to select for wood density in Austrian pine but that it is essential to assess the effects on density from selections made for other characteristics.

Inheritance studies in percent latewood, which sometimes are equated with density determination, are rare. Those available were discussed in Chapter 4.1.1.1. The studies that have been made show relatively high heritabilities, nearly as large or even larger than for wood density (van Buijtenen 1962). The narrow sense heritability for latewood percent for loblolly pine varied from 0.25 to 0.92, which is a somewhat similar range in values to that of specific gravity (Goggans 1962).

5.2 The Genetic Control of Wood Density in the Conifers

A whole chapter could be written on the controls that affect the physiology of the tree which in turn result in the production of woods of differing density; some of these were mentioned in Chapters 1 and 2. These are not the subject of this book, but the simplistic statement can be made that: "... anything that causes a change in growth pattern of a tree can cause differing wood densities to be formed" (Zobel and van Buijtenen 1989). Date of bud break is a physiological characteristic affecting wood density that has been studied in considerable depth.

In general, the early flushing families of Norway spruce (*Picea abies*) tend to have lower wood density (Mergen et al. 1964, Birot and Nepveu 1979). However, an opposite result was reported for Douglas-fir, (McKimmy 1959, Kennedy 1970), where the early flushing families had higher wood density.

Of the many papers that could be cited for genetic control of wood density in the conifers, a few are listed in Table 5.1. Most reports are for the pines, but other genera, like *Pseudotsuga* and *Picea*, have also been extensively studied.

Table 5.1. The variability and/or inheritance of wood density (specific gravity) in several conifer genera – some general examples

Genera and species	Reference	Comments
Genus Abies		
A. alba	Nepveu 1976	Genetic correlations are high for juvenile-mature wood density
Genus Araucaria		
A. cunninghamia	Eisemann et al. 1990	Heritability estimates for basic wood density on 15 year-old trees was very high
A. cunninghamia	Harding and Woolaston 1991	Heritability of specific gravity was found to be high for 15-year-old trees
Genus Cryptomeria		
C. japonica	Kurinobu et al. 1990	Specific gravity is strongly inherited
Genus Larix		
L. leptolepsis	Isebrands and Hunt 1975	Improvement of the wood of Japanese larch through a selection program is very feasible; this includes uniformity of wood density within an annual ring
Larix spp.	Lewark 1980	Although the heritability values were rather high for wood density, the actual differences were too small for obtaining meaningful gains from genetic manipulation
Genus Picea		
P. abies	Ericson 1960a	There is a large variability in wood density among clones and a good correlation between ortet and ramet specific gravities
P. abies	Parrot 1960	Wood density is variable among provenances; it was uniform among seven progenies
P. abies	Kennedy 1966	There were great differences among clones in specific gravity, caused by both latewood and earlywood variations. The range of specific gravity differences within clones was dwarfed by the differences among clones

Table 5.1. (*contd.*)

Genera and species	Reference	Comments
P. abies	Rone 1970	Only a low heritability was found for wood density
P. abies	Lewark 1982b	Strong inheritance was indicated for specific gravity in a young clonal test
P. mariana	Khalil 1985	Wall thickness had only a small amount of genetic control compared to other wood characteristics
P. mariana	Boyle et al. 1987	There was a high heritability for specific gravity
P. sitchensis	Brazier 1967	There was a good potential for genetic improvement based upon the great variation among trees
Genus Pinus		
P. caribaea	Brown 1971	For trees grown in Jamaica, significant differences in wood density were found among trees. There was potential for selection of trees for high density offspring
P. caribaea var. *hondurensis*	Harris 1977	There was a huge difference in wood density among individual trees within an area as well as among differing areas
P. caribaea	Plumptre 1977	The variation in wood density between sites and among trees within sites was large
P. caribaea var. *hondurensis*	Burley et al. 1983	The ranges of family means (13 to 25%) and individual tree variances indicated potential gain from selective breeding
P. caribaea var. *hondurensis*	Kanowski 1986	A high heritability was obtained for wood density in the species
P. densiflora	Iwakawa et al. 1967	Parent progeny correlations for specific gravity were significant
P. densiflora	Sudo 1983	In akamatsu pine pulpwood programs, selection was made for wood density and latewood percent
P. echinata	Wiselogel and Tauer 1982	Heritability estimates were high enough to make it possible to obtain substantial genetic gain in wood density
P. elliottii	Echols and Dorman 1962	Modest gains in specific gravity are possible through selection and breeding
P. elliottii *P. taeda*	Jackson and Warren 1962	The specific gravities of the progeny were significantly correlated with those of the parents

Table 5.1. (*contd.*)

Genera and species	Reference	Comments
P. elliottii	Zobel et al. 1962a	Inheritance of specific gravity from parent to offspring was good; it was weak from donor tree to graft
P. elliottii	Sohn and Goddard 1975	Inheritance of specific gravity was high
P. elliottii	Allen 1985	Wood density has good potential for a breeding program as shown by its high heritability
P. elliottii var. *densa*	White and Saucier 1965	Specific gravities among individual trees from the same plot were statistically highly different
P. kesiya	Geary 1967	Family differences in wood density were relatively large. The family with the heaviest wood was 13% denser than the family with the lightest wood
P. kesiya	Guldager 1972	Density investigations should have a high priority in genetic programs
P. kesiya	Ladrach 1986	A tree improvement program will give good gains in wood density and wood uniformity
P. nigra	Iwakawa et al. 1967	There were good parent to progeny correlations in specific gravity
P. nigra	Nanson et al. 1975	Specific gravity was under rather strict genetic control
P. oocarpa	Lima 1987	Specific gravity was under strong genetic control compared to height, diameter and straightness
P. patula	Kromhout and Toon 1978	Large selection differentials makes a good breeding program possible
P. pinaster	Nicholls 1967a	Basic density has a reasonably strong heritability but it can be drastically modified by the environment
P. pinaster	Polge and Illy 1968	A high specific gravity inheritance was shown by young progeny
P. pinaster	Keller 1973	Parent to progeny correlations were strong for specific gravity and heritability is high
P. pinaster	Nepveu 1976	The juvenile-mature genetic correlations for wood density were high
P. ponderosa	McKimmy and King 1980	Genetic selection of families with denser wood will produce increased strength of wood
P. radiata	Dadswell et al. 1961	Percentage latewood is under a considerable degree of genetic control

Table 5.1. (*contd.*)

Genera and species	Reference	Comments
P. radiata	Harris 1965a	Wood density, despite its complexity, shows strong genetic control and values may increase with age of wood so h^2 approaches H^2
P. radiata	Nicholls et al. 1964	Narrow sense heritability of wood density was relatively low compared to other studies
P. radiata	Nicholls 1967b	Inheritance of specific gravity was strong but it changes with age and ring number from the pith
P. radiata	Burdon 1970	Clonal repeatability was high, about 0.7, showing genetic constancy within and among clones.
P. radiata	Nicholls et al. 1976	Ramets and ortets had similar specific gravity
P. radiata	Nicholls et al. 1980	Good gains are possible by selecting for juvenile wood density
P. radiata	Bannister and Vine 1981	There were large differences in wood density among families
P. radiata	Johnson and Colley 1991	Juvenile basic density had high individual heritability
P. radiata	Burdon et al. 1992b	Eleven trees were sampled in each of 30 open-pollinated progenies in New Zealand. Results showed that wood density could be readily manipulated by selective breeding
P. sylvestris	Ericson 1960a	The wood density of the parents can be used for improvement in genetic programs
P. sylvestris	Ericson 1960b	There was a strong, positive correlation between the basic density of the original trees and their grafts grown in different environments. Basic density depends on genotypic differences among the trees
P. sylvestris	Ericson 1961	The variance in wood density among clones was due to genetic differences
P. sylvestris	Persson 1972	Inheritance of wood density, based on numerous families, was strong
P. sylvestris	Velling 1974	Results of crossing tests gave good support to the idea that the density of wood is a property under strong genetic control
P. taeda	Brown and Klein 1961	Wood specific gravity is under genetic control in young trees

Table 5.1. (*contd.*)

Genera and species	Reference	Comments
P. taeda	Goggans 1962	Percentage of latewood and specific gravity, along with tracheid length, are the easiest wood properties to use in a tree improvement program
P. taeda	van Buijtenen 1963a	Wood density differences among 5-year-old grafted clones was strong; among 6-year-old open-pollinated progenies, moderately strong
P. taeda	Stonecypher et al. 1964	Inheritance of specific gravity was strong for seedlings
P. taeda	van Buijtenen 1965	Wood specific gravity and percent latewood are strongly inherited
P. taeda	Boyd 1968	There were large variations among clones in specific gravity
P. taeda	Shelbourne et al. 1969	Specific gravity had a very high additive genetic variance
P. taeda	Blair et al. 1975	Use of trees with high density juvenile wood will increase the overall density of trees in the plantation
P. taeda	Zobel et al. 1978	Wood density of juvenile wood can be significantly improved using genetics
P. taeda	Pearson et al. 1980	Specific gravity of juvenile wood can be changed by breeding trees with high specific gravity
P. taeda	Kellison 1981	Progeny of parent trees selected for high juvenile wood density yielded 32 kg/m^3 higher density than progeny from unselected trees
P. taeda	Jett and Talbert 1982	Wood density should be included in advanced tree improvement programs. One generation of selection increased wood density by 0.75 lbs/ft^3 in 12-year-old trees
P. taeda	McKinley et al. 1982	Estimates of genetic gain for specific gravity ranged from 0.99 to 2.99%, after adjustment for growth differences
P. taeda	Talbert et al. 1982b	Wood specific gravity inheritance of 20-year-old progenies and parents was strong. Good gain can be made by selection
P. taeda	Zobel 1984	Genetic improvement in juvenile wood will partially offset poorer quality wood in young trees

for species in the US like spruce pine (*P. glabra*). Separate sections on wood genetics of the hard and on the soft pines have been developed, with *P. radiata* included with the hard pines.

To assess the worth of genetic changes in wood, it is necessary to know the value of the wood properties per se. The bulk of the operational research on wood properties of the hard pines has been done on their use for fiber products. However, in the last few years this has been much expanded to include also solid wood products. Over many years, numerous basic studies on the strength properties of wood were made, but often these were not translated into economic effects on boards, plywood, or other finished products. One can obtain implications from these test results, but it is not always possible to estimate monetary values of genetic changes on the final products from them.

There are major problems related to wood density determination in the pines (see Chap. 3.5). If these are not recognized and corrected, results obtained that are supposedly genetic can be incorrect and lead to erroneous estimates of gains and false conclusions. For example, if accuracy is to be obtained in genetic estimates of the wood density of pines, *all wood samples should have their resin extracted* before weighing. Areas of wood near branches, wounds, ring shakes, compression wood, and areas of resin concentration should be avoided when obtaining samples for wood density assessments. In older trees that contain heartwood, a thorough extraction of resinous materials is mandatory because production of heartwood varies greatly in individual trees. In young trees, the absolute values for specific gravity are changed by extraction, but the rankings are about the same. Perhaps the most common error, resulting in contradictory results and poor genetic assessments of wood density, is the failure to take ring age or heartwood resin deposition in older trees into account.

Most of the genetics work on wood properties has been done on the hard pines and Douglas-fir, especially with loblolly pine and slash pine along with radiata pine. More recently, studies have been established with the tropical pines, especially *P. caribaea*, but results are not generally available except those related to provenance (see Chap. 9). Results on the inheritance of wood density in the hard pines have been dramatic indeed and are having a major influence on operational forestry programs and even on mill establishment locations.

A short but simple table (Table 5.2) shows the results obtained in the progeny when loblolly pine parent trees of known overall specific gravity are crossed. The differences when crossing high density with high density, low with low, and high with low are rather dramatic. Although these data are only for 2-year-old progeny, similar results are found at later ages (Talbert et al. 1982a).

Many pages could be filled listing all of the studies done on the genetics of wood density in the hard pines. Some of these are summarized in Tables 5.3 and 5.4. Some studies are quite complete and complex – a recent one on slash pine by Hodge et al. (1992) reported in the 34th Annual Progress Report of the Florida Cooperative Forest Genetics Research Program illustrates this. It included genetic analyses of the following:

Table 5.2. Inheritance of wood specific gravity in two-year old *Pinus taeda* seedlings[a]

Type of cross (parental sp. gr.)	Progeny mean of seedlings	Number of seedlings
High (0.65) × high (0.58)	0.389	150
High (0.65) × low (0.47)	0.369	100
Low (0.49) × low (0.47)	0.350	150
High (0.65) × wind	0.385	95
Low (0.47) × wind	0.356	150

[a] Adapted from Brown and Klein 1961. Based upon crosses of specific loblolly pine parents of known wood specific gravity, (in parenthesis).

1. Change in inheritance of wood density over time.
2. Inheritance of wood density of the whole tree, earlywood and latewood.
3. Differences in wood density related to progeny test locations.
4. Assessment of wood uniformity.
5. The transition age from juvenile to mature wood.
6. Growth rate related to density.

Some of their heritability results are shown below for 13-year-old trees.

Characteristic	Specific gravity	Narrow sense heritability
Earlywood	0.32	0.13
Latewood	0.74	0.22
Mature wood	0.55	0.33
Juvenile wood	0.45	0.15
Whole core	0.48	0.16

In comparison with most of the values shown in Table 5.3, the above heritabilities are relatively low; but the heritabilities appear to be increasing as the trees become older or in rings further from the pith to about the 9th ring. The same pattern of increase was reported by Nicholls (1966).

Another detailed inheritance study of hard pines was done by Nicholls et al. (1980) on juvenile *Pinus radiata*. They determined both the additive and nonadditive variances for the fourth annual ring of the same crosses growing in two places. The heritabilities were determined for ring width, latewood ratio, maximum density, minimum density, and average density. Heritabilities were quite high, especially for the maximum density ($h^2 = 0.66$). They found the nonadditive genetic variance to be negligible.

The inheritance of wood density in the hard pines, clearly the most important wood characteristic, has been intensively studied genetically. It is obvious (see Tables 5.3 and 5.4) that the heritabilities reported vary greatly, even within species and sometimes with different studies done by the same researchers. Age differences are largely responsible for this, as well as the conditions under which

Table 5.3. Narrow sense heritability values (h^2) for wood density in the hard pines

Species	Age (years)	Heritability	Kind[b]	Reference
P. banksiana	4	0.72	Indiv.	Okwuagwu and Guries 1980
P. banksiana	10–20	0.55–0.73	Family	Villeneuve et al. 1987
Pinus caribaea	7	0.71 (mean sp. gr.)	Indiv.	Burley et al. 1983
		0.50 (min. sp. gr.)		
		0.50 (max. sp. gr.)		
P. caribaea var. *hondurensis*	11	0.62	Indiv.	Harding et al. 1991
P. contorta	9	0.33 (provenance #1)	Indiv.	Fries 1986
		0.34 (provenance #2)		
		0.76 (provenance #3)		
P. elliottii	14	0.21 (open-pollinated)	Indiv.	Squillace et al. 1962
		0.56 (control-pollinated)		
P. elliottii	6	0.64	Indiv.	van Buijtenen 1964
P. elliottii	6	0.43	Indiv.	Goddard and Cole 1966
P. elliottii	9	0.47	Indiv.	Sohn and Goddard 1975
P. elliottii	14	0.69	Indiv.	Allen 1985
P. elliottii	13	0.48	Indiv.	Hodge et al. 1992
P. nigra	10	0.67	Indiv.	Arbez and Millier 1972
P. nigra	22	0.60–0.70	Indiv.	Nanson et al. 1975
P. oocarpa	4	0.82	Family	Lima 1987
		0.58	Indiv.	
P. pinaster	Seedlings	0.75	Indiv.	Polge and Illy 1968
P. pinaster	15	0.44	Indiv.	Chaperon et al. 1988
P. radiata	13	0.43–0.57	Indiv.	Dadswell et al. 1961
P. radiata	7	0.20	Indiv.	Nicholls et al. 1964
P. radiata	15	0.50	Indiv.	Bannister 1970
P. radiata	5	0.55–1.00	Indiv.	Burdon et al. 1972
P. radiata	12	0.55	Indiv.	Shelbourne et al. 1972
P. radiata	4	0.47	Indiv.	Nicholls et al. 1980
P. radiata	15	0.51–0.72	Indiv.	Bannister and Vine 1981
P. radiata	5	0.50	Indiv.	Burley et al. 1983

P. radiata	20	1.00	Indiv.	Burdon and Young 1991b
P. radiata	13	0.50	Indiv.	Dean et al. 1991
P. radiata	7	0.65	Indiv.	Burdon and Low 1992
P. radiata	20	near 1.00	Indiv.	Burdon et al. 1992b
P. sylvestris	20	0.56	Indiv.	Persson 1972
	18	0.46	Indiv.	
P. sylvestris	Young	0.57	Indiv.	Velling 1974
P. taeda	9	0.76–0.87	Indiv.	Goggans 1962
		0.04–0.35 (earlywood)		
		0.50–0.79 (latewood)		
P. taeda	2	0.37–0.49	Indiv.	van Buijtenen
	6	0.64–1.00	Indiv.	et al. 1962
P. taeda	6	0.64	Indiv.	van Buijtenen 1964
P. taeda	9	0.90	Indiv.	Goggans 1964
P. taeda	2	0.56	Indiv.	Stonecypher
	3	0.72	Indiv.	et al. 1964
P. taeda	5	0.73 (unextracted)	Indiv.	Stonecypher
	6	0.64 (extracted)		and Zobel 1966
P. taeda	5	0.75	Indiv.	Shelbourne et al. 1969
P. taeda	10	0.76	Indiv.	Barker 1973
P. taeda	5	0.47	Indiv.	Matziris and Zobel 1973
P. taeda	6	0.52	Indiv.	Stonecypher et al. 1973
P. taeda	10	0.42	Indiv.	Jett and
	8	0.57 (second gen.)	Indiv.	Talbert 1982
P. taeda	20	0.44 (Juvenile wood)	Indiv.	Talbert et al.
		0.45 (mature wood)		1982a
P. taeda	10	0.84 (Juvenile wood)	Indiv.	Talbert et al. 1982b
P. taeda	7	0.20	Indiv.	Bridgwater et al. 1983
P. taeda	General	0.50	Indiv.	Megraw 1985
P. virginiana	6	0.13–0.19	Indiv.	Rink and Thor 1973
P. virginiana	8	0.38–0.41	Indiv.	Meier and Goggans 1977

[a] Standard errors are not shown, but some are quite large.
[b] Based on individual tree or family assessments.

Table 5.4. Broad sense and parent-progeny heritability values for wood density in conifers

Species	Age (years)	Heritability	Kind	Reference
Abies alba	4	0.33	Parent to progeny	Polge et al. 1971
Picea abies	Young	0.56–0.86	Rooted cuttings	Kennedy 1966
Picea abies	7	0.80	Control crosses	Nilsson 1963
Pinus elliottii	14	0.73	Control crosses	Squillace et al. 1962
Pinus elliottii	5	0.54–0.63	Grafts	Zobel et al. 1962a
Pinus elliottii	5	0.93	Ortet to ramet	Einspahr et al. 1964
		0.44–0.49	Grafts	
Pinus elliottii	6	0.64–0.84	Grafts	van Buijtenen 1964
Pinus elliottii	6	0.43	Parent to progeny	Goddard and Cole 1966
Pinus nigra	22	0.50–0.80	Parent to progeny	Nanson et al. 1975
Pinus radiata	20	0.45–0.75	Rooted cuttings	Dadswell et al. 1961
Pinus radiata	10	0.75	Rooted cuttings	Burdon and Harris 1973
Pinus sylvestris	14	0.46–0.56	Parent to progeny	Persson 1972
Pinus sylvestris	Young	0.72–0.88	Graft	Velling 1974
Pinus taeda	9 (earlywood)	0.26	Parent to progeny	Goggans 1962
	9 (latewood)	0.18		
Pinus	1	0.17–0.58	Graft	van Buijtenen 1962
taeda	5	0.64–0.84		
Pseudotsuga menziesii	12	0.89	Rooted cuttings	Megraw 1985

the studies were made and the method of determining density and calculation methodology. Such heritability estimates depend greatly on the environmental variances present. In general, however, it can be clearly stated that *inheritance of wood density in the hard pines is high, indicating that good gains will be obtained by selective breeding.* There are exceptions such as *Pinus resinosa* which has a very uniform wood (Fowler and Morris 1977). It has been estimated that *Pinus resinosa* has remained unchanged for the last 5 million years!

5.2.2 Soft Pines

In comparison to the hard pines, only little work has been done on wood density of the soft pines. The soft pine species are less valuable to industry though their wood is especially in demand for certain decorative or finishing lumber rather than standard construction uses. In the standard kraft pulping process, soft pines are less used for pulp and paper because of their low density and small amount of thick-walled latewood. Currently, with TMP (thermomechanical pulp) white pine wood will have a wider use in pulping but still will not be a generally desired wood. Early growth of most soft pine species is slow, making them silviculturally difficult to establish in plantations. However, some species like *Pinus strobus* and *Pinus chiapensis* grow very rapidly later in life.

5.2.3 Other Conifers of Major Importance

A special section (Chap. 5.2.3.1) has been devoted to the wood of spruces and firs, as well as one for Douglas-fir and larch (Chap. 5.2.3.2). A great deal is known about the specific gravity of the wood of Douglas-fir. Considerable research has been done with the spruces and larches. Studies on the wood density of other conifer species are scattered, and often the genetics of their wood properties is unknown. In general, trends for Douglas-fir and the larches are similar to those of the hard pines, but the spruces and firs have quite different wood variation patterns and their response to genetic manipulation sometimes differs. This is also true of some of the other conifers, such as the Taxodiaceae and Cupressacae.

5.2.3.1 The Spruces and Firs

There has been some very successful genetic manipulation of the wood of the spruces (*Picea* spp.). The genetics of the wood of the firs (*Abies spp.*) is less well known. These two are the really important genera, along with the pines and Douglas-fir, in the northern latitudes and at the higher elevations. Overall, spruce has low wood density and even though its long thin tracheids are ideal for paper products, especially for fine papers and newsprint, wall thickness is frequently deficient. Both genera make good to excellent high-quality lumber, with the spruces having a wood that is exceptionally easy to handle. The firs have wood suitable for pulp manufacture but are not highly desired for solid wood products because of the instability of the wood upon drying; sometimes a bad odor and often much defect is present even in relatively young trees.

Both genera should respond as favorably as Douglas-fir to wood property manipulation by genetics, but much more effort is needed to definitely prove this. The one common characteristic that all spruces and firs have is large tree-to-tree variation in wood density. This, along with the high heritability of this property, enables a quick change in wood density by breeding. This has already been proven for Norway spruce (*Picea abies*) and currently available information indicates a similar response in white spruce (*P. glauca*) and black spruce (*P. mariana*). Changes in wood density of the spruces and firs will not be as dramatic as for the larches and Douglas-fir because their overall wood density averages are lower and there is less variance among trees. However, worthwhile gains in wood density from breeding are possible, and intensive breeding in that direction should not be ignored.

Actual inheritance values were shown for clones of *Picea abies* grown in Canada by Kennedy (1966). The gross heritability for specific gravity of $H^2 = 0.84$ is very high. In his study, high heritability estimates for specific gravity were found at similar positions within the ring. Similarly, Lewark (1982c) found the broad sense heritability for Norway spruce in Saxony to be $H^2 = 0.70$ to 0.86 for maximum density while $H^2 = 0.87$ for mean density. Such high values indicate that good gains are possible from a clonal selection program. Other studies on *Picea abies* showed $H^2 = 0.84$ (Kleinschmit and Knigge 1967), $H^2 = 0.50$ to 0.71 (Worrall 1975) and $H^2 = 0.29$ to 0.49 (Birot and Nepveu 1979).

In a study of interior white spruce in British Columbia, Yanchuk and Kiss (1992) reported that the heritability based on individual trees was $h^2 = 0.47$ and based upon family mean was 0.67. These values are in the same range as that for most of the hard pines. Provenance differences with $h^2 = 0.40$ to 0.60 were found by Mothe (1983) for *P. abies*.

There are only a few heritability studies on the firs (*Abies* spp). One, based on 4-year-old seedlings of silver fir (*Abies alba*) (from 20 open-pollinated families) reported values varying by method from $h^2 = 0.33$ to $h^2 = 0.22$ (Polge et al. 1971). The authors made the point that these heritability values were low because of the young age of the trees.

5.2.3.2 Douglas-Fir and Larch

Overall, Douglas-fir wood has a relatively high basic density and is usually quite strong, but there is much variation by locality and often huge tree-to-tree differences within a locality. However, Douglas-fir east of the Cascade Mountain Range has wood of lower density.

A recent study by Vargas-Hernandez and Adams (1991) showed the magnitude of inheritance for several aspects of wood density form 60 open pollinated families 15 years of age from seed. Heritabilities were obtained based on individual trees and on a family mean basis. They were subdivided into latewood and earlywood densities, as well as overall densities. In addition, the intraring density variations were determined. The results are summarized below.

Wood type	Average specific gravity	Individual heritability	Family heritability
Earlywood density	0.342	0.47	0.51
Latewood density	0.824	0.36	0.46
Overall density	0.455	0.59	0.55
Intraring density variation	0.198	0.25	0.39

The heritabilities were moderate to strong and in the range of those reported in other studies and for other species. The variances were quite large, as is common to all heritability studies. In a 12-year-old controlled pollination test of

Douglas-fir, King et al. (1988) reported individual tree heritability for wood density as $h^2 = 0.90$, while individual tree heritability for diameter was $h^2 = 0.23$. A high heritability of 0.75 for wood density was reported for 7-year-old Douglas-fir by Megraw (1985). The results from the available studies show that wood density is inherited strongly enough in Douglas-fir so that gains can be made in a breeding program despite some negative correlation with volume (see Chap. 10).

Larch has relatively thick cell walls, but in general larches have been shunned as highest quality sawn timber because of splitting and splinters. However, if care is used in sawing and drying, a good product can be made. A few studies have been made on the genetics of larch wood, mostly related to provenance. In general, heritability values have been small, and Lewark (1980) stated that heritability for wood density in larch was too small to use for an effective breeding program. In a general statement, Isebrands and Hunt (1975) reported that the large inherent differences among trees that was found, including trees having consistently high or low density wood, make for a good possibility of selecting for wood properties in a tree improvement program of Japanese larch (*Larix leptolepsis*). In European larch (*Larix decidua* var. *polonica*), Wodzicki (1961) found that 42% of the variation in wood density of seedlings was associated with phenological variables, which are themselves frequently strongly inherited. For 20 different seed sources of 17-year-old Japanese larch grown in Michigan, specific gravity varied from 0.39 to 0.42, indicating some potential for genetic control (Lee 1975).

5.2.4 Other Conifers of Minor Importance

Only a relatively few sound, basic genetic studies on wood have been made on other than the pines, Douglas-fir, and spruces.

In *Araucaria cunninghamia* (hoop pine) in Queensland, Australia, the heritability for basic density varied from $h^2 = 0.40$ to 0.83, a very high value which indicates that a large change in density would be possible through a selection and breeding program (Harding 1990). In another study on hoop pine, Harding and Woolaston (1991) report $h^2 = 0.60$. Specific gravity was also reported to be highly heritable in sugi (*Cryptomeria japonica*) (Kurinobu et al. 1990).

5.3 The Genetic Control of Wood Density in Hardwoods

Specific gravity of the hardwoods has received far more study than any other hardwood property because of its relation to strength, workability, pulpability, and ease of measurement. As emphasized earlier, specific gravity in the hardwoods is related to a combination of fibers, vessels, rays, and parenchyma cells. Usually the volumes of the different cell types are considered the key to specific gravity determinations. For example, Taylor (1965) reported a direct relationship between specific gravity and vessel volume in yellow poplar (*Liriodendron tulipifera*). Another example is Zhang and Zhong (1992), who found that wood specific gravity in oak (*Quercus liaotungensis*) is closely related to the proportion of vessels present. This is illustrated by the two annual rings of a ring porous species shown in Fig. 5.2 illustrating the prominence of the vessels.

The genetics of specific gravity in the hardwoods was reviewed by Zobel (1965a) but since then considerably more information has become available, especially in the *Eucalyptus* and *Populus* genera. (Some of this is summarized in Tables 5.5 and 5.6.)

Interest in the genetics of wood density in the hard-woods is increasing. For example, in 1965, Zobel summarized: "Breeding for improved wood in hardwoods has barely begun ... breeding for desired wood in hardwoods will probably be more difficult than in conifers." For paper manufacture, Dadswell and Wardrop (1959) listed seven desired hardwood wood qualities and stress how the diversity of products, and the complexity of the wood structure, all cause difficulties and interfere with genetic work on wood from this group of trees.

Numerous studies have been made of geographic variation as well as individual tree differences in specific gravity in hardwoods. Geographic tests nearly all show large variability, but there is no way to determine whether the reported variation results primarily from environmental or genetic control (see

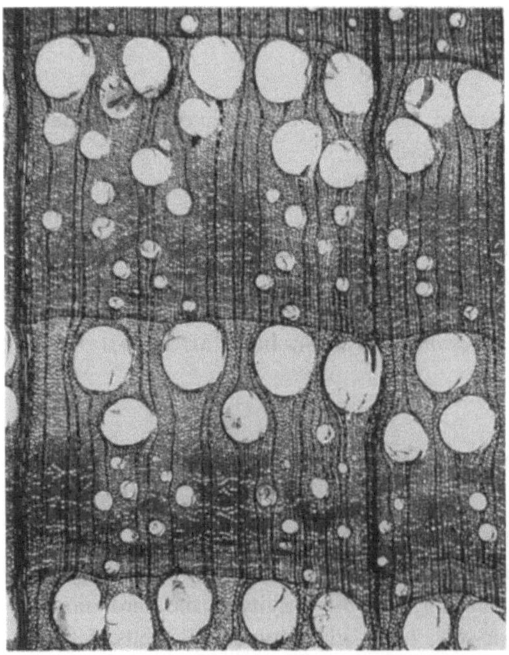

Fig. 5.2. The ring-porous hardwoods produce most of their vessels early in the season's growth. Shown are two annual rings in a ring-porous hardwood. It is evident how the volume of vessels can affect specific gravity. (Courtesy of Elisabeth Wheeler, NC State University)

Table 5.5. The strength of inheritance and variation of wood density in hardwoods

Genera and species	Reference	Comments
Acer rubrum	Saucier and Taras 1966a	Specific gravity was definitely different between clones
Betula pendula	Nepveu and Velling 1983	Wood density was inherited strongly enough to make gains possible in a breeding program
Betula pendula	Velling 1983	Inheritance of basic density was strong
Eucalyptus camaldulensis	Siddiqui et al. 1979	In a 10-year-old plantation, large differences were evident in specific gravity among provenances
Eucalyptus globulus	Dillner et al. 1970	Wood density and cellulose yields varied greatly among the 260 trees tested
Eucalyptus grandis and E. tereticornis	Hans 1976	There were significant differences in specific gravity among open-pollinated families
Eucalyptus grandis	Wang et al. 1984	Variations in wood density and moisture content were largely due to genetics
Eucalyptus spp.	Harding et al. 1989	The range of densities within and between species and provenances make breeding for wood quality improvement possible
Fraxinus pennsylvanica	Lowe and Greene 1990	Wood density of 42 families at 10 years of age was affected by plantation, provenance, and family within provenance. Family heritability was 0.80.
Gmelina arborea	Akachuku 1984	Significant differences were observed in wood density among trees grown under identical conditions in Nigeria
Platanus occidentalis	McCutchan 1982	Gains of 29 to 37% in dry weight would occur from selection for wood density
Platanus occidentalis	Jourdain and Olson 1982	Direct selection for specific gravity yielded only small gains
Populus spp.	Paul 1956	Wood density was uniform despite differences in site
Populus spp.	von Pechman 1958	The effect of heredity on wood was evident for several clones
Populus spp.	Göhre 1960	Individual tree wood densities varied greatly, independent of environment

Table 5.5. (*contd.*)

Genera and species	Reference	Comments
Populus spp. (hybrids)	Marton et al. 1967	Percent summerwood was not under strong genetic control
Populus spp. (hybrids)	Hamilton and Wendel 1967	There was a difference of 0.09 in specific gravity (24%) between the heaviest and lightest wood of 13 clones
Populus spp. (hybrids)	Mohrdiek 1979	Offspring specific gravity depended more on individual parents than on species or hybrid groups
Populus spp. (hybrids)	Stevens et al. 1983	Specific gravity can be improved by genetics in a selection program in the poplars
Populus deltoides and P. trichocarpa	Gabriel 1956	In both species, clonal differences in specific gravity were statistically significant
Populus deltoides	Farmer 1970	Clonal means for specific gravity ranged from 0.29 to 0.42
Populus deltoides and P. nigra	Avanzo 1974	Strong genetic differences in specific gravity occurred within and between species
Populus deltoides and P. euroamericana	Nepveu et al. 1978	Juvenile to mature correlations in wood density were 0.72 and 0.60 for the two species
Populus deltoides	Herpka 1979	In Europe, a high heritability was found among 3-year-old clones
Populus deltoides	Olson et al. 1985	Differences in specific gravity were large among clones and highly heritable
Populus × *euroamericana*	Scaramuzzi 1960	Wood density is closely related to the fiber wall thickness and fiber radius
Populus nigra	Nepveu and Tessier du Cross 1976	The density of wood in Euramerican hybrids was a variable character and subject to genetic control
Populus nigra P. euroamericana	Nepveu et al. 1978	There were fairly good juvenile to mature wood correlations showing genetic control
Populus tremuloides	van Buijtenen et al. 1959	There was only a small difference among clones compared to within clones
Populus tremuloides	Einspahr et al. 1967	Considerable natural variation occurred in wood properties including wood density of triploid aspen

Table 5.5. (*contd.*)

Genera and species	Reference	Comments
Populus tremuloides	Yanchuk et al. 1982a	There were large clonal differences in wood density. Density varied from 0.35 g/ml to 0.40 g/ml
Populus trichocarpa	Cech et al. 1960	There was a large clonal influence on specific gravity on the wood of black cottonwood
Robinia pseudoacacia	Stringer et al. 1982	No significant specific gravity differences were found among nine 20-year-old clones
Swietenia macrophylla	Chudnoff and Geary 1973	There was no relationship between the density of mother trees and their progenies when grown in several sites in Puerto Rico. Density variation among mother trees was large
Terminalia ivorensis	Nepveu 1976	The juvenile to mature genetic correlations for wood density were high

Table 5.6. Heritability values for wood density in hardwoods

Species	Age (years)	Heritability[a]	Kind[b]	Reference
Eucalyptus citriodora	0.5	0.91	Family	Almedia et al. 1981
E. deglupta	0.5	0.68	Family	Davidson 1972
	7.0	0.46	Indiv.	Davidson 1983
E. globulus	4.0	0.50	Family	Celbi 1979
	7.0	0.60	Family	
	12.0	0.50	Family	
	8.0	0.80	Family	Dean et al. 1990
	8.0	0.78	Indiv.	
	9.0	0.45	Indiv.	Clarke 1990
	8.0	0.80	Indiv.	Anonymous 1991
	9.0	0.50	Indiv.	
	8.0	0.65	Indiv.	Borralho et al. 1992
E. grandis	3.0	0.65	Coppice clones	King 1980
	1.0	0.83	Family	Wang et al. 1984
		0.80	Indiv.	
	9.0	0.45	Family	Malan 1988
	9.0	0.45	Indiv.	
	6.0	0.90	Clonal	Bertolucci et al. 1992
E. nitens	8.0	0.62	Family	Dean 1990
	8.0	0.69	Indiv.	
E. obliqua	7.0	0.84	Family	Dean et al. 1990
E. regnans	1.0	0.46	Indiv.	Rudman et al. 1969

Table 5.6. (*contd.*)

Species	Age (years)	Heritability[a]	Kind[b]	Reference
E. saligna	3.0	0.65	Clonal	King 1980
E. viminalis	3.0	0.61	Family	Otegbeye and Kellison 1980
Platanus occidentalis	4	0.78	Indiv.	Webb et al. 1973
	5	0.64	Family	Jourdain and Olson 1982
	7	0.73	Indiv.	Nebgen and Lowe 1982
	6	0.68	Indiv.	Land et al. 1983
Populus spp. (*Triploids*)	51.0	0.38	Indiv.	Einspahr et al. 1963
P. deltoides	Juvenile	0.62	Indiv.	Farmer and Wilcox 1966
	Mature	0.70	Indiv.	Farmer and Wilcox 1968
(Favorable envir.)	0.5	0.64	Indiv.	Farmer 1970
(Stress envir.)	0.5	0.76	Indiv.	
P. tremuloides	Old	0.43	Indiv.	van Buijtenen et al. 1962
	25.0	0.74	Indiv.	Einspahr et al. 1967
	25.0	0.42	Parent-progeny regression	
P. trichocarpa	—	0.56	Clonal	Reck 1974
Robinia pseudoacacia	20.0	0.10	Clonal	Stringer et al. 1982
Terminalia ivorensis	20.0	0.56	Parent-progeny regression	Nepveu 1976
Quercus petrea,	6.0	0.56	Clonal	Nepveu 1984a
Q. robur		0.59		
Q. rubra		0.37		

[a] Standard errors are not shown, but some are quite large.
[b] Based on individual (Indiv.), family, or clonal analyses.

Chap. 9). Therefore, these will not be cited here, with the exception of the tew studies listed in the next paragraph to indicate why researchers initially assumed that there was strong genetic control of wood density in hardwoods.

As an example, Valentine (1962) found the following specific gravity values for 26 different stands of *Populus tremuloides* sampled in the Adirondack Mountains of New York.

Individual tree sp gr. 0.29 to 0.47
Area mean sp gr. 0.34 to 0.43.

Wood density of yellow poplar (*Liriodendron tulipifera*) in the eastern United States from the mountains is lower than in trees from the piedmont (Barefoot

1963, Taylor 1965, Kellison 1967). In contrast, Webb (1964) found no differences in specific gravity for sweetgum (*Liquidambar styraciflua*) from different areas, but he obtained huge tree-to-tree differences on each site, jut as was found for yellow poplar. Although the large individual tree differences would indicate potential genetic differences, no positive statement can be made until specific tests are made. One can only state that site differences are involved, but probably genetic differences also (see Zobel and van Buijtenen 1989, Chap. 2 if greater detail is desired).

The most extensive studies of inheritance of specific gravity have been done with poplars where vegetative propagation is often the standard method of regeneration. In a clonal study on the wood differences between *Populus deltoides* and *P. trichocarpa*, Gabriel (1956) concluded that many of the large clonal differences in specific gravity within species (*P. trichocarpa*, 0.40 to 0.70; *P. deltoides*, 0.47 to 0.66) must be genetic. Similar differences between the species *P. trichocarpa* and the hybrid *P. regenerata* and clones among species led Kennedy and Smith (1959) to suggest that significant improvement can be made by selecting clones of high specific gravity.

An actual study of broad sense heritabilities for wood density of *P. tremuloides* by van Buijtenen et al. (1959) showed only a weak $H^2 = 0.17$. They suggested that the results indicated that a breeding and selection program would be of little value. However, in a follow-up study, van Buijtenen et al. (1962) found H^2 from 0.17 to 0.43. In a similar study, Einspahr et al. (1963) reported the broad sense heritability for specific gravity to be 0.38, while for the same species Yanchuk et al. (1983a) reported H^2 to be 0.35. All studies indicate relatively low heritabilities; despite this, all authors felt that specific gravity was under strong enough genetic control to warrant breeding programs.

Another group of hardwood trees in which genetic studies of wood density have been made are in the genus *Eucalyptus*. In 1956, Pryor et al. discussed the considerable work on the genetics of wood of the eucalypts and showed a strong inheritance pattern with some hybrids. In clonal forestry, inheritance of wood density in the eucalypts has been amazingly strong. Large clone-to-clone differences exist, while the members of a clone have good uniformity, with the result that the wood of all the trees of a clone is nearly identical. It is the use of the clonal uniformity, large clonal differences, and strong inheritance that makes possible the superiority of *Eucalyptus* wood in the pulp industry. Some results with the eucalyptus are reported in Chapter 5.4.

5.4 Genetic Gains in Wood Density Using Vegetative Propagation and Coppice

The best way to improve wood density on an operational scale is to use vegetative propagation, which in forestry usually means rooted cuttings. As mentioned in Chapter 2, one of the advantages of rooted cuttings is that essentially all the genetic variation in wood density, which includes both the additive as well as the nonadditive genetic variance, is transferred from the donor tree to the rooted cutting. Since wood density has a high additive component, gains will be good through seed regeneration but they will be even better when using rooted cuttings. Another reason for large gains is the high selection differential that can be applied. For example, Wright and Endo (1993) found that selection for wood density in the ortets of *E. grandis* will result in good gains in the clones. Relative wood density of the clones was maintained on different sites. In Colombia, the mean wood density of all clones was 0.41. Also for *E. grandis*, Jesus and Vital (1986) found significant correlations between wood density of rooted cuttings and that of the ortet in Brazil.

Wood uniformity is another benefit of genetic improvement which adds greatly to the value of rooted cuttings. All trees of a given clone growing on similar sites usually have essentially the same wood density but there are some minor exceptions (see Chap. 10.5). If the clones used in the plantation are chosen carefully, the wood produced in a plantation will have a narrow density range which is of great value to the wood user. Inheritance values for wood density in rooted cuttings are high because they represent broad sense heritabilities, giving genetic values about as high as for any other tree characteristic with the exception of resin yield. As a summary statement, if one is to attain the best genetic returns for wood density, vegetative propagation should be used whenever feasible because of the gain in both density and uniformity of the wood produced. For the occasional wood property that has very low heritability, such as cellulose yield (Jett et al. 1977), vegetative propagation is the best way to make genetic gain.

It must be remembered that gain = selection differential × heritability. It is of little value to have high heritability if the selection differential is low. Therefore, it can be stated that genetic gain using vegetative propagation can *only be satisfactory* if it is accompanied by an *intensive selection program.* This need to enhance genetic gain is frequently overlooked in breeding for wood density.

A few operational programs are using vegetative propagation to its fullest to obtain dramatic gains in wood properties. The example cited below is for Aracruz Florestal in Espirito Santo, Brazil (Brandáo 1984). The company started its use of rooted cuttings about 1975 and for several years have planted essentially only rooted cuttings. In 1990, nearly 40 million rooted cuttings were planted. In addition to improving growth, form, adaptability, and disease tolerance, an objective of the Aracruz program is the improvement of both the properties per se and the uniformity of the wood. The wood properties concentrated upon were wood density and cellulose yield per unit dry weight of wood. Listed in Table 5.7 is a summary of a table from the eucalypt report (Brandáo 1984).

In addition to the huge volume increase, the excellent current forest and mill yields are a combination of the improved wood density and the fact that more cellulose is obtained from pulping a ton of dry wood from genetically improved trees. Although not directly indicated, the much improved uniformity of the wood (500 to 600 kg/m^3) compared to the original (300 to 900 kg/m^3) results in a much more uniform pulp and much greater manufacturing efficiency. This is an

Table 5.7. Changes obtained by wood manipulation in *Eucalyptus* 7 years of age

	Original plantations	Rooted cutting plantations
Volume yield (m^3/ha/yr)	33	70
Density range (kg/m^3)	300 to 900	500 to 600
Average basic density (kg/m^3)	460	575
Percent pulp yield	48	51
Pulp yield (kg pulp/m^3 wood)	238[a]	293[b]
Mill consumption (m^3/ton pulp)	4.20[a]	3.41[b]
Forest productivity (tons pulp/ha/yr)	7.85	18.45

[a] Pulped with bark.
[b] With bark removed.

outstanding example of breeding for wood properties combined with maximum use of the genetic potential through vegetative propagation.

Along with the use of vegetative propagation is the question as to the wood quality produced in coppice forestry. After adjusting for the age of the wood formed, will the wood of the coppiced trees be similar to the wood of the tree from which the coppice arose? Has there been a genetic change? In general, the answer is that the wood from coppice is essentially the same as wood of the same age that came from the tree which produced the coppice. In working with *Eucalyptus saligna*, King (1980) found: "The specific gravity of coppice wood was similar in value to similarly aged tree wood of this species." He reported a variation in specific gravity among clones to be 0.34 to 0.52 (see Chap. 10.5). Similarly, the basic density of the coppice of *E. camaldulensis* was similar to the original crop, although the investigators did report that the first cutting decreased basic density a little, while it increased pulp yield and tracheid length (Sesbou and Nepveu 1991).

5.5 Inheritance of Within-Tree Variation in Wood Density

There is a definite pattern of development within a ring from the earlywood, through the transition zone and to the latewood. The earlywood has thin cell walls, the latewood thick walls with those in the transition zone being intermediate.

However, the differences between the two often vary from tree to tree. A study on this was made on Norway spruce clones by Kennedy (1966). His investigation centered on a study of within-ring variation in several wood properties and the degree to which these properties might be heritable. Four Norway spruce clones were available for this purpose. There were highly significant differences among the clones. Latewood percentage was highly correlated with specific gravity ($r = 0.90$). Although the quality of earlywood varied between clones, the range of variation within clones appeared to be dwarfed by differences among clones. The results showed high heritability estimates for specific gravity at similar positions within the ring; clone II was superior in specific gravity not only because of its higher latewood percentage but also because of its heavier earlywood. The within tree genetic variation and inheritance patterns are related to juvenile wood and its inheritance. This is covered in Chapter 8.3.

In radiata pine, Nicholls (1965) found a heritability trend as follows: near the pith, $H^2 = 0.60$; 9th growth ring from the pith, $H^2 = 0.24$: to the 25th ring, $H^2 = 0.60$. Nicholls hypothesized that the expression of different genes could become evident at different times. Although not shown as heritabilities the wood density at year 7 and later for several species in the Northwestern United States had strong genetic correlations to that produced at 15 years (Anonymous 1990).

When a comparison was made in inheritance of wood density between different sections of loblolly pine trees, Talbert et al. (1982b) found similar heritabilities for juvenile wood, mature wood, and the weighted specific gravity of the whole tree, but these were a little lower than the h^2 at age 10. Thus, any of these three (genetic correlation of $r = 0.88$) can be used to estimate inheritance of specific gravity.

In a detailed study of slash pine, Hodge and Purnell found that inheritance of earlywood density across all rings was $h^2 = 0.13$, ring density was $h^2 = 0.17$, and latewood density was $h^2 = 0.22$. These values are lower than many reported, but the authors give a detailed explanation as to why inheritance data from several sites would be lower than if obtained from a single site. Despite the lower values, the authors estimate that, with a 25% selection, wood density of juvenile wood can be changed by 0.01 and mature wood 0.025 g/cm^3, which certainly are economically meaningful. These gains are similar to those found for 10-year-old loblolly pine of 0.016 g/cm^3 (Zobel et al. 1978).

5.6 Summary

The inheritance of wood density is moderate to strong in nearly all tree species; only a couple of papers were found that indicated otherwise. This is rather surprising considering the complexity of wood density, which is determined by several traits, each of which, in itself, shows a weak to strong genetic component. Heritability values of wood density frequently change relative to the number of annual rings from the tree center, or pith. There is a mass of data on the strength of inheritance of wood properties, which are shown in several tables in this chapter. These are summarized as:

1. Of the seven genera included in 84 different papers, (61 of which dealt with pine in Table 5.1), all but a few reported that wood density is inherited strongly enough, and variation patterns are large enough, so that good gains can be obtained in a tree breeding program. This is especially true of the 17 papers on *Pinus taeda* and the 12 papers on *Pinus radiata*. There can be no question about the intensity of inheritance of specific gravity in these species. Several papers emphasized that wood density was among the strongest of inherited characteristics in forest trees.
2. Heritabilities were very large for the hard pines, with many narrow sense heritabilities being greater than $h^2 = 0.40$. Values for the 20 *Pinus taeda* reports cited were somewhat variable, but generally high, while many of the results for *Pinus radiata* were around $h^2 = 0.50$. These results indicate that in a tree improvement program based upon seed regeneration, very useful changes in wood density can be achieved by breeding.
3. As expected, the 18 broad-sense and parent-progeny heritabilities shown in Table 5.4 are high to very high, indicating the large gains possible with a control-pollinated program or with the use of vegetative propagation.
4. No information was available relative to inheritance of wood density in the soft pines.
5. Heritability values for the spruces, firs, and Douglas-fir vary considerably but are generally high, especially the broad sense values. Density of latewood has a higher inheritance value than that of earlywood. The inheritance values for Norway spruce were especially strong.
6. Considerable work has been done on the inheritance of wood density in the hardwoods, but nearly all of this is concerned with *Eucalyptus* and *Populus* (Table 5.5). Of the heritabilities in the 24 studies on the eucalypts, many were reported as family values. These are remarkably high, showing the potential for good gains in a genetics program (Table 5.6). Values in the eight studies for *Populus* are equally high. From the data available, one can conclude that the inheritance of wood density in the hardwoods is stronger than in the conifers.
7. The wood density of sprouts from coppice is generally the same as that of the original tree at the same age although there are reports of differences between the two. Of greater importance, the wood properties of a tree from rooted cuttings or grafts are similar to the donor tree at the same age. The uniformity of the wood of ramets within a clone is very high, a result that enables the production of large amounts of uniform wood in a vegetative propagation program. The gains possible from genetic manipulation and use of rooted cuttings, based upon the Aracruz operation with *Eucalyptus grandis* in Brazil, are shown in Table 5.7.

Chapter 6

Inheritance of the Cellular Components of Wood, Cellulose Yield and Pulp and Paper Products

6.1 General Concepts

There are many properties of wood that are under genetic control, most of which are essentially independent except for those that are actually dependent on each other, like cell wall thickness and specific gravity. Most show a relatively strong inheritance pattern. A few are of importance in determining product quality, but others play only a minor role. Frequently, it is not the cell component itself that is of importance, but ratios, such as cell length to wall thickness or to lumen size, that are important.

Before genetic gain can become meaningful, variability must be large and have a significant genetic base. Such variability is usually present, and even a short-fiber species like *Populus deltoides* has fiber length variation from tree to tree ranging from 0.70 to 1.38 mm (Boyce and Kaeiser 1961).

The literature related to the genetics of cell components is very scattered, and the results cited are often inconclusive because of design and analytical deficiences with different references sometimes showing conflicting results. When such divergent results are encountered, both opinions will be cited and finally the opinions of the authors of this book will be stated. The extent of the genetic base and the mating structure, as well as the number of samples and replications used in studies, will affect the scope of inference for different studies. When possible, those with narrow or very specific applicability are noted so the reader may assess the value of the reference. This is not always possible, and when a point of special use is encountered, the reader must check the original publication for breadth of applicability. Some early studies on the genetics of cell morphology were summarized and discussed by Zobel (1965a).

6.1.1 Variability and its Causes

The variability in cell dimensions within and between trees has been quite thoroughly studied and, in some instances, documented. One example of this was the 1965 paper of Schultz-Dewitz on variation in cell diameter and wall thickness of northern conifers. He listed numerous authors reporting variability of different cell types for numerous species, but often only the authors were listed and no details were given of the studies. Variability does not prove that there is genetic diversity but it usually indicates that such a diversity probably exists.

Generally, cell variation is considered to be of only minimal importance in its effects on the final products compared to wood density. This was expressed by Kibblewhite and Lloyd (1983) as: "The inclusion of pulp fibre length in the regression analysis did not significantly improve wood property/handsheet property relationships." Yet as more information from sophisticated studies is obtained, cell differences of even small magnitude may become important. For example, an intensive study of fiber length in which a large amount of variation was found was made by Campinhos and Claudio-da-Silva (1990) with relatively short-fibered *Eucalyptus*. They stated regarding cell length: "... its importance in a tree breeding program may be higher than earlier anticipated ... particularly where combined with the selection based on fiber wall thickness ...". They found that actual length differences from 0.64 mm to 0.85 mm in fiber length produced a pulp that was bulkier and with lower opacity from the short fibers compared to that obtained from the longer fibers. Even fibers that differed in length only from 0.60 mm to 0.66 mm produced papers with noticeably different properties.

Small differences in cell morphology will probably be of greater importance in hardwoods than in the conifers. This results from the complexity of the cell types in the wood (vessels, rays, fibers) and because small differences in cell length of 0.1 to 0.2 mm in the hardwoods can sometimes have an effect on the final product.

The physiological functions that influence cell structure and which can also be altered by genetics are many and will only be mentioned here. The relation of growth rate to the number of anticlinal cambial cell divisions affects cell length (Bailey 1920, Bannan 1967). Faster growth results in the formation of shorter cells because the anticlinal divisions take place before full cell length has been obtained. In his treatment of physiological control of wood properties, Larson (1960) stated: "Any factor that causes terminal elongation growth to begin prematurely or to cease will bring about a respective increase or decrease in cell size. The increase in summerwood cell wall thickness is due to a physiological process separate from that causing a decrease in cell diameter." He explains that the two phenomena occur at the same time when normal latewood is developed and that control of cell diameter is probably caused by an auxin produced in the bud or growing shoot. Auxin variation is believed to be of less importance in determining cell wall thickness. The physiological control of wood formation has been covered in detail by several publications such as by Roberts et al. (1988).

6.2 Cells of the Hardwoods

The cell components of the hardwoods are more complex than in the conifers. Not only is cell morphology itself important, but the arrangement of the cells in the gross structure of the wood also affects the ultimate product. The proportions of different cell types such as vessel elements, tracheids, libriform fibers, and medullary ray parenchyma are important factors, as reported by Mottet

(1965) (Fig. 6.1). This was well illustrated by Dadswell and Wardrop (1960) who showed that it was not the specific gravity per se of hardwoods that was of value, but of even more critical importance was the arrangement of the thick-walled cells in the annual ring. In assessing studies of the inheritance of cell characteristics, the genetic control of the proportion and arrangement of the cell elements means that one must remember that the results apply only to the species or provenances used in the test and only under the specific conditions of the experiment.

Since cell characteristics of the hardwoods can be altered by either environmental or genetic manipulation, a good assessment of the genetic pattern is difficult. Also, the great complexity of hardwood morphology makes the problem of assessing the inheritance pattern of the cells as well as relating a given fiber characteristic to product quality very difficult (Lange 1959).

There has been much discussion about strength properties of individual cells and their effect upon the final product. This was applied to the genetics of triploid aspen by Einspahr et al. (1963), who stated: "Fiber strength, as measured by zero span tensile strength, had variance ratios of 0.60 and 0.82 and appears to be under fairly strong genetic control."

6.2.1 Fiber Length

Except for wood density, the inheritance of fiber length has been the most studied wood property in the hardwoods because it is often considered to be the limiting factor in their use. The fibers of most hardwoods are short, making them useful for certain products and undesirable for others. For example, fiber length is considered to be the most important characteristic (along with the wall thickness) for making high quality groundwood in the eucalypts (Campinhos and Claudio-da-Silva 1990).

Fiber length has proven to be quite strongly inherited but the value of an improvement in length is usually questioned, since most fibers are less than 2 mm in length and more commonly about 1 mm. According to Hale (1959), the average fiber length for 70 North American hardwoods is 1.4 mm. The species with the longest fibers is black gum (*Nyssa sylvatica*) with an average length of 2.4 mm, but many species have fibers less than 1.0 mm in length. Fibers are frequently even shorter in some tropical species. Thus, even if percentage variation in cell length within and between species is considerable, in a practical sense it is small.

There are many reports of a large variation in fiber length among individual families and individual trees grown on similar sites. These are indica-

Fig. 6.1. The arrangements of the vessels, rays, and fibers in hardwoods greatly affect the utility of the wood. All three cell types respond to genetic manipulation. Shown are cell types of the wood of *Gmelina arborea* with 44× magnification (*top*), 88× magnification (*bottom*). (Courtesy Companhia Florestal, Monte Dourado, Brazil)

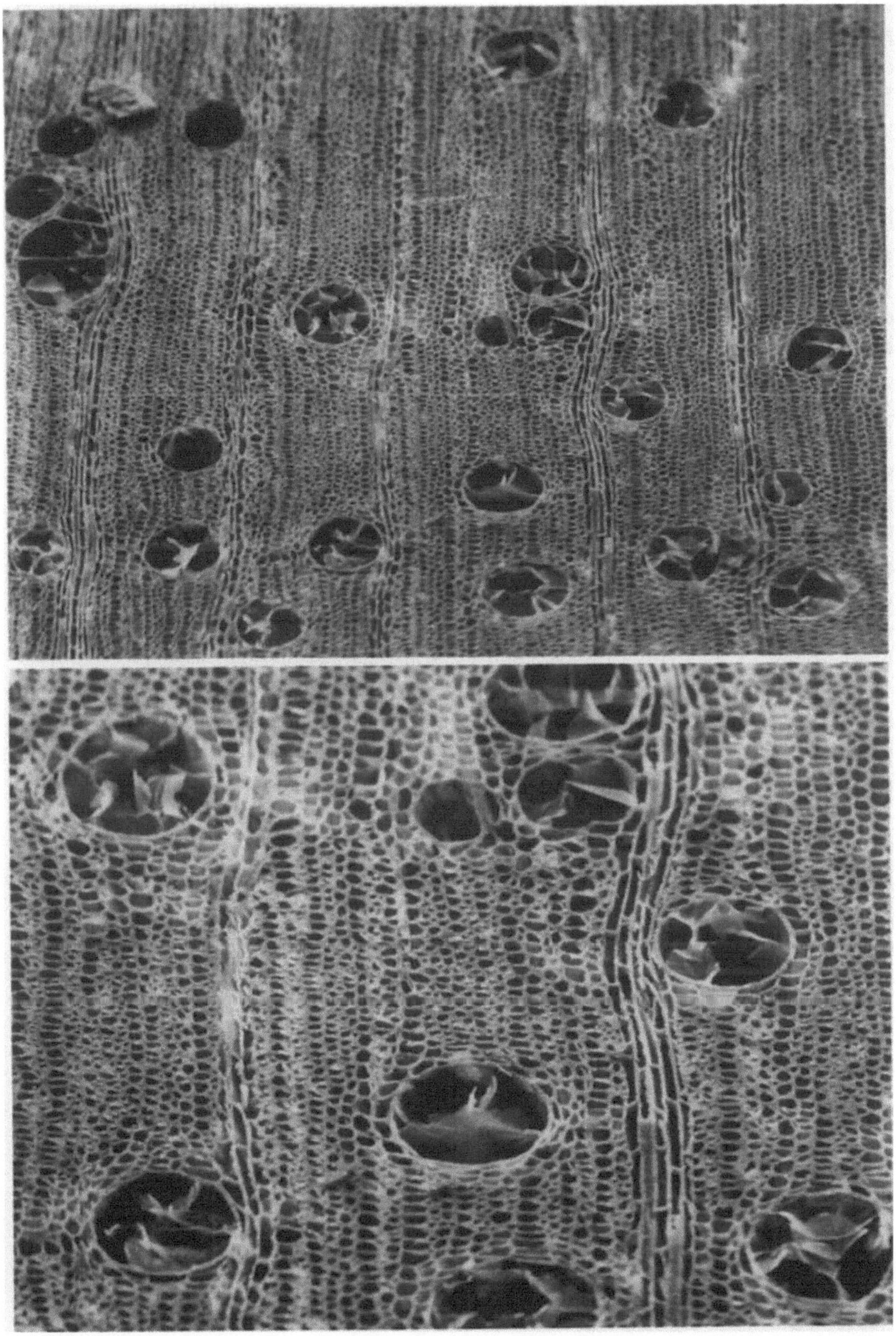

tive of possible genetic control but do not prove that such control is involved. A few examples of variability in fiber length are given by Taylor (1965), who found large differences in fiber length for yellow poplar (*Liriodendron tulipifera*), as did Webb (1964) in sweetgum (*Liquidambar styraciflua*). Webb concluded that the wide tree-to-tree variation observed in every stand or population of sweetgum resulted from a large measure of genetic control. The same conclusion was reached by Akachuku (1984) for the tropical hardwood *Gmelina arborea*, when he found large differences in Nigeria among trees of the same age grown under the same conditions. This differs somewhat from what was reported by Meyer-Uhlenreid (1959) for the genus *Populus*, where fiber length was constant for different trees within a species, irrespective of site. However, Meyer-Uhlenreid did find distinct differences in fiber length between the sections of poplars, namely *Aigeiros* and *Tacamahaca*.

Sampling for fiber length must be intensive because of the relatively modest size differences of fibers among species and among individual trees within a species (van Buijtenen 1960). When obtaining wood samples, for example, boring for increment cores cuts may fibers and care must be taken to measure only the uncut ones. Of special importance, there sometimes appears to be much variation in fiber length within an annual ring (Bosshard 1951, Liese and Ammer 1958), and the location where the fibers are sampled for measurement can therefore be critical. The usual patterns of change in radial and vertical stem directions have been reviewed by many authors such as Spurr and Hyvärinen (1954).

Fiber length variation is usually measured between and within trees, but there also may be differences depending upon the geographic source (see Chap. 9). However, the large fiber length differences among trees, as well as the relatively large variability within annual rings, makes geographic source differences difficult to detect. The result appears to be, as reported by Thorbjornsen (1961) for *Liriodendron tulipifera*, that geographic differences from varied test sites are site effects, not true genetic differences related to place of origin of the seed.

Several papers cited in Table 6.1 indicate the strength of inheritance of cell length in the hardwoods. In almost all instances, the inheritances pattern is from relatively strong to very strong. While the value of this in a breeding program is debated, some foresters have an interest in increasing fiber length in the hardwoods. There has been particular interest in breeding for longer fibers in the genus *Populus*. The heritabilities are relatively high; a few have been indicated in Table 6.2. These will enable a high percentage of improvement. However, the basic lengths of poplar cells are so small that even large percentage improvements may not be of major importance to improving the utility of the wood.

As for all wood properties, cell length is affected rather strongly by both genetics and environment. It is generally well accepted, for example, that short cells result from fast growth because the mother cells do not have a chance for full elongation before another transverse division takes place (Bailey 1920). The idea is that this relationship exists, no matter what the cause of the fast growth. However, this hypothesis was not accepted by Armstrong and Funk (1979) for

Table 6.1. Inheritance of fiber length in the hardwoods

Species	Reference	Comments
Eucalyptus globulus	Clarke 1990	A narrow sense heritability of $h^2=0.63$ was obtained for fiber length
E. grandis	Malan 1988	The h^2 of fiber length based upon family means was 0.54 for 9-year-old trees
E. nitens	Dean 1990	Family $h^2 = 0.38$ and individual tree heritability was $h^2 = 0.32$ for 8-year-old trees grown in Australia
E. regnans	Rudman et al. 1969	On an individual tree basis, $h^2=0.52$ for 1-year-old trees
E. viminalis	Otegbeye and Kellison 1980	The family h^2 for fiber length was 0.42; that based on individual trees was a low 0.12 when grown in the Southern USA
Fraxinus pennsylvanica	Lowe and Greene 1990	42 open-pollinated families in three 10-year-old tests showed strong family inheritance ($h^2 = 0.73$)
Liriodendron tulipifera	Thorbjorensen 1961	There were large fiber length differences among trees but these were statistically non-significant due probably to large variability within the annual rings
Platanus occidentalis	Jourdain and Olson 1982	There was family H^2 of 0.50 for fiber length. Direct selection for fiber length would give gains of about 3%
Populus deltoides	Boyce and Kaeiser 1961	Fiber length varied from 0.85 to 1.28 mm among natural cotton-wood trees. with a genetic variation of 30%
	Gabriel 1956	Statistically significant differences in fiber length were found among clones.
	Posey 1969	Strong clonal differences occurred for fiber length in Oklahoma
Populus hybrids	Hamilton and Wendel 1967	Within the 13 clones studied, no significant differences in fiber length were found
	Mohrdiek 1979	A number of hybrid progenies were studied; there were no significant differences in fiber length among them
P. tremuloides	van Buijtenen et al. 1959	There was a good inheritance of fiber length with $h^2=0.35$ and 0.51
	Einspahr et al. 1963	Triploid aspen had H^2 of 0.86 at age 30 for fiber length; $H^2=0.50$ for tree average fiber length
P. tremuloides	Einspahr et al. 1967	Five-year-old clones had fiber lengths with moderate narrow sense heritability
	Yanchuk et al. 1983b	Broad sense heritability of fiber length was $H^2=0.43$

Table 6.1. (*contd.*)

Species	Reference	Comments
P. trichocarpa	Gabriel 1956	The within-species differences were large enough to be statistically significant
Triplochiton scleroxylon	Oteng-Amoako et al. 1983	There was an increase in fiber length of less than 30% from juvenile to mature wood. This is usual for trees with storied structure

Table 6.2. Heritability values for cell length in the genus *Populus*

Species	Age (years)	Heritability	Kind	Reference
Populus deltoides	0.5	0.48	H^2 (good site)	Farmer 1970
	0.5	0.43	H^2 (stress site)	Farmer 1970
P. tremuloides	Mature	0.51	H^2	van Buijtenen et al. 1959
	30.0	0.86	H^2	Einspahr et al. 1967
	25.0	0.74	h^2	Einspahr et al. 1967
	25.0	0.42	Parent to progeny regression	Einspahr et al. 1967
	1.0	0.36	H^2	Farmer and Wilcox 1968
P. trichocarpa	Young	0.71	H^2	Reck 1974

seedlings of white ash (*Fraxinus americana*), who stated: "There was no correlation between the rate of cambial divisions and cambial derivative lengths."

A number of studies on the inheritance of cell length have been made in the eucalypts. Five of these were listed in Table 6.1. A number of general references could also be cited, like Rudman et al. (1969), who reported marked fiber length differences among clones for *E. camaldulensis*. Some eucalypt species have relatively long fibers. Although it is usually considered that the differences found will have little effect on the final product, Campinhos and Claudio-da-Silva (1990) found that even small differences had an effect on the end product.

6.2.2 Fiber Diameter, Wall Thickness, and Percentage of Cell Types

Fiber diameter and wall thickness are both closely tied to specific gravity. Each has its own inheritance pattern and the two can be genetically independent. In general, despite the inheritance patterns, fiber diameter and wall thickness are rarely included in a tree breeding program because of their overall influence on the more easily measured specific gravity of which they are a part. An exception was the study by Otegbeye and Kellison (1980) on 3-year-old *Eucalyptus viminalis*, where they reported $h^2 = 0.82$ for fiber diameter and $h^2 = 0.94$ for wall thickness. These values are unusually high but they have a high standard error. A similar high value of $H^2 = 0.85$ was obtained by Bertolucci et al. (1992) for fiber wall thickness in *Eucalyptus grandis*.

The diameter of the cells can play a major role in wood quality, as shown for oak by Zhang and Zhong (1992). They reported that both radial and tangential shrinkage in oak was related to fiber diameter. Fiber wall thickness, and its effect on pulp and paper of the eucalypts was discussed in some detail by Campinhos and Claudio-da-Silva (1990). As is well known, fibers from high density woods with thick walls have a low flexibility and are more resistant to conforming during papermaking. Fiber coarseness, a commonly used assessment in pulping, relates strongly to wall thickness and cell diameter. For *Eucalyptus grandis* in Brazil, Bertolucci et al. (1992) found a high value of $H^2 = 0.86$ for coarseness, which could be of great value in a vegetative propagation program when less coarse fibers are desired. In the same study they obtained $H^2 = 0.76$ for fiber diameter. Another frequently used measurement related to fiber diameter is the number of fibers per gram. Many papermakers consider that a key wood property is the fiber number, essentially a reflection of cell diameter which, however, varies due to changes in cell length. Inheritance of number of fibers per gram has rarely been studied, but it was found by Bertolucci et al. (1992) to have a $H^2 = 0.74$. With this high heritability and the use of a rooted cutting program, it would be possible to readily develop clones with a more desirable number of fibers per gram.

Usually pulp and paper quality discussions relate to the percentage of vessels or rays in the wood, but occasionally studies are made on the percentage of fibers. This was done by Nepveu (1984b) on English oak (*Quercus robur*) where he found only slight clonal effects. He stated that most variation was the result of environmental differences. Similarly, Scaramuzzi (1958) found fiber volume to have very little variability between clones in the hybrid poplars.

6.2.3 Vessels and Rays

There is considerable variability in the amount and arrangement of vessels in the hardwoods, and this diversity may have a very important effect on wood quality.

For example, vessels contribute little toward paper strength but can cause trouble in printing (Dadswell and Watson 1961). Vessel amount and arrangement are difficult to study and remarkably few genetic studies have been carried out, even though authors like Kanowski et al. (1991) state that the vessel characteristics are strongly controlled genetically (Fig. 6.2). It is widely accepted among those who work with hardwoods, and especially with ring-porous species, that there is a strong genetic component related to the vessel characteristics.

Generally, it is believed that vessel area is related to the silviculture applied to the tree, as well as location within the tree. This was the theme of the report by Phelps and Workman (1992) on black walnut (*Juglans nigra*). They concluded that, in addition to management and its effect on growth rate and vessel volume, there were within-tree variations that must also be considered. These, presumably, are under some degree of genetic control. In hybrid poplars, the area occupied by vessels was found not to be associated with clones, and therefore displayed little genetic control (Marton et al. 1967). However, in English oak, Nepveu (1984b) reported that genetic control was moderate for vessel percentage in earlywood, even though the environment strongly influenced the percentage of fibers. The result is a reasonably strong genetic control of the vessel volume and width of the earlywood. Also, in two different poplar hybrids, Crist and Dawson (1975) found considerable differences in the percent of vessels; the clone with the smaller amount of vessels had a higher wood density, as one would expect. For burr oak (*Quercus macrocarpa*) in Nebraska, latewood vessel diameter varied directly with ring width but vessel density varied inversely with ring width. Latewood vessel diameter was closely related to precipitation (Woodcock 1989).

Vessel size has a marked effect on the final product, as grain patterns in solid wood products and in effects on paper quality. Although vessel diameter was less strongly inherited than fiber length, Bertolucci et al. (1992) did find that it had a $H^2 = 0.72$ in *Eucalyptus grandis*. This means that if special clones were desired with small vessels, it would be possible to select for this.

For poplars in general, Meyer-Uhlenreid (1958) reported that those of the *Aigeiros* Section had vessels with different tangential diameters than did those in the *Tacamahaca* Section. In a study of the wood, including the vessels of cultivars of the hybrid *Populus* × *euramericana*, Scaramuzzi (1960) found remarkable differences among the various poplar types and between one provenance and another in most cell characteristics, including wall thickness and diameter. He stated: "The latter are believed to be linked, at least in part, to genetical factors. ..." In *Eucalyptus globulus* from South Africa, vessel diameter had a $h^2 = 0.61$ in a study by Clarke (1990).

Information on inheritance of ray volume is very sparse. An indication that ray volume showed a large amount of variation was given by Scaramuzzi (1960) for *Populus* x *euramericana*. In beech (*Fagus sylvatica*), Tellerup (1953) found distinct individual differences in the shape of wood rays from several different clones.

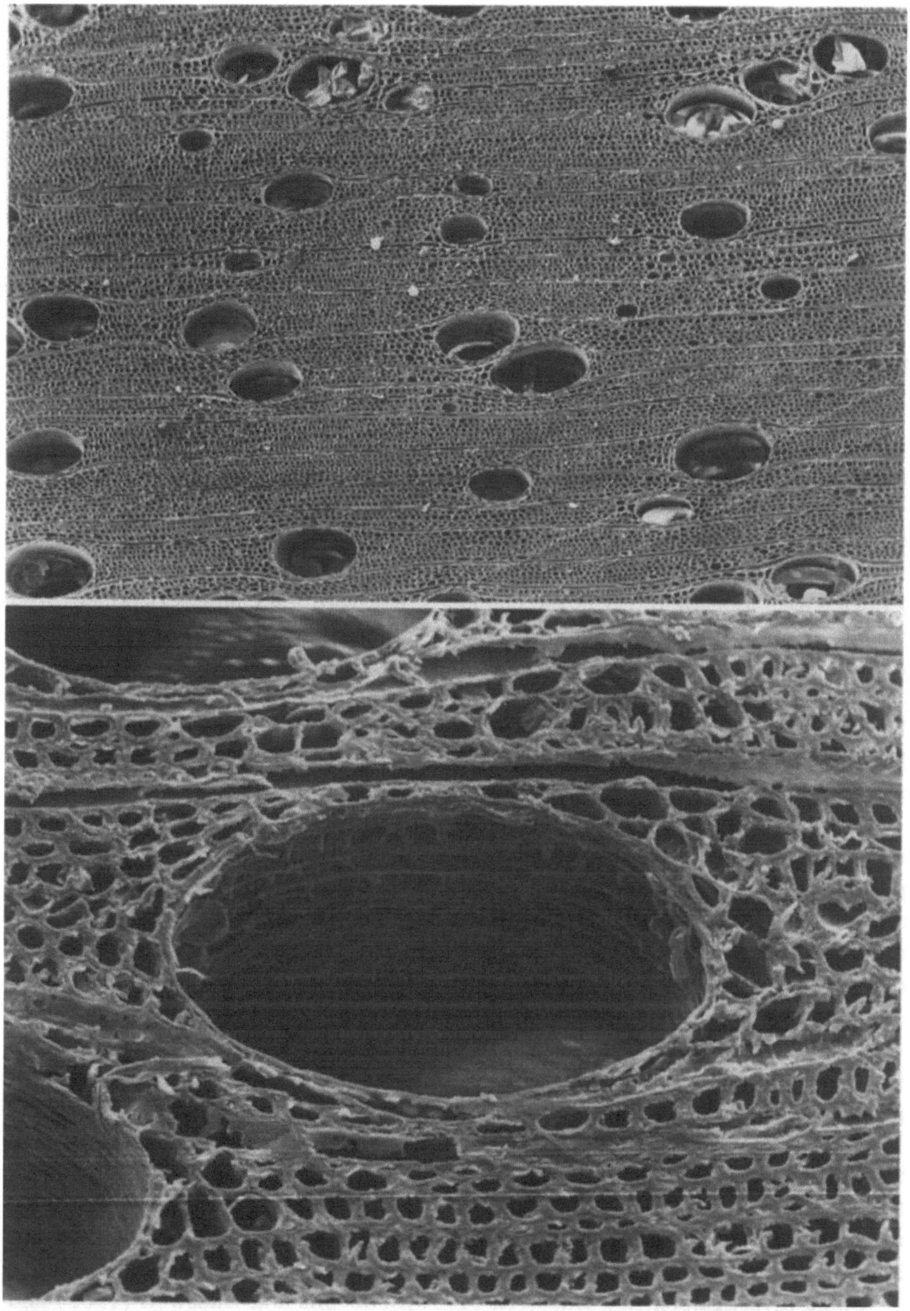

Fig. 6.2. The inheritance of vessel characteristics is reported to be strong. Shown here is a magnification of the vessels and fibers in the wood of *Eucalyptus deglupta*, *top* magnification 33×, *bottom* magnification 242×. It is evident why the amount, location, and size of vessels in wood have a strong influence on the final product. (Courtesy Companhia Florestal, Monte Dourado, Brazil)

6.3 Cells of the Conifers

The conifer cells, the bulk of which are tracheids, are relatively simple compared to the hardwoods. Ray cells and other minor cell types are present but tracheids are the cell components of conifers that dominate the effect on wood usage. Most of the tracheid characteristics are relatively strongly inherited; this is particularly true for cell length.

6.3.1 Tracheid Length

While the inheritance and importance of tracheid length have been widely studied in the pines, it is also widely debated. Heritabilities are high enough and variation large enough that good genetic gains are possible. Tracheid length of slash (*Pinus elliottii*) and loblolly (*P. taeda*) pines is strongly enough controlled genetically so that it is feasible to breed for long or short tracheids (Jackson and Greene 1958). The major unknown is the economic worth of the changes obtained from genetic manipulation of tracheid length. When tracheids are very short, such as in juvenile wood, a moderate increase in length will have a significant effect on pulp and paper quality. An increase from 2.0 mm to 2.5 mm can be important. However, in mature wood, an achievable increase from 3.5 mm to 4.5 mm will have a very minor effect on paper properties, since 2.5 to 3.0 mm are considered the minimum necessary to produce good paper from conifers.

The point at issue is whether the gains from breeding for longer cells are large enough, as expressed in product value, to warrant the effort. For example, Nicholls (1967b) states that cell length is of major importance for tearing strength of paper. In Australia, he suggests that any genetics program should include tracheid length in its selection program to ensure that current lengths are maintained. This is because they are now satisfactory, and gains in tear cannot be detected from a further increase in tracheid length. Greater cell length is favorable to both tensile and tear strength of paper, as stated by Britt (1967). Tracheid length has an effect on bursting strength, tearing strength, and tensile strength of paper according to van Buijtenen (1965). In fact, the value of increasing tracheid length varies with the product. For example, van Buijtenen (1965) points out that tracheid length generally has a favorable influence on pulp and paper properties, especially when there is a large amount of juvenile wood present. The same idea was expressed by Wiselogel and Tauer in 1982 for shortleaf pine (*Pinus echinata*) from which they expect substantial genetic gains in tracheid length.

Some investigators, like Dadswell (1957), Nicholls and Dadswell (1961) and Strickland and Goddard (1966), put particular stress on tracheid length in softwoods. Dadswell stated that average lengths of 2.0 mm and over for the wood of the first growth ring are most desirable for pulp production (the usual average is between 1.0 and 1.5 mm). However, there is no universal agreement on the importance of increasing tracheid length in radiata pine (*Pinus radiata*). For example, Burdon and Thulin (1965) were not certain of the value of increased tracheid length and they considered the tracheid length of radiata pine to be adequate for pulp. Different arguments were expressed by others that longer tracheids would be desirable for mixing with short fibered hardwoods. Cell length differences among trees within a species can be large but they can also be very large between species (Fig. 6.3).

A number of reviews of the genetics of tracheid length have been published, such as those by Dinwoodie (1961) and van Buijtenen (1965). A summary of the findings to that time is that cell length is strongly inherited. Currently, it is considered that there is moderate genetic control of cell length (Zobel and van Buijtenen 1989). However, this is sometimes not found, as reported by Keller (1973), who stated that heritability of cell length in maritime pine (*P. pinaster*) is low based upon a study of half-sibs with one mother tree. A similar result was found by Loo-Dinkins et al. (1984), where heritability of tracheid length of loblolly pine was low for both juvenile and mature woods. Both broad and narrow sense heritabilities for tracheid length in the hard pines are quite variable but some are high. Overall, good gains can be made from a breeding program. Reported heritabilities range from $h^2 = 0.01$ to $h^2 = 0.97$ and H^2 from 0.28 to 0.84 (Table 6.3).

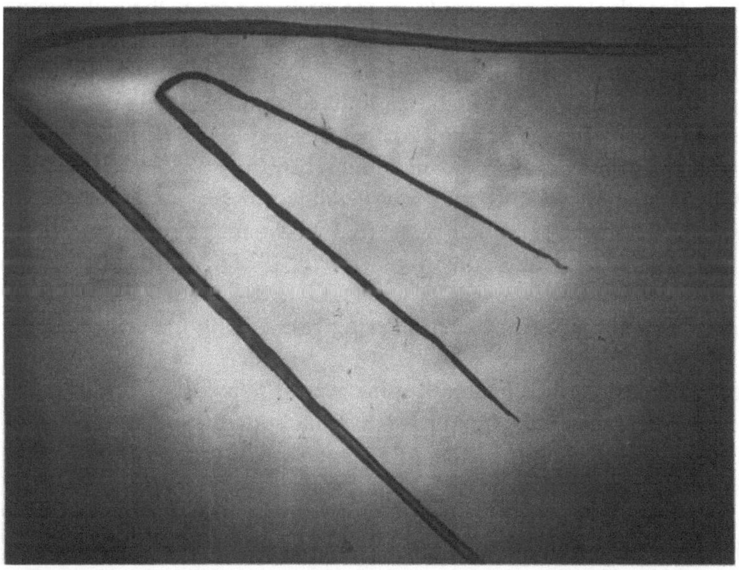

Fig. 6.3. The huge differences that can be found in tracheid length and size by species are shown for *Pinus taeda* (the short cells) and *Pinus oocarpa* (the long cells). Increasing the length of either of these by breeding would have very little effect on the final product. (Magnification is 10×)

Table 6.3. Heritability values for tracheid length in the hard pines

Species	Heritability[a] Value	Type[b]	Reference
Pinus elliottii	0.73	Broad	Einspahr et al. 1964
	0.56	Broad	Zobel et al. 1962a
	0.25	Narrow	Allen 1985
P. pinaster	0.01	Narrow	Chaperon et al. 1988
	0.14	Narrow	Keller 1973
P. radiata	0.66	Broad	Dadswell 1960
	0.73	Broad	Dadswell et al. 1961
	0.80	Broad	Nilsson 1963
(age 4)	0.44	Broad	Nicholls et al. 1964
(age 7)	0.28		
P. taeda	0.85to 0.97	Narrow	Goggans 1962
(latewood tracheids)	0.54to 0.77	Narrow	Goggans 1962
(earlywood tracheids)	0.44	Narrow	Stonecypher et al. 1973

[a] Standard errors are not shown but some are quite large.
[b] Broad or narrow sense heritability.

If the results reported by Greene (1966) for loblolly pine apply that: "... trees with significantly longer tracheids in the first ring from the pith maintain relatively long lengths throughout the succeeding rings ...", obtaining longer tracheids would not be difficult. However, in some of our studies on loblolly pine, we found that tracheid length did not stabilize until after the 10th ring from the pith. In radiata pine, Nicholls (1965) showed that heritability of tracheid length increased from the pith outward until a maximum of $H^2 = 0.55$ was reached for the 5th to 9th rings from the pith and then declined to $H^2 = 0.20$. None of the tracheid length differences of radiata pine (3.4 to 4.6 mm in mature wood) could be attributed to environmental conditions (Harris 1961), while Cown et al. (1992) did not find significant differences in tracheid length among 15 families of 20-year-old progenies of this species. Such differing results with the same species are puzzling.

Variations of tracheid lengths of loblolly pine were studied by Wheeler et al. (1965). Significant differences were found for lengths among trees at all height levels for both juvenile and mature wood (3.38 versus 4.44 mm average within a stand). The size and consistency of these differences would appear to indicate a degree of genetic control and the potential to make genetic gain. In addition, small stand differences were found which suggest some degree of genetic control. An unusual result of tracheid inheritance was reported by Jackson and Greene (1958). Rather than the usual intermediacy of the progeny between the parents, they found that the progeny were influenced more by female than by the male parent. No other reports of this pattern are known.

For a small crossing experiment of a limited number of young trees of slash pine grown in Queensland, Allen (1985) broadly generalized that tracheid length has no appreciable genetic variation. This finding was similar to those of Keller (1973) and Chaperon et al. (1988) for seedlings from a limited series of crosses on *Pinus pinaster*, who found heritability of cell length was low. Considering all the information available from a large number of studies, it is evident that tracheid diameter, wall thickness, and fibril angle are inherited less strongly than tracheid length, as expressed by van Buijtenen (1965). This is true despite the isolated reports of low inheritance of tracheid length reported for seedlings.

The genetic response to increasing tracheid length is also encouraging for many conifer species other than the pines. For example, in Norway spruce (*Picea abies*), Ujvári and Szönyi (1973), expected to increase tracheid length 20% above normal. Based on a series of studies of tracheid length, a special grafted seed orchard has been established to produce wood with longer cells. The major objective was to reach the same tracheid length in 40 years that normally would take 60 years, and to do this in 20 years when using vegetative propagation. In black spruce (*Picea mariana*), Boyle et al. (1987) found tracheid length to be predominantly controlled by additive variance, meaning that response to selection would be good. Other studies on *Picea abies* indicated useful heritability values for cell length. As an example, Kennedy (1966) obtained an $H^2 = 0.38$ while Rone (1970) stated that there was a high heritability for cell length in this species. Working with Chinese fir (*Cunninghamia lanceolata*), Yitai et al. (1992) found good general combining ability for tracheid length, indicating the predominance of additive gene action for this characteristic.

It is not sufficient to speak of inheritance of tracheid length in general. For example, Goggans (1962) and Smith (1967) report for conifers that latewood tracheid length appears to be more highly heritable than earlywood tracheid length. Also, the position in the growth ring and within the tree where the tracheids are sampled is of importance. For example, in Norway spruce, Kennedy (1966) found latewood tracheids to be 15% longer than the earlywood tracheids. He summarized: "This change agrees well with the 12–25% increase generally found across a coniferous ring."

6.3.2 Other Tracheid Characteristics

Essentially all tracheid characteristics measured have shown moderate to strong genetic control although there are some exceptions like cell width (Fig. 6.4). In most instances considerable variability exists, as was reported for tracheid tangential diameters of both mature and juvenile wood by Wheeler et al. (1965). In his study to determine characteristics of tracheids, van Buijtenen (1965) feels that many variations in tracheid characteristics are related to percent latewood and stated: "Percent summerwood ... is one of the primary factors controlling the average cross-section tracheid dimensions." He further pointed out the close

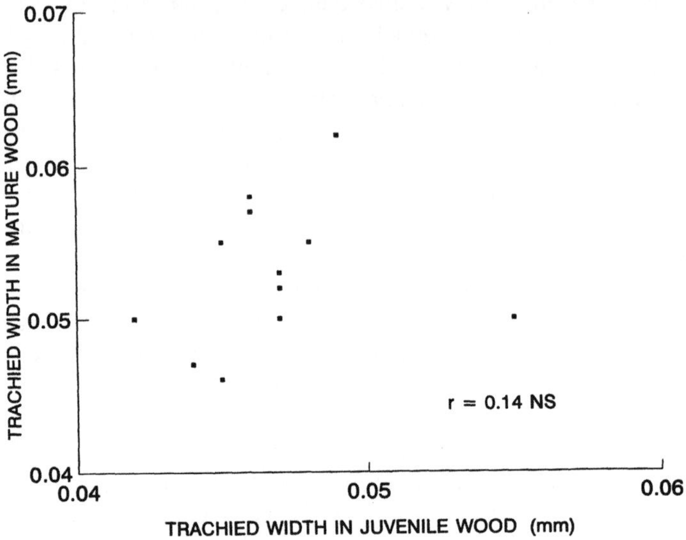

Fig. 6.4. Cell width has a genetic component but it is weak. Even within a tree, the cell width of the tracheids in the juvenile wood is not well correlated with those of the mature wood, as shown. Breeding for cell width has not proven to be very productive

relationship between cell wall thickness and wood specific gravity. Increasing wall thickness increased tear strength but reduced tensile strength and burst. Smaller tracheid diameters usually improve major strength properties of paper.

In contrast to tracheid length, van Buijtenen (1965) provides evidence for more environmental control of tracheid diameter and wall thickness, although some genetic control appears to be evident. This was quantified for wall thickness, where Goggans (1962) reported $h^2 = 0.84$ for wall thickness in latewood and $h^2 = 0.13$ for earlywood. However, in a study of young Caribbean pine (*Pinus caribaea* v. *hondurensis*) in South Africa, Barnes et al. (1983) showed similar heritabilities for earlywood and latewood, with h^2 about 0.50.

On a species basis, Malan (1993) found that tecunuman pine (*Pinus tecunumanii*) grown in South Africa had higher wood density than did *P. taeda, P. elliottii,* and patula pine (*P. patula*). This resulted even though *P. tecunumanii* had only half the latewood percentage of *P. patula* and a third of that of *P. taeda* and *P. elliottii.* The earlywood zones of *P. tecunumanii* were substantially denser than in the other species while the latewood densities were similar.

Tracheid diameter has considerable variability in Norway spruce (Olesen 1977) and appears to be under strong additive genetic control in black spruce (Boyle et al. 1987). In his 1965 summary, van Buijtenen suggested selecting for small diameter cells and thinner cell walls to improve both quality and yield for many types of paper. In fact, in 8-year-old loblolly pine clones, van Buijtenen reported in 1969 that there were definite differences in tracheid lumen diameter and wall thickness in the wood of trees selected for high and low specific gravity. In studies of inheritance and strength of different clones of *Picea abies* grown in Canada, major clonal differences were found by Kennedy (1966), who indicated that the differences among clones were the result of variation in cell wall architecture. This affected not only specific gravity but also tension strength and stiffness of the wood.

Cell characteristics and their effects often cannot be handled as discreet units. For example, the use of the Runkel ratio (sometimes called Double Wall Thickness ratio) was emphasized by van Buijtenen (1965).

$$\text{Runkel ratio} = \frac{(2 \times \text{tracheid wall thickness})}{(\text{tracheid lumen diameter})} .$$

Changes in cell width or cell wall thickness caused by genetics will change this ratio, and van Buijtenen warns that it should be prevented from becoming too high. This agrees with Barefoot et al. (1964), who reported for pine that: "... the Runkel Ratio was the best single predictor of paper properties, accounting for at least 58% of the variation ...". In their 1994 summary, van Buijtenen and Zobel stated that cell diameter and wall thickness appear to be moderately inherited, based upon limited samples, but they are not as strongly inherited as cell length.

6.4 Cellulose Yield and Pulp and Paper Products

Sometimes changes in wood properties have been studied with regard to their effect on paper properties. This different way of assessing the results of genetic changes in wood properties has led to the conclusion that there must be a strong genetic control over fundamental fiber properties which, in turn, influences paper quality and strength properties. For example, Palmer et al. (1984) showed the variation of pulp made from *Gmelina arborea* in the Solomon Islands. Occasionally, as in Clarke (1990), the term "fiber yield" is included with the pulp and paper properties. It is the product of tree volume, wood density, and pulp yield. It was found to have a generally rather high heritability.

In a paper on breeding eucalypts for pulp and paper, Dean et al. (1991) state that the inheritance of wood and pulping properties is highly heritable and that resultant variations have a high impact on the overall cost of production. Similarly, in a paper written when there was only limited early information on the inheritance of wood properties that affect paper, McElwee (1963) suggested that tearing strength, bulk density, bursting strength, and fold and tensile strength could all be improved by genetic manipulation of wood cells. For a number of hybrid poplars, Marton et al. (1967) showed that burst, breaking length, tear, and brightness were all significantly related to clonal origin and that consequently selection for clones could have an effect on the paper produced. For forest trees in general, Schreiner (1935) predicted that in the final analysis, it is the inherited characteristics of the fibrous raw material that determines paper quality, even when differing manufacturing processes are used. The genetic potential relative to utilization was expounded on by Schreiner 22 years later, after many of his early predictions had been found to be true (Schreiner 1958).

Although the chemical properties of wood are of importance in papermaking, it appears that the morphology of the cells is of greater importance. This was specifically reported by Byrd (1964), who found that in *Pinus taeda* differences in chemical constituents were not as closely related to paper properties as tracheid morphology, both within and among trees. It has been frequently stated that latewood is of special importance. As one example, Gladstone et al. (1970) found that latewood yielded 2 to 7% more pulp than earlywood. Much of the yield difference can be related to the greater resistance of latewood cellulose to degradative pulping reactions. They also reported that between tree rankings for lignin, holocellulose and alphacellulose are the same for earlywood as for latewood. Cellulose yields vary greatly from tree to tree, as shown in Fig. 6.5.

Although not commonly determined, the genetic control of papermaking itself has been occasionally assessed. One study did this in considerable detail for triploid aspen (Einspahr et al. 1963). The inheritance values were remarkably high, indicating that there must be strong genetic control over fundamental fiber properties that influence handsheet strength properties. Table 6.4 illustrates this. Actual value differences in this kind of study are of minor consequence. The important thing is their relative sizes, which are high even for broad sense

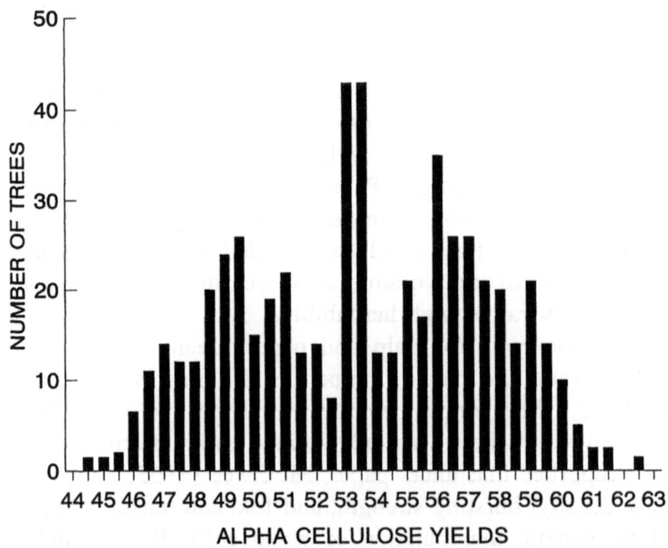

Fig. 6.5. As shown for this group of loblolly pine trees, alpha-cellulose content varies greatly by individual tree. The trees are all of the same age growing on the same site. The cause of the bimodal distribution is unknown. Much of the tree-to-tree difference results from nonadditive genetic variation.

heritabilities. These results are a strong indication that the cell properties are inherited strongly enough to have a real effect on the pulp properties. Despite the rigors and changes with the pulping treatment, cell properties appear to have strong enough genetic control to alter the final product. In a later paper, Einspahr et al. (1967) showed pulp yields to have a heritability of $H^2 = 0.63$ and zero span tensile $H^2=0.29$ for quaking aspen (*Populus tremuloides*).

In earlier studies of normal diploid quaking aspen, van Buijtenen et al. (1959, 1962) reported variance ratios (the upper limit for broad sense heritability) as follows:

Tear factor, 500 ml freeness	0.52
Breaking length, 500 ml freeness	0.71
Burst factor, 500 ml freeness	0.84
Zero span tensile, 500 ml freeness	0.19.

All these values were lower than those reported for the triploid aspen shown in Table 6.4, indicating stronger inheritance within the population of polyploids. In a study assessing the effects of provenance on pulping in pine, Wright (1987) found some effect in *Pinus tecunumanii*. A summary is shown in Table 6.5.

Working with *Pinus pinaster*, Chaperon et al. (1988) found the inheritance of pulping quality to be very low ($h^2=0.04$). However, for slash pine, Einspahr et al. (1964) found an inheritance for pulp yield to be $h^2=0.59$ and zero span tensile strength 0.84. Fiber yield inheritance was $h^2=0.38$ for 30-year-old *Pinus pinaster* (Keller 1973), based on parent to progeny correlations.

Although it is not measured or used in selection programs directly, the inheritance of compression wood, assessed mostly through the genetic control of straightness and limb form, grossly affects the quality of paper produced (see Chap. 7.5). Some of these effects were listed by Pillow and Bray (1935) in a comparison of pulp made from normal and compression wood. A summary indicates that compression wood gives a lower yield of pulp (42% for compression wood, 49% for normal wood), a decrease in tearing and folding strength

Table 6.4. Broad sense heritability estimates for handsheet strength properties of triploid aspen

Zero-span tensile (beaten)	0.82
Zero-span tensile (unbeaten)	0.60
Yield, unscreened	0.87
Yield, lignin-free basic	0.88
Tear factor, 500 ml freeness	0.90
Tear factor, 750 ml freenesss	0.88
Bursting strength, 500 ml freeness	0.83
Bursting strength, 750 ml freeness	0.76
Tensile strength, 500 ml freeness	0.57
Tensile strength, 750 ml freeness	0.54

Table 6.5. Pulp and papermaking characteristics for two provenances of *Pinus tecunumanii* grown in the Eastern Transvaal and Zululand, South Africa. (Wright 1987)

Trait	Eastern Transvaal		Zululand	
	CAM[a]	MPR[b]	CAM[a]	MPR[b]
Total yield (%)	40.8	41.4	46.6	43.2
Burst index	5.5	5.6	6.1	6.1
Tear index	9.6	9.0	8.2	8.7
Specific gravity	0.463	0.509	0.494	0.507

[a] CAM = Camelia Provenance.
[b] MPR = Mountain Pine Ridge Provenance.
The same results were evident for pulping as for wood properties.

properties, a decrease in chemical composition with 8% less cellulose, and a marked increase in bleach requirement. In addition, more severe cooking conditions are needed. The physical characteristics of loblolly pine compression wood that most affect pulp are the shorter fibers and the flatter slope of the microfibrils. This brief summary of compression wood of loblolly pine clearly illustrates how vitally important it is to breed for both morphological characteristics of the tree and internal tree qualities to improve paper properties. This dual breeding emphasis is currently especially important for the tropical pines to improve their pulp and paper properties as well as the characteristics of the solid wood products produced.

In *Eucalyptus globulus*, the heritability for pulp yields was $h^2 = 0.43$ for individual tree analyses in Tasmania, Australia (Dean et al. 1991); the pulping quality varied from different provenances (Turner et al. 1983). Pulp yield family inheritance values have been reported for 7-year-old *E. obligua*, $h^2 = 0.48$ (Matheson et al. 1986), $h^2 = 0.62$ for 8-year-old *E. globulus* (Dean et al. 1991) and $h^2 = 0.18$ and $h^2 = 0.26$ for 10- and 12-year-old *E. globulus* in Portugal (CELBI 1979). These all indicate that improvement in pulp yields can be obtained through selection and breeding. For the eucalypts, Dean et al. (1991) stated in summary: "... studies of wood and pulping properties have demonstrated that these are generally highly heritable and that variations have a high impact on the overall cost of production ... breeding is very worthwhile. ..." They summarized that pulp and paper improvement by breeding is important, quoting heritabilities of 0.72 for yield and 0.25 for tensile index of paper.

Although not inherited directly, in South Africa SAPPI reported large tree variations (over 8%) in pulp yields for *E. dunnii* (Anonymous 1988). There were large within family differences for pulp yield and kappa number. It was suggested that some of the differences are due to genetic variations in wood, making it possible to change pulp yields and quality by breeding.

In a detailed study on *Eucalyptus grandis* Bertolucci et al. (1992) found that clonal eucalypts gave the following data when pulped to 20 Kappa:

Variable	Broad sense heritability (H^2)
Production gain	0.46
Net yield	0.45
Dry solids	0.64
Alkaline charge	0.74
Pentosans	0.95
Roughness	0.91
Apparent density	0.89
Viscosity	0.87
Tear factor	0.85
Tensile strength	0.84.

These values are all remarkably high. They indicate that using clonal forestry with *E. grandis* gives worthwhile improvements for some particularly important pulping characteristics. Of course, all factors cannot be improved at one time because of their relationships and interactions. Nevertheless, the results are encouraging for the development of trees with specific desired character(s) as they may be needed for specialty products.

Working with various categories of pulping of *Eucalyptus grandis*, Clarke (1990) made an intensive study of the inheritance of pulp and paper making properties of the species. Some of his results are summarized below based upon a macropulping technique:

Property	Heritability (h^2)
Pulp yield	0.19
Fiber yield per tree	0.71
Kappa number	0.30
Brightness	0.21
Burst (2250 revs)	0.21
Tear (2250 revs)	0.33
Breaking length (2250 revs)	0.25.

Some people question the value of these types of assessments, believing that they are controlled by a number of different wood characteristics and influenced by conditions of assessment. This is true, but the remarkable fact is that such high heritabilities are obtained with so many variable and unknown factors. The important thing is that despite all these factors (wood, growth, tree form, manufacturing process, etc.), it is possible to select and reproduce individuals with certain kinds of pulp characteristics. Especially after vegetative propagation is operationally used, it will be possible to develop strains of trees whose wood will have better burst or better tear or other desired pulping characteristics through direct selection by pulp tests.

For another, completely unrelated, hardwood, namely European white birch (*Betula pendula*), Velling (1983) found that pulp yields had a low heritability.

6.5 Summary

The literature related to the genetics of cell components and their chemistry is scattered and sometimes contradictory. In contrast, cell morphology variation is quite well documented. Although the inheritance patterns are often strong, the variations among cells have generally not been considered highly important in the determination of qualities of the final product. However, newer studies indicate that they are, in fact, quite important. Small differences in cell morphology appear to be of greater importance in the hardwoods than in the conifers. In hardwoods, cell morphology is complex but the arrangement of the cells in the wood has the largest effect on the final product. Fiber length of hardwoods has been the most studied property, and moderate to strong heritabilities are reported. Unfortunately, because of the basically short cells in hardwoods, changes in length resulting from genetic manipulation generally have only a limited effect on the properties of the final product. Although fast growth within a tree stimulated by the environment usually results in short cells (for both hardwoods and conifers) there is no connection between inherently fast-growing genotypes and cell length.

Fiber diameter and wall thickness are reasonably strongly inherited, but since they are closely allied to specific gravity, separate genetic breeding is rarely applied for them. If it were done, cell diameter and wall thickness could have a major role in product quality because of their effect on flexibility and paper conformation. This is sometimes reported in the form of coarseness, which, in itself, has a strong heritability.

The vessels and rays of hardwoods can have a strong inheritance pattern but are considered by some to have little effect on paper strength although they can adversely affect printing. Others state categorically that vessel volume has an influence on the final product, whether solid wood or paper.

Tracheids are the main cell type in the conifers and their length is moderately to strongly inherited with sufficient variability for it to be feasible to develop long- or short- fibered landraces. Despite this, little breeding is done for cell length because in mature wood an increase of even 1 mm will have little effect on the final product. However, some of the earlier researchers put special emphasis on increasing tracheid length in the conifers. This is especially true for juvenile wood with its short tracheids. Despite the controversy about the value of increasing tracheid length, nearly everyone is in agreement that good genetic gains are possible. There are reports that the tracheid length of latewood cells is more strongly inherited than those of the earlywood.

Tracheid characteristics other than length also show a strong inheritance pattern. The latewood percent, and therefore the percent of thick-walled cells, is commonly listed as being a major determinant of product quality. Small-diameter cells are preferred for most papers but essentially no breeding has been done for that objective. In papermaking, cell ratios (like the Runkel Ratio) are often considered key indicators of quality. Breeding for changes in wall thickness or cell diameter will affect this ratio. The genetics of yield and quality of paper products have received a surprisingly large amount of emphasis. Despite their complexity, and all of the interrelations involved, there is considerable genetic influence on paper properties. Of course, the inheritance of paper quality results from inheritance of the basic wood properties, but the sum product has been shown to have a good inheritance pattern. Several genetic studies of this nature report very high heritabilities. When the chemical properties of the wood are compared with the morphology of the cells, it is found that morphology has the greatest effect on paper properties.

The broad consensus, especially in the eucalypts, is that pulp yield can be improved by direct selection and testing of clones. Except for a couple of organizations, this use of genetics is not being employed in operational programs.

Chapter 7

Grain, Fibril Patterns, and Internal Defects

7.1 General

Differing grain patterns, fibril angles, internal defects, and reaction wood are of great importance both for the strength and quality of the final product and for esthetic uses of wood. There are many patterns in this context; for the purpose of simplicity, they have been discussed as six types in this chapter. Certainly the divisions could be much finer but in most cases genetic studies have either not been carried out, or are just now underway.

The general statement can be made that, from observation and experience, it appears that most grain patterns are reasonably strongly inherited while reaction wood is weakly inherited. As usual, there are differences from tree to tree as well as differing ideas about inheritance. For example, in Chapter 7.2 a paper is cited stating that large gains are possible by genetic selection against spiral grain; another paper reports that, in effect, spiral grain is not inherited strongly enough to expend the effort to genetically improve it.

Microfibrillar angle is so important to strength properties, and is so closely related to compression wood and juvenile wood, that it has been included even though there are only suggestions that there is significant genetic control. The defects resulting from adverse microfibrillar angles certainly are so serious that if genetic control should be found important, genetic studies on fibril angle certainly should be given a high priority.

7.2 Spiral Grain[3]

Spiral grain is a common defect of wood. Spiral grain and related wave phenomena in wood were both treated in great detail in a recent monograph by Harris (1989). Most reports relative to spiral grain concern the conifers, but some work has been done with hardwoods. It is of primary interest to the production of solid wood products and relates directly to timber strength and stability (Fig. 7.1). According to McBride (1967), losses from spiral grain are heavy. He presents dollar values for Douglas-fir (*Pseudotsuga menziesii*) and Engelmann spruce (*Picea*

[3] It is essential to differentiate spiral grain from spiral bole. In the latter the whole bole develops like a corkscrew. Spiral grain is an orientation of the grain from the vertical which often occurs in a tree with a perfectly straight bole.

Fig. 7.1. Spiral grain is commonly recognized as a severe defect in lumber. As shown by the herring-bone pattern on the veneer, spiral grain is also adverse for plywood. Spiral grain, especially near the stem center, often has a strong genetic component. (Courtesy Division of Forest Products, CSIRO, Australia)

engelmanii). There is a reduction of 7.4% in grade for Douglas-fir and 1.6% for spruce. He explained the problem as: "A spiral angle of one degree will result in a change of 48 degrees in the angle of the heart shake at the opposite end of a 16 foot log ... it is evident ... spiral grain in trees is a defect which results in serious devaluation of lumber and logs."

The size of the grain angle is important for the value of the final product. Several investigators, such as Elliott (1958), summarized the value lost by spiral grain. He showed for Douglas-fir and western hemlock (*Tsuga heterophylla*) that grain angles varied from 1.0° to 10.0°.

The impact on the wood product produced is related to the angle; for Douglas-fir and western hemlock, a grain angle of 5.8° reduced resistance to impact bending by 38%. As a general directive, Hancock (1962) states that a 6.0° spiral, which is 1 inch to 10 inches, is used as a limit in lumber grading rules for construction lumber grades. In radiata pine, (*Pinus radiata*) Cown et al. (1991b) showed that a left hand spiral of 4.7° in the inner ten growth rings caused significant problems in processing and marketing of lumber because of drying degrade, strength loss, and movement of the wood after it has been put into service. Any tree with an angle of grain greater than 1 in 16 should be rejected in a breeding program, according to Harris (1969). Since younger trees are now being harvested, and utilization is becoming more intensive with the use of the juvenile core, spiral grain in wood has recently become of much greater concern. This is especially true when mass-sawing techniques are used, like the Swedish gang saw or chip-n-saw, where the central part of the tree is included in the sawn boards without any attempt to saw for quality.

The angle of spiral grain can vary from 0.6° to 6.9° in Japanese larch (*Larix leptolepis*) (Mikami 1973). For this species, the limit for usefulness is a maximum grain angle of 3.0°. Only about 5% of the 358 trees Mikami sampled satisfied the 3.0° criterion. The grain angle in the first ten rings gives a good estimate of the utility of the wood of a tree.

Among the numerous papers attesting to the importance of spiral grain, Banks (1967) stated: "Twist is shown to be the most significant reason for degrade of spiral grained softwood timber produced in South Africa ... foresters ... have long been aware of the presence of spiral grain in species of pine and the possibility that this is inheritable ... excessive spiral grain is a major cause of twist in poles and loss of strength and warping of wood." Similarly, Brazier (1965) emphasized the adverse effects of distortion and twist in lumber. For some species, like the southern pines, spiral grain has not been considered of much importance in mature trees (Zobel 1965b), but for other species it is considered a major defect.

Each species has its own characteristic pattern of spirality; some change from left rotation in youth to right rotation in maturity, others do the reverse, and in some the initial spiral in youth is continued into maturity, especially if it is a right hand spirality.

In nearly every study, the size of grain angle at the tree center has been emphasized, such as by Zobel et al. (1968b) in loblolly pine (*Pinus taeda*) or Taiwan incense cedar (*Calocedrus formosana*) (Chiu and Lee 1992). Typically, the large spirality near the tree center changes towards the bark. However, a few exceptions occur, such as in some hardwoods or individual conifer trees of several species where there is no spiral change from pith to bark. The accepted and usual pattern is a large left angle near the pith, changing to the right toward the bark, which usually shows a smaller angle. In assessing the seriousness of spiral grain, Harris (1984, 1989) emphasized that the problem is greater with fast-grown trees. For example, the highly spiraled core wood of *Pinus radiata* is often 10 inches in diameter and thus constitutes a large portion of the log. Development of a "spiral grain index" to truly assess the effect on the tree was recommended by Brazier (1965), since the worst spiral is near the center of the tree, where it represents only a small portion of the volume of the log. Spiral grain can be serious in mature wood (15 to 20 years of age); Cown et al. (1983) recorded angles in excess of 10.0° in Fijian-grown *Pinus caribaea* var. *hondurensis*. The grain spiral in Taiwan incense cedar apparently can be reasonably well estimated from bark grain spirality, where the wood grain angle is about 1/3 of the bark grain angle (Chiu and Lee 1992).

In South Africa, Banks (1967) reported the greatest spirality near the tree center, decreasing towards the bark. He stated: "The core wood of *Pinus elliottii* (slash pine) has the least spiral, followed by *P. taeda, P. patula* (patula pine), *P. radiata* with *P. pinaster* (maritime pine) having the most spiral." The latter two species have the rings closest to the pith with average spirality of 4.5° to 5.8° but at the tenth ring it is only 1.5° to 2.0°. Banks showed that pieces sawn a short distance from the pith are less likely to twist than pieces containing pith. Similarly, Cown et al. (1991b) showed that a spirality of 4.7° in the first ten rings causes degrade.

For maritime pine and Calabrian pine (*Pinus nigra* var. *calabrica*), spiral grain was always to the left in both grafts and seedings, and it was greater in the grafts than in the seedlings (Baradat et al. 1978). Spiral grain was less in Douglas-fir than in Corsican pine, but for both, the spiral increased from the base to the top of the tree (Birot et al. 1979). This would be expected, of course, since the greatest average spirality usually occurs near the center of the tree, regardless of height, in the wood that is overall closest to the tree center.

Because of the change in spirality and its random appearance, Krempl (1965) came to the conclusion in Norway spruce (*Picea abies*): "It is impossible to derive with certainty the degree of spiral grain in the interior of the stem from external characteristics...." Krempl emphasized the control of spiral grain by silvicultural methods. This contrasts with Nicholls (1967f), who states that it is not possible to control spiral grain by influencing environmental factors. Yet Lowery (1966) studied Douglas-fir in four states, as well as larch, lodgepole pine (*Pinus contorta*), and Engelmann spruce, and concluded that there was genetic control of spiral grain with little geographic difference. Heritability of spirality changes with age of the annual ring and is maximum in the early life of the tree. The high spiral grain with the large variability among trees indicates strong genetic control (and also the need for more efficient sampling methods) according to Cown et al. (1991b).

A classification and measurement of spiral grain was made by Lowery (1966), who took 1089 samples for lodegepole pine, western larch (*Larix occidentalis*), Douglas-fir, and Engelmann spruce. He reported that spiral grain was common in these species, but each had a characteristic distribution pattern of spirality. There was a left-handed spiral in 48% of the trees and right-handed in 47%, with very few being straight-grained. He found that initial right-hand spiral is less likely to change direction than initial left-hand spiral. Even though different species had strong left- or right-hand spiral tendencies, all species contained trees that spiraled in both directions.

The inheritance of spiral grain and its genetic control has been evident for a long time, beginning with the pioneering studies by Champion in 1927, 1929, and 1930; he studied open-pollinated progenies of trees with and without spiral grain and concluded that spiral grain is under genetic control and follows the Mendelian laws of inheritance. In other studies in India, Kadambi and Dabrall (1955) found that twist in seedlings of *Pinus longifolia* (chir pine) appeared to be a dominant genetic characteristic (Fig. 7.2).

Numerous papers deal with the inheritance of spiral grain; a few of these are listed in Table 7.1. One species (*Pinus radiata*) has been intensively studied, as reported by Fielding (1967) and Harris (1989). In addition to describing a simple methodology and the change in grain from left to right with age of ring, Fielding

Table 7.1. Genetic strength of spiral grain in conifers

Species	Reference	Location	Strength of inheritance
Araucaria cunninghamia	Eisemann et al. 1990	Center to bark	Narrow sense heritabilities were $h^2 = 0.33$ for the whole tree. Heritability of spiral grain was highest near the pith
	Harding and Woolaston 1991	Near tree center	The heritability of spiral grain was $h^2 = 0.20$
Calocedrus formosana	Chiu and Lee 1992	Overall	Genetic improvement of spiral grain through selection is effective in Taiwan incense cedar
Larix leptolepis	Mikami 1973	Overall	Eight and 9-year-old clonal Japanese larch had $H^2 = 0.35$ to 0.42 by growth ring and 0.49 for maximum grain angle
Pinus brutea	Baradat et al. 1978	Overall	The heritabilities obtained were high at $h^2 = 0.61$
P. caribaea	Harding et al. 1991	3rd ring	Heritability was $h^2 = 0.12$
		9th ring	Heritability was $h^2 = 0.46$
P. elliottii	Allen 1985	Overall	A $H^2 = 0.15$ was found for the whole tree in 14-year-old trees
P. longifolia	Kadambi and Dabrall 1955	Near tree center	Twist in seedlings appears to be a dominant character. When twisted trees were crossed, 68 to 82% of the resultant seedlings were twisted

P. pinaster *P. nigra* var. *calabrica*	Arbez et al. 1978	Overall	Heritabilities of maritime pine were $h^2 = 0.16$ and $H^2 = 0.43$ and $h^2 = 0.38$ and $H^2 = 0.61$ for Calabrian pine
Pinus radiata	Dadswell et al. 1961	Second ring	A H^2 of 0.66 was found for rooted cuttings
		7th and 8th ring	Heritabilities dropped to $H^2 = 0.28$ on the outer rings
	Fielding 1967	7th ring	In 9-year-old open pollinated trees $H^2 = 0.55$ for individual trees. Differences also occurred among families
	Nicholls et al. 1964	4th ring	In 10-year-old trees, $H^2 = 0.44$ was found. Values are large enough to make gains in a breeding program
		7th ring	Heritability was reduced to $H^2 = 0.24$ in the 7th ring from tree center
	Pederick 1971	2nd ring	$H^2 = 0.40$ with most of the variance being additive
	Burdon and Low 1992	5th ring	h^2 was generally over 0.50 for spirality
P. taeda	Zobel 1964	Overall	A decrease in heritability with age of ring was found

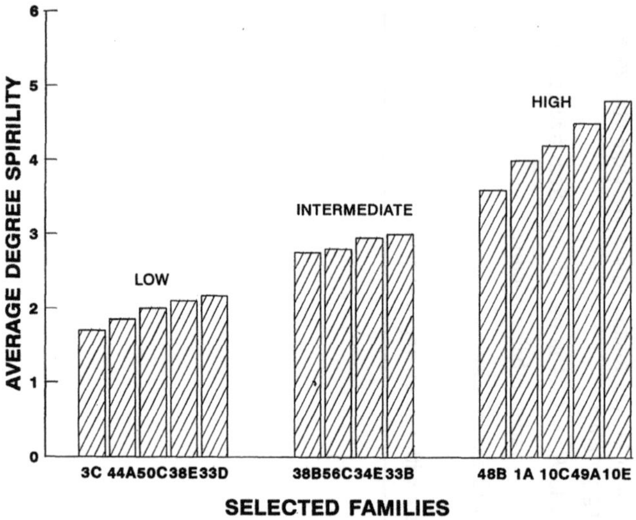

Fig. 7.2. Spiral grain is one of the more strongly inherited wood characteristics. Shown are groupings of trees by severity of spirality for *Pinus taeda*. This species generally has a low frequency of spirality. (After Zobel and van Buijtenen 1989)

opines: "The possibilities of reducing spiral grain by breeding are good; variances are large and heritabilities are high ... angles as great as 10.0° are not uncommon in the corewood." Spirality of adjacent rings within a tree is similar. However, the inheritance of spiral grain often changes as spirality changes direction. A reading at ring seven was found to be nearly as good as measuring the other rings (Harding et al. 1991).

For spiral grain in general, a number of researchers report that reasonable gains can be obtained from breeding. For example, Pederick (1971) stated for radiata pine: "It therefore appears that progress can be made to reduce spirality in corewood through standard seed orchard breeding procedures." Pederick recommends a selection intensity of 50%; using this, the improvement would be 0.73° and the average angle would be reduced from 3.8° to 3.1° in ring two. For slash pine in Queensland, Allen (1985) found that spiral grain values, especially near tree center, are quite large, and the genetic approach should be used to reduce them. In radiata pine, 25-year-old rooted cuttings had a $H^2 = 0.55$ near the pith with a rapid decrease to 0.08 at the ninth annual ring. Fielding (1967) summarized, also for *P. radiata*,: "The possibilities of reducing spiral grain by breeding are good; variances are large and heritabilities are high." In contrast, Zobel et al. (1968b) made the following summary based upon 4-year-old loblolly pine: "... genetic variation in this trait is mostly of the non-additive type. The study involved 1,043 trees ... mass selection to reduce spiral grain will not be very effective ... spirality extensive enough to cause serious degrade in sawn or veneer products rarely occurred; this raises some question as to whether spiral grain in loblolly pine is a serious problem".

Reference to Table 7.1 shows that spiral grain can have from low to high heritabilities depending on the species. Nearly all researchers find higher heritabilities for this characteristic near the tree center than near the bark but there are some major exceptions. Much of the variation in spiral gain within a species can be accounted for by the various methods used to assess this characteristic (see Chap. 3.6.1). The important thing is that spiral grain often has genetic control strong enough to use in a breeding program.

In addition to concern about the genetics of spiral grain is the amount of variability associated with it (Dadswell et al. 1961). Northcott (1957) suggested that spiral grain is a normal wood phenomenon and is not abnormal. He reported that of 594 mature trees, representing six conifer and six hardwood species, only three specimens showed no spiral grain at all. Of 709 seedlings, representing nine conifers and eight broadleaf species, 81% had spiral grain. The large variability in spiral grain tendency of different species, and trees of the same species growing together, is well documented (Northcott 1958, Krahl-Urban 1960).

A major question relates as to how much spiral grain is a result of growing conditions of the trees and how much to genetics (Krahl-Urban 1960, Ohkura 1960). In a study specifically designed to measure the effect of growth rate on spiral grain, Sachsse (1965) came to the conclusion that there was confusion, and some researchers found greater spirality with fast growth, others greater with slow growth, and others, such as Paterson (1967a), found no correlation. In an intensive study of *Pinus sylvestris* Liese and Ammer (1962) could find no relationship between the extent of spiral grain and the width of the annual ring. They also reported that a fiber deviation of less than 30° did not influence the normal pattern of cell length with advancing age, but has a major effect resulting in extreme twist.

It seems quite obvious that some of the confusion is related to a lack of appreciation of the effects of ring width, juvenile wood qualities, and closeness to the center of the tree. Despite Sachsse's report (1965) and others, many foresters still feel that spiral grain is more pronounced in the faster-growing tress.

Although the details are not of major concern here, there have been several explanations of the growth phenomena that result in the formation of spiral grain. One rather complete coverage of this subject was by Harris (1989), who reported that the polarity of xylem development is determined by the direction of flow of the growth substances acting on the cambium needle traces in the corewood of radiata pine. The traces obstruct the flow of growth substances and by their spiral array induce left-hand spiral flow. Later, flow becomes more normal and spirality becomes less.

Although recent studies on spiral grain have been few with the exception of the inclusive monograph by Harris 1989, during the 1960s there was considerable interest, as evidenced by the review paper by Noskowiak (1963). Upon close examination, however, it is evident that many of the papers contained only ideas and opinions about inheritance, without research results. The reason is obvious; spiral grain is so difficult to assess and to understand physiologically that few good studies have been conducted. Difficulties involved are evident from the measurement of spiral grain as explained in Chapter 3.6.1. Although there are several ways of assessing this characteristic, answers tend to be different and there is no clearcut, standard, simple method of measuring it. Several of those outlined in Chapter 3 are said to be simple, yet colleagues of the researchers did not seem to think they were, and often developed new methods for themselves. Differences in methodology of assessment have introduced many discrepancies among reported results relative to the genetic control of this characteristic.

Most emphasis relative to spiral grain has been on the pines (Noskowiak 1963, Harris 1989) and very little information is available on the spiral grain of hardwoods. One for beech (*Fagus sylvatica*) by Teissier du Cross et al. (1980) reported a good correlation between spiral angles at three different tree heights showing uniformity within the tree. The narrow sense heritability was 0.66. Another on beech in Germany by Krahl-Urban (1960) concluded that spiral grain was a dominant characteristic and is also present in the branches. It is of special interest that these, and several other researchers, later comment that the spiral tendency can be observed in very young seedlings. Spiral angle was also tested in white ash (*Fraxinus americana*) and was found to be constant radially for 50 growth rings without the large changes observed in many conifers (Hiller 1968). In *Eucalypstus dalrympliana*, Birot et al. (1980) found spiral grain to be under moderate genetic control. In an 11-year-old *Eucalyptus nitens* provenance trial, there were no significant differences between seed lots for spirality (Purnell 1988).

7.2.1 Interlocked Grain

Although considered separately by most authors, some treat interlocked grain only as a severe form of spiral grain (Webb 1967) (Fig. 7.3). In his article about spiral grain, Northcott (1957) also refers to interlocked grain as an extreme type of spiral grain with rapid and large changes in grain direction, often in adjacent annual rings. Although it occurs wherever trees are grown, interlocked grain appears to be more common and severe in tropical climates (Harzmann 1965, Harris 1989). It is very common in some tropical hardwood species and is so severe in some *Eucalyptus* species that they are nearly useless for solid wood products. Even the splitting of blocks for firewood can be very difficult when severe interlocked grain occurs, such as in *Eucalyptus globulus* from parts of California.

There are few studies that show interlocked grain to be under strong genetic control. One example is Webb (1969), who states that its severity in sweetgum (*Liquidambar styraciflua*) can be reduced by selection and breeding. He suggests selection in a tree improvement program to reduce interlocked grain because of its lack of relationship with growth rate, straightness or self-pruning ability. From 180 trees he found 5 with straight grain (index value less than 5.9), while the tree with most severely interlocked grain had an index on 52.0 caused by a maximum

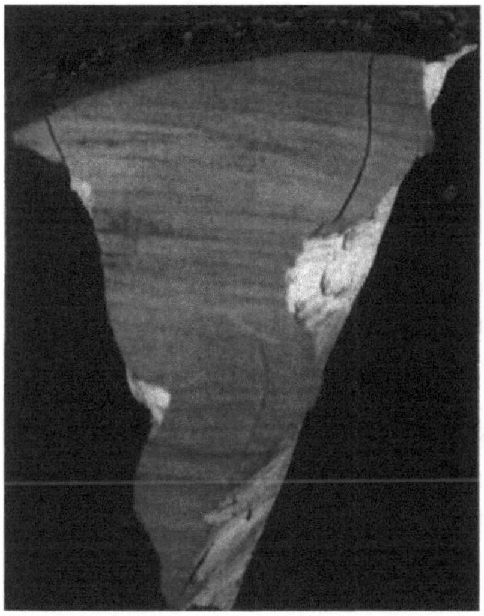

Fig. 7.3. Interlocked grain is common in many hardwoods, especially tropical hardwoods. It also occurs in species such as *Liquidambar styraciflua* shown here. There is a strong enough inheritance of interlocked grain so that straight-grained landraces can be developed

left spiral of 31° and a maximum right spiral of 28°. The average index value for all trees was 14.3 (12°).

The direction and angle of spiral varies greatly and changes frequently. There is a substantial range of variation in severity of interlocked grain among trees in the same stand as well as between different stands (Webb 1969). Webb found no geographical differences of interlocked grain in sweetgum, some differences among stands within an area, but huge differences among individual trees in the same stand. Occasional trees occur with essentially straight grain. The direction of spirality in interlocked grain is set by the direction of pseudo-transverse divisions in the cambium. The direction of flow of growth regulators and direction of mechanical pressures on the cambium influence the orientation of pseudo-transverse divisions in the cambium (Webb 1967). Although nothing was stated about the part that genetics would play, Harzmann (1965) reports that the wood of *Liquidambar formosana* (Taiwan sweetgum) has "crossed spiral grain", which causes "fractious surface" of the wood. For the tropical hardwoods *Entandrophragma* spp., Hejnowicz and Zagorska-Marek (1974) describe the mechanism of intrusive growth which produces an interlocked type of grain. This also occurs in the cambium of *Tilia* spp. (basswood).

Although interlocked grain is very common and dramatically restricts the usage of some species for quality products, little has been done with it genetically. All indications from its pattern of occurrence and the few studies made are that it has a strong genetic background that would respond to breeding.

7.3 Microfibrillar Angle

The walls of wood cells are not solid but are made up of small units called microfibrils. The terms used in wood anatomy (morphology) today are all standardized and agreed upon by organizations, such as the International Association of Wood Anatomists. Dictionaries are available. Microfibril (or microfibrillar) angle is the accepted term, not fibril, micellar, or fibrovascular angle. The microfibrils are the basic building units of all wood elements. They vary in width from 1 micrometer for the primary walls to 10 micrometer in the secondary walls. The "angle" usually referred to, and certainly in this book, is the microfibril angle in the dominating, secondary wall, S_2. "Fibril" is a term that has never been defined. The orientation of these microfibrils has a major effect upon wood quality and stability, especially longitudinal and tangential shrinkage (Harris and Meylan 1965). A generalized statement was made by Cave and Walker (1994): "There appears to be sufficient variation in microfibril angle between trees to justify selection of clones to yield stiffer timber." When the microfibril angle is steep, longitudinal shrinkage of the wood upon drying is small, but when it is flat, such as in reaction wood, juvenile wood, or around knots, longitudinal shrinkage can be large, as much as nine times that of normal wood. Therefore, when boards contain wood with differing microfibril angles they are unstable and warp, check, split, and twist. In addition, board stiffness is not closely related to wood density but appears to be more related to fibril angle (Cave and Walker 1994). In fact, one of the worst features of reaction and juvenile woods is the flat microfibril angle; resulting defect and degrade are very high. Even paper and plywood can be unstable from differential shrinkage, and paper properties can be affected. (Kellogg et al. 1975). This was also shown by Meylan and Probine (1969), who felt that large microfibril angles had a greater effect on corewood properties than does density or other factors. The importance of the microfibril angle is becoming greater as larger amounts of juvenile wood are being harvested in conjunction with younger trees or more intensive utilization of harvested trees. Meylan and Probine (1969) suggest that breeding is necessary to improve microfibrillar angle. Mergen and Furnival (1960) found that the microfibril angle was under some genetic control in the hybrid between *Pinus thunbergii* and *P. densiflora*.

Microfibril angle is not constant and varies between earlywood and latewood. For example, McMillan (1973) found in loblolly pine that microfibril angles were 33° for latewood and 27° for earlywood. The angle was constant with varying growth rates when wood specific gravity was low but it did vary when it was high. In three different genetic groups of trees, Donaldson (1993) found that all groups had similar pith to bark trends. The microfibril angle at breast height was found to be similar by Donaldson (1992) but in the 1993 study, a decline in microfibril angle was evident above breast height, as had been found in several softwood species. The differences in the angles of individual genotypes are large enough that the properties of solid wood are significantly affected (Donaldson 1992).

For a wood defect of such great importance as adverse microfibril angle, one would expect it to have been the subject of much study. This is true (Boyd and Foster 1975), but it is interesting that very little has been genetic research. This serious wood problem is closely related to stiffness of boards. When studied, grossly different patterns are evident; in loblolly pine, the microfibril angle decreases consistently and markedly from pith to bark, while in ash (*Fraxinus americana*), the angle varied little over a span of 50 growth rings (Hiller 1968). Such genetic control from species to species is a good indication that there is genetic variation within species. In slash pine, the ring number from the pith was correlated ($r^2 = 0.72$) with microfibril angle (Jackson 1964). In another study of the microfibril angle, Hiller (1954) stated: "Trees of the same age and similar vigor of growth differ in the size of the fibril angles in comparative annual rings from the pith. The size of the fibril angles in all consecutive annual rings seems to be dependent on the size of the fibril angles of the very first annual rings. This suggests that genetic factors may directly or indirectly influence the size of fibril angles in a tree."

The microfibril angle often is not constant within an annual ring and the microfibril angle varies from pith to bark with the largest angles occurring in the first five to ten growth rings, according to many investigators (Donaldson 1993). As an example, Smith (1967) reported that the heritability of microfibril angle in the last formed latewood is high compared to the rest of the ring in the third and subsequent rings. There was an increase in the heritability of microfibril angle with age from the pith. Heritability was highest for the first 10-year increment.

Although the variation of microfibril angle from tree to tree appears to be genetically controlled, the degree of inheritance is not well known. There are strong suggestions, such as found by Jackson (1964), that the microfibril angle of the open-pollinated progeny of slash and loblolly pines was significantly correlated with the microfibril angle of their respective parents. In control-pollinated families, Jackson (1964) found that the progeny were related to the female parent or mid-parent value, but not to the male parent. He also reported that the microfibril angles in the first-year wood in the stems and branches were closely correlated. He concluded that branch wood can be used to evaluate the variation of the microfibril angle of both progeny and parents. However, in *Pinus pinaster* Keller (1973) found nonsignificant values for heritability of the microfibril angle.

7.4 Miscellaneous Wood Grain Patterns, Figured Wood

One of the attractions of wood is the varied grains it exhibits. Wood from the genus *Populus* or *Gmelina* show a very minimal grain and these white woods are often stained to simulate desired wood like black cherry (*Prunus serotina*), black walnut (*Juglans nigra*), or any one of many highly desired species. Other species, especially the ring-porous hardwoods, as well as some of the diffuse-porous tropical woods, have very distinctive and sometime dramatic grains. Some are rare and very valuable, as described for some hardwoods by Harris (1989).

Very little work has been done on the genetics of desired wood grain patterns, although the inheritance is definitely related to species, and sometimes by provenance within species. Whether the differences are caused by genetics or environment is usually not known. For example, walnut wood from the midwestern USA is supposed to be of better quality than walnut wood from the south but this is disputed. Black cherry from Pennsylvania is touted as being superior to that from the Blue Ridge mountains further southwest. There is no question that teak (*Tectona*) wood varies considerably by source as do the mahoganies (*Swietenia*). However, the causes of the variations are not well understood.

Work on the genetics of figured wood was discouraged by early failures. For example, Bailey (1948) reported that vegetative propagation of figured trees of red maple (*Acer rubrum*), yellow poplar (*Liriodendron tulipifera*), and black walnut did not produce trees with figured wood. These results were similar to those obtained by Lamb, who in 1940 grafted curley grained walnut trees that after 18 years had not yet developed curly grain. To counter this, however, Tellerup (1953) proved that there were distinct individual differences in the shape of the wood rays of *Fagus sylvatica* (beech) from several different clones. Also, Walters found in 1951 that the figured characteristics of *Juglans nigra* were transmitted by grafting.

Proof of genetic control of desired grain variations through seed regeneration are few. One of the early and best known was curly grained birch (*Betula verrucosa*) reported by Heikinheimo in 1951. He found that when curly-grained parent trees were crossed, about 50% of the progeny produced curly-grained wood, but many of these did not grow to tree size, remaining as shrubs. It has been said that over the years this cross was so productive that curly-grained wood became much less valuable. In Sweden, Johnsson (1950) concluded that wavy grain in birch is conditioned by a hereditary disturbance in the function of the cambium. Control crosses showed a high frequency, but less than 100%, curly progeny. He even sometimes found that open-pollinated trees produced a high proportion of curly progeny. Based on his own research and that of others, Ruden (1954) proposed the varietal name *Betula verrucose* v. *maserica* for curly birch produced by genetic peculiarities in bark growth. Currently, in Finland, there is one silver birch seed orchard (*Betula pendula* f. *carelica*) to produce curly grain. Inherited grain deviations were reported in sycamore maple (*Acer pseudoplatanus*) (Conrad 1977). Another type of wavy grain (associated with spiral grain) is found in *Aesculus hippocastanum* (buckeye) (Pyszynki 1977). It probably is more a defect than a decorative wood. Two kinds of waves occur – long ones, about 500 mm, and short ones of 6 mm.

Variations in the amount and location of wood rays, along with the arrangements of vessels, give striking examples of differing wood grains. One example is the large wood rays sometimes found in beech which give the wood an unusual appearance along with high tangential shrinkage (Keller and Thiercelin 1975).

7.5 Reaction Wood

Compression wood in conifers has been summarized by Timell (1986). Usually the inheritance patterns of reaction wood (tension and compression woods) are not calculated because reaction wood is the result of so many different environmental factors and tree form characteristics, and especially with the straightness of the tree. Shelbourne et al. (1969) (Fig. 7.4) found the severe form of compression wood of loblolly pine to be closely related to straightness, but the mild form actually was not related to straightness. The authors concluded that other factors, in addition to tree straightness, control formation of reaction wood. There sometimes seems to be a genetic pattern to reaction wood production. Apparently, the potential of some trees to produce reaction wood more strongly than others under a given set of environments is genetically controlled. A difficulty in determining inheritance in reaction wood relates to the problems in measuring it.

Although some inheritance patterns for reaction wood are small, there have been reports on a high heritability of reaction wood ($h^2 = 0.95$) by Shelbourne et al. (1969) for severe compression wood in loblolly pine. In *Pinus radiata*, Burdon (1975) confirmed the strong inheritance of compression wood. For slash pine, Einspahr et al. (1964) reported $H^2 = 0.34$ for compression wood. In a study of 5-year-old quaking aspen (*Populus tremuloides*) Einspahr et al. (1967)

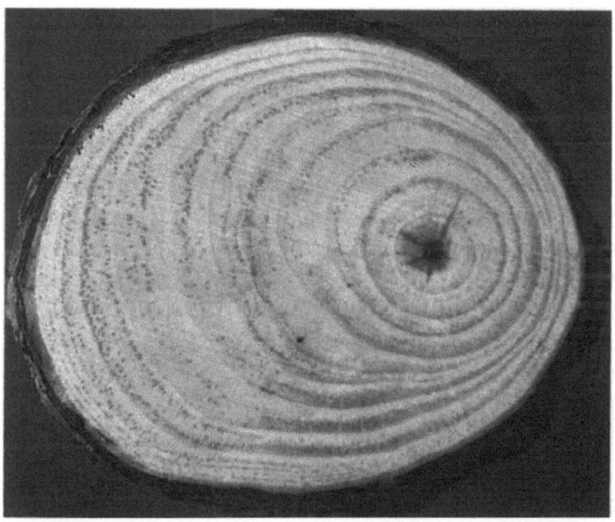

Fig. 7.4. When a tree is not straight it produces reaction wood. Shown is compression wood from the underside of a leaning pine. Surprisingly, reaction wood is not only related to tree straightness (and limbs) but there is a moderate genetic control of the propensity to produce reaction wood. Two trees, equally crooked, often produce differing amounts of reaction wood

found that there was essentially no additive genetic control of tension wood. This was similar to studies of others such as Harding et al. (1991) with compression wood of *Pinus caribaea* where an h^2 of only 0.02 was obtained. For *Araucaria cunninghamia* (hoop pine), Harding and Woolaston (1991) reported that compression wood had a heritability of 0.19. However, Eisemann et al. (1990) obtained a value of $h^2 = 0.44$ for compression wood in the same species.

For 20-year-old progeny of *Pinus radiata*, Burdon and Young (1991a) report an h^2 of 0.61 for compression wood while Cown et al. (1992) found $h^2 = 0.64$ for 20-year-old radiata pine. They summarized: "The heritability estimate ... confirms that there is a strong genetic influence on the incidence of compression wood in *Pinus radiata*." These are relatively high values for genetic control of this characteristic; they are similar to those reported by Shelbourne et al. (1969) ($h^2 = 0.95$) in loblolly pine, based upon both open and control pollinated progenies. Such high heritabilities were completely unexpected.

The heritability of reaction wood would be expected to follow the inheritance of tree straightness since the two are closely related. This however, is often not the case, no doubt because of the complexity of the causes and plant responses which result in reaction wood formation. However, when reaction wood is present it can have a pronounced effect on the final product, both for solid wood and pulp products. For the latter, Wangaard (1958) reported that pulp yields of loblolly pine were lowered by 5 to 6% in the sulfite process and 3 to 5% in sulfate pulping when compression wood was used. Its negative effect on solid wood products is widely recognized.

7.6 Cracks, Shake, and Other Internal Defects

There are numerous defects that occur internally within a log. Most of these are associated with environmental factors, branches, lean, or other external forces. Yet, directly and indirectly there is some genetic control. An example of indirect control was cited for *Quercus robur* and *Q. petrea* in which shake and longitudinal fissures were related to vessel area (Kanowski et al. 1991). There was genetic control of vessel area with a large narrow sense heritability of $h^2 = 0.60$ and broad sense of $H^2 = 0.93$.

The presence of heartshake, splitting, and collapse near the center of the tree can be a major cause of lumber degrade (Purnell 1988). Heartshake is generally considered to be the result of wounding in the nursery or planting stages. The damage at the root collar zone causes a point of weakness that enlarges longitudinally and radially into heartshake as the tree grows. However, based on his studies of *Pinus elliottii*, Darrow (1992) ended his discussion: "There is a possibility that there are genetic differences in the incidence of heartshake among families of *Pinus elliottii*." He pointed out that there are very large differences among trees in the severity of heartshake and the degree of resinosis associated with heartshake that cannot be presently explained.

In wood manufacture and seasoning, all types of internal anomalies occur within the wood as the result of many differing factors. These are usually not considered from a genetic viewpoint, but sometimes they are, as was done by Sesbou (1981) regarding shrinkage with collapse in *Eucalyptus camaldulensis* (Fig. 7.5). He reported a large variability among trees, and 40% of the shrinkage with collapse had a genetic relationship. In 11-year-old *E. nitens*, no significant differences among seedlots could be found. Collapse was most evident between the heartwood and sapwood; it was not related to splitting (Purnell 1988). Collapse is related to the percentage moisture content saturation and percentage cell cavity volume that contains water (Chafe 1985).

Another serious degrade is splitting of the log, at or after harvesting. This is especially prevalent in many tropical hardwoods and in species such as *Eucalyptus cloeziana*. Strains of trees, or clones, have developed in the eucalypts, where splitting is minimal and where genetic control is obvious (Kromhout and Toon 1978). In an 11-year-old provenance test of *Eucalyptus nitens*, no differences in splitting between seedlots could be detected (Purnell 1988). Purnell stated that this was the result of huge differences among trees. The sawing quality of *Eucalyptus saligna* can be improved by breeding and selection. Interestingly, Barr (1983) found the tendency to split to be more associated with stands than with

Fig. 7.5. Many trees, especially the tropical hardwoods, have internal stresses and strains that cause degrade, as shown by this log from the Amazon. The tendency to develop such stresses is suspected to have a genetic foundation but there are few solid data about its genetic control. (After Zobel and van Buijtenen 1989)

individual trees. This is opposite to our experience with *E. grandis* and *E. globulus*, where the nonsplitting characteristic varies greatly by individual tree. This enables the development of clones that produce wood suitable for high quality solid wood products when rooted cuttings are used. A method of assessing splitting, called triangular splitting percentage, is used for *E. grandis* but it is not accurate for *E. nitens* because the elliptical splitting is large and the two must be assessed together (Purnell 1988).

7.7 Summary

There are several different grain and fibril patterns and internal defects that can influence wood utilization. Most of the variation patterns are greatest from tree to tree.

1. *Spiral grain* has been widely studied. It has a major effect on the utilization of solid wood products. The grain angle is important for some species (Douglas-fir, larch, and some pines) while of lesser importance for others (the southern pines). The greatest spiral is nearly always near tree center and usually decreases and even reverses towards the bark. There are many different ways to measure spiral grain and results can differ considerably. Yet, despite this, a moderate to high heritability is usually found. Heritability values are listed in Table 7.1.
   a) Highest heritabilities are near tree center. An example is radiata pine with $H^2 = 0.66$ for the second ring but only $H^2 = 0.28$ for the eighth ring.
   b) Spiral grain is common and found in most species, but it is difficult to decide whether the cause is environmental or genetic.
   c) Most emphasis has been on the inheritance of spiral grain in the conifers but its presence is also evident in some hardwoods.
2. *Interlocked grain* is an extreme and complex form of spiral grain that, when severe, makes wood essentially useless for some products. The studies that have been made, mostly on the hardwoods (including the eucalypts) have found interlocked grain to be moderately inherited with great tree-to-tree differences.
3. Little genetic work has been done on the genetics of *microfibrillar* angle, despite its importance. There are a number of modern methods now available for measuring microfibril angle, no more difficult to apply than those used for determining wood density or tracheid/fiber length. They are also quick. Large tree to tree variations exist. Most (of the few) studies made on its genetic control show a moderate strength of inheritance.
4. There is much interest in the cause and development of *grain patterns and figured wood* because of their value and rarity. Early tests mostly showed that figure was closely related to environment but later ones indicate that strong inheritance patterns sometimes occur.
5. *Reaction wood* and especially compression wood in conifers is generally believed to be highly related to tree form, straightness, and limb characteristics.

This is generally true, but sometimes strong genetic control is found in which certain genotypes respond to the environment different than do others. Although some papers report little additive variance (therefore low heritabilities), others find strong h^2.

6. Cracks, heart shakes, splitting, and collapse are internal defects, usually occurring near the tree center. They are commonly thought to be caused by environmental factors but some also have a reasonably strong genetic component. Heartshake and splitting are common; for the latter, in certain eucalypt species, nonsplitting genotypes have been found and are used in vegetative propagation programs to produce straight-grained lumber.

Chapter 8

Tree Form and Internal Tree Characteristics

8.1 Introductory Comments

Previous chapters have dealt with the most common wood characteristics and their inheritance patterns. This chapter, and Chapter 11, cover a miscellany of wood properties that did not fit into the previous chapters; but that does not imply that this chapter does not contain important information. It does! For example, improving tree form is the fastest and easiest way to improve wood properties and needs major consideration. Although a limited amount can be done with it genetically, an alteration of the pattern of juvenile wood production will have a major effect on wood utilization. Although wood chemistry often shows a strong inheritance, its effect on the final product is usually less than that of anatomical changes, and research on genetic aspects of wood chemistry has been limited. Moisture content, bark percentage, and wood color all have important inheritance patterns.

8.2 Stem Form and Branching

Genetic improvement of wood quality is usually related to internal characteristics such as cell length or wood density. However, great improvements in wood properties can also be obtained by altering the shape of the tree and its branches by genetic manipulation. The basic phases of stem form development have been discussed by Larson (1963). Changing tree form and branches is frequently the fastest way to improve wood because of their heritability patterns and ease of measurement. This relatively rapid and easy way of improving wood by genetic manipulation is too often overlooked.

8.2.1 Stem Straightness and Sinuosity

This section deals specifically with the genetic control of stem straightness or, as it is sometimes called, stem sinuosity. Reaction wood, and especially compression wood, which is commonly associated with stem crook or sinuosity, is exceedingly important in all wood products, and is a major determinant of wood quality (see Timell 1969, 1986 and Chap. 4 of Zobel and van Buijtenen 1989 which deal extensively with this subject). The importance of stem straightness is recognized by the lumber industry. For example, Wagner and Taylor (1993) reported that the 20% lower lumber recovery in pines from the southern USA (when compared to that from the Rocky Mountains) can only be accounted for by log shape, including roundness, straightness, and taper. These can all be influenced by genetic manipulation of tree shape.

Stem straightness and sinuosity can be measured in many ways. It is not the place here to go into details other than to point out that simple methods work well. For loblolly (*Pinus taeda*) and slash (*P. elliottii*) pines, we have found that a simple subjective method using a scale of 1 to 6 works as well in categorizing the trees as does actual measurement. In Douglas-fir (*Pseudotsuga menziesii*), the method of scoring by maximum displacement for stem sinuosity is relatively simple and usable according to Adams and Howe (1985), who found a family heritability for stem displacement of $h^2 = 0.59$. To rank trees and families, the grosser and simpler measures are usually satisfactory.

Tree straightness has a moderately strong inheritance pattern. Because a coverage of the genetics of stem straightness occurs in many other publications, such as Timell (1986), who lists dozens of publications as early as 1905, it will be dealt with only peripherally here. Table 8.1 covers some of the older references. However, many more recent references could have been added to the table, such as Allen (1985), who found the narrow sense heritability of straightness for slash pine to be $h^2 = 0.14$ and the broad sense $H^2 = 0.46$ or Ladrach and Lambeth (1991) for *Pinus patula* where $h^2 = 0.24$. For noble fir (*Abies procera*) Doede (1993) reported $h^2 = 0.33$ for stem sinuosity. Because of its moderate herita-

Table 8.1. Early examples of the inheritance of tree straightness. (After Zobel and van Buitjenen 1989, Table 4.2)

Species	Reference	Inheritance pattern
Larix decidua	Engler 1905	Differences in stem straightness were great from seed of different origins
Pinus taeda	Perry 1960	Straight parents produced straight progeny
Pinus taeda	Goddard and Strickland 1964	Large differences occur in the inheritance of crookedness
Pinus elliottii	Gansel 1965	The number of crooks/unit length had $H^2 = 0.38$; degree of crook had $H^2 = 0.34$
Larix spp.	Keiding and Olsen 1965	Large differences in straightness were observed among clones
Pinus taeda	Shelbourne 1969	$h^2 = 0.39$ for stem straightness, a heritability high enough to obtain good gains quickly
Pinus sylvestris	Ehrenberg 1970	There was a good inheritance for straightness
Dalbergia sisoo	Vidacovic and Ahsan 1970	Heritability of crookedness varied from 0.42 to 0.65

bility, stem straightness can be effectively used to reduce the frequency of compression wood. In a meeting to assess the major needs for better forests and wood in New Zealand, stem straightness was listed as being of primary importance (Burdon and Thulin 1965).

The reason stem straightness is a major breeding objective is because of its importance to the final product along with its moderately strong inheritance. Several studies have been made to quantify this importance, indicating how strongly straightness should be emphasized as a breeding objective. This was done for paper production in loblolly pine (Blair et al. 1974), where it was found that straight trees produced significantly better yields and better tear strength than did the crooked trees (Fig. 8.1). Zobel et al. (1977) found that a 13% greater dollar value was obtained for plywood from straight compared to crooked loblolly pine trees. These authors summarized: "These results give emphasis to the importance of tree quality ... in a well balanced tree improvement program." A similar study on the effects of tree straightness on lumber production was done by Bridgwater (1984). Again, the importance of stem straightness was confirmed.

In maritime pine (*P. pinaster*), Polge and Illy (1967) investigated the anisotrophy of trees (i.e., the tree is not round but wider on one side than the other) with its resultant compression-like wood. In nine half-sib families (237 trees) they found that $h^2 = 0.21$ for the tendency to east-west anisotrophy. Thus, some improvement in anisotrophy could result by selecting for roundness of the tree stem.

There is no question that the best and easiest way to improve wood quality of a tree is to breed for better form; this was summarized by Martin (1984): "... genetic selection will result in straighter trees and more disease-free trees with fewer and smaller knots...." Straightness is among the easiest of traits to control genetically. This was proven in the selection program of the North Carolina State University-Industry Cooperative, where in one generation of intensive selection for tree straightness, enough gain was made that this characteristic did not need to be further emphasized in advanced generations. Similar ideas of the importance of stem form were expressed by Blair and Olson (1984) when they emphasized the opportunity to obtain better wood in forest trees by improving either stem straightness or crown form. As outlined above, there perhaps *can be no greater pay-off from genetics in a wood quality program than strong selection and breeding for tree straightness.* Stem deformities related to compression wood, the effect of compression wood, and the genetics of compression wood formation were thoroughly covered in the three-volume treatise by Timell (1986).

8.2.2 Stem Taper

Stem taper has a major effect on wood quality and wood utilization. The few studies reported on genetic control of taper have basically been inconclusive. For

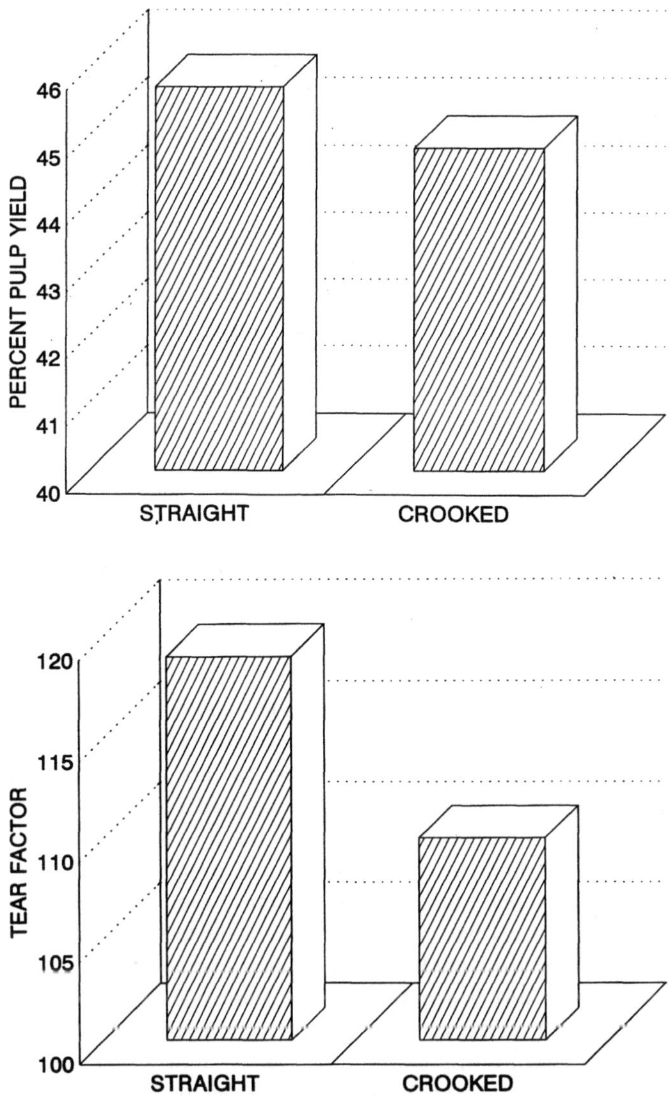

Fig. 8.1. Tree straightness is related to the amount of compression wood in conifers, and compression wood is related to yield and properties of paper. Shown is the effect of tree straightness in loblolly pine on both yield (*above*) and quality (tear factor), (*below*) of pulp

example, the taper assessments on loblolly pine by Pederick (1970) indicated that the differences in form class between families were too small to materially affect volume determination. Similarly, for noble fir (*Abies procera*) only a $h^2 = 0.09$ was obtained by Doede (1993). Pederick summarized that: "... evidence for strong genetic control of differences in stem form has as yet not been clearly demonstrated." This was also reported for Douglas-fir by King et al. (1992), where the low heritability of about $h^2 = 0.10$ was found for bole taper. In contrast, a broad-based report by Johnson (1960) suggested that stem form was under genetic control, even within individual populations. He felt that it would be easier to work genetically with taper than with growth rate. Provenance differences in stem form were reported for ponderosa pine (*Pinus ponderosa*) by Callaham and Liddicoet (1961) and Squillace and Silen (1962).

The easiest way to improve stem form of pine is to use rooted cuttings from 2- or 3-year-old trees or hedges. It has been generally observed by many foresters, and is widely accepted, that trees from such cuttings are more cylindrical with less butt swell than is present in trees grown from seedlings. This information is used widely in New Zealand and elsewhere to produce trees best suited for solid wood products. Further, the limbs from the rooted cuttings are smaller and have flatter branch angles than those from seedlings. However, no genetic studies have been reported on this form of control of bole characteristics. It is, however, an excellent example of a way to obtain improvement in tree quality using a nongenetic methodology.

8.2.3 Branching Characteristics

The presence and absence, as well as the sizes, angles, and locations of branches, has a major effect on wood properties. Most limb characteristics are only moderately inherited (Zobel and van Buijtenen 1989). Despite this, New Zealand foresters made smaller limbs and dispersed knots a major goal in improvement of *Pinus radiata* (Burdon and Thulin 1965). Breeding for smaller limbs with flatter branch angles is quite common in tree improvement programs, but published results are not plentiful. A few comparisons of limbs from trees where a mild selection for limb characteristics have been made with those from commercial planting stock were revealed in a study of the branch characteristics in an 8-year-old loblolly pine progeny test is summarized in Table 8.2 (Zobel 1973).

It is obvious from Table 8.2 that even the low selection differential used for branch traits in this study with limited families resulted in useful gains, mainly in branch diameter, branch size, and knot volume. A detailed study on the inheritance and value of limb characteristics was made with Douglas-fir by King et al. (1992). They found that significant amounts of additive genetic variance were evident for all eight crown traits studied except for branch thickness, and three traits had significant nonadditive genetic variance.

Branch size has a major effect on paper quality. Large differences were found in tearing strength of paper made from kraft pulp of young loblolly pines (Blair et al. 1974); large-limbed trees produced paper with considerably lower tear strength than did small-limbed ones. It was not possible to estimate exactly the value of limb size in making plywood, but Zobel et al. (1977) found that large knots required patching, which considerably reduced the amount of valuable sur-

Table 8.2. Branch characteristics of young improved and unimproved loblolly pine in an 8-year-old progeny test

Characteristic	Trees from commercial seed	Trees from seed orchard seed
Branch diameter (mm)	19.0	15.5
Total number of branches	30	32
Basal area of branches (cm^2)	103	71
Branch angle from vertical (degrees)	56	61
Knot volume (%)	1.3	0.9
Branches over 2.5 cm diameter (no)	4.5	0.7

face plies that were produced. The conclusion is that even though limb size is not as strongly inherited as bole straightness, its effects on plywood are significant enough for branch size and angle to be included in a tree breeding program if quality plywood is desired.

Degrade caused by branches is of two types: that due to the knot-wood itself and that due to the reaction wood associated with the branches. In young (8-year-old) loblolly pine, von Wedel et al. (1968) found 1% of the wood of the merchantable tree to be actual knotwood while 7% of the wood was compression wood associated with the branches. In the last (top) bolt in the tree there was 12% knot and associated compression woods.

Most foresters consider size of branches to be closely related to tree size. However, in Douglas-fir, King et al. (1992) found that a positive association existed between yield and a fine branching form type. They stated that multitrait index selection can be used to improve both yield and fine branching in Douglas-fir. For Norway spruce (*Picea abies*), Lewark (1981) warns against selecting for trees with large volumes because this will result in low-quality trees with large branches. This idea was also expressed by Bamber and Burley (1983) for radiata pine when they stated that spacing and stocking are the chief determiners of branch size and that branch size is not strongly inherited. In *Poplus tremuloides*, Jefferson and Yanchuk (1985) reported only a low heritability for branch size. However, the inheritance of branch size can be fairly high (Fig. 8.2) as illustrated in Fig. 8.3. Broad-sense heritability for branch size of slash pine was $H^2 = 0.40$, according to Gansel (1965). Good inheritance of branch fineness in *Picea abies* was reported by Karki (1980). A heritability of $h^2 = 0.26$ was obtained for the ratio of branch thickness to stem diameter for Douglas-fir (King et al. 1992) and $h^2 = 0.15$ for branch diameter in noble fir (Doede 1993).

Although the relationship of branch length to wood properties is not as clear as branch size per se, length is related to wood properties by its association with branch size and branch angle. There is controversy concerning the genetic control of limb length, which is often referred to as crown slenderness. There is no question that this characteristic is strongly controlled by stocking, as mentioned by Bamber and Burley (1983) for radiata pine. However, in *Picea abies*,

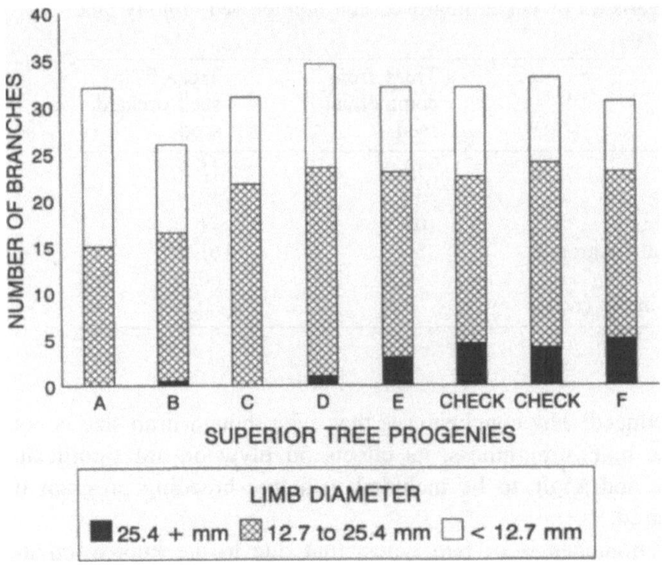

Fig. 8.2. In addition to the size of branches, the number of branches are important in determining wood properties. Shown are the results for breeding for limb size and number. Note that trees that have more limbs also generally have smaller limbs

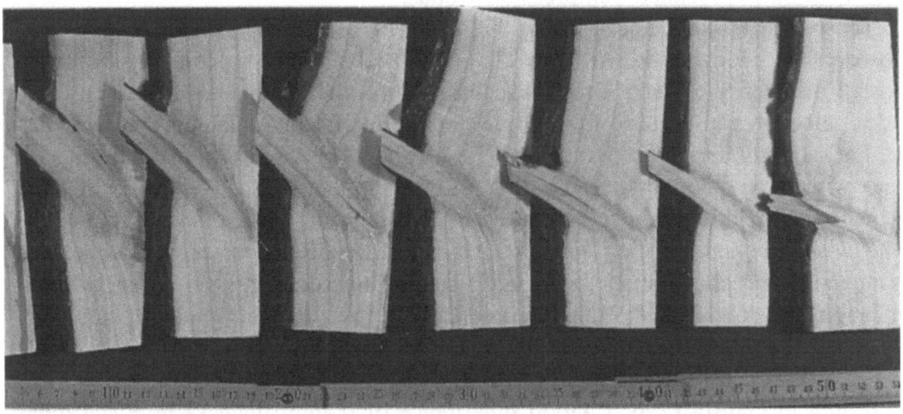

Fig. 8.3. Branch size is inherited strongly enough to warrant breeding for improvement. Shown are representatives of seven open-pollinated families of *Pinus taeda*, indicating differences in both branch size and angle. Part of the differences can be utilized in a genetics program. Note the differences in branch size in the different progenies

Karki (1980) reported that there was a reasonable inheritance for narrow tree crowns, as does Arita (1967) for *Cryptomeria japonica* (sugi). The Japanese use crown slenderness as a criterion in their vegetative propagation programs with (*Cryptomeria japonica*). In *Pinus elliotti*, Gansel (1965) reported $H^2 = 0.43$ for limb length, a high value for this branch characteristic. In noble fir, branch length had a high heritability ($h^2 = 0.49$) (Doede 1993). In contrast, branch length in *Pinus elliotti* was reported to be under very little genetic control (Strickland and Goddard 1965). The same finding was obtained for *P. radiata* by Bannister and Burdon (1969), while for Douglas-fir, King et al. (1992) found only $h^2 = 0.11$ for branch length.

Branch angle usually has the strongest inheritance pattern of any limb characteristic, and this angle has a major effect on wood quality. This has been reported (Ehrenberg 1970, Bailey et al. 1974) mostly for conifers. However, van Buijtenen et al. (1962) found branch angle inheritance to be high in trembling aspen (*Populus tremuloides*) with broad sense heritabilities varying from $H^2 = 0.68$ to 0.78; similarly, Jefferson and Yanchuck (1985) also reported a high heritability for branch angle in *Populus tremuloides*. Another example is Holst (1960), who observed that an acute branch angle is strongly inherited in black spruce (*Picea mariana*), as has also been found for maritime pine (Polge and Illy 1967). For lodgepole pine (*Pinus contorta*) branching habit was found to be strongly inherited by Polk (1972) and this included branch angle. Bamber and Burley (1983) reported that branch angle in radiata pine has a reasonably strong genetic component, and Jefferson and Yanchuk (1985) found that genetic gains for branch angle breeding are the highest in the branch category. For Douglas-fir, branch angle was found to have the highest heritability of any limb characteristic by King et al. (1992). Yet in noble fir, branch angle $h^2 = 0.23$ was quite low (Doede 1993). Branch angle is the only branching characteristic where there appears to be unanimity about the existence of a strong genetic control.

Another branch characteristic that is of importance in affecting wood properties is the location or formation of the branches as single whorls, clustered or scattered on the stem. This is particularly critical for some species like radiata pine, where there is controversy as to whether one should breed for uninodal or multinodal branching type trees. Studies on the genetics of internode length have been made (Carson and Ingliss 1988). The exact genetic control was difficult to define because of the effects of several environmental factors. The uninodal trees produce long sections of knot-free wood but they have heavy concentrations of large knots at the nodes, which greatly weakens boards cut through that location. The multinodal trees have smaller knots scattered throughout the board. The broad sense heritability for number of limb whorls in radiata pine was strong with a heritability around $H^2 = 0.70$ according to Fielding (1960). In contrast, Jefferson and Yanchuk (1985) reported for Canadian species that expected genetic gains would be least for the number of branches per whorl; this is similar to the $h^2 = 0.10$ for branch number in noble fir (Doede 1993). The number of branches was found to be associated with 42% of the family differences in Douglas-fir (King et al. 1992).

Closely related to branch number is internode length and number of internodes. These characteristics were analyzed in provenances of *Pinus oocarpa* and *P. tecunumanii* (Corea 1989). Considerable differences in internode length were found by provenance (1.00 vs 0.75 m) in comparing *P. tecunumanii* sources.

One source had significantly more branches per unit stem length than the other 15 sources. This study indicated that, under certain circumstances, provenance selection can be beneficial to improve wood as it is affected by limb numbers and dispersion.

The inheritance for branch number and location appears to be only moderate. Data have been reported for inheritance of branch number in loblolly pine by von Wedel et al. (1967), who found significant differences in the number of limbs produced by different families. Similarly, Franklin and Callaham (1970) found that branch numbers and location were genetically controlled in lodgepole pine. Others, however, like Kristinic (1968), working with white willow (*Salix alba*) have reported that branch number is controlled mostly by the environment.

Forking is another type of branching that has a genetic component, although it usually is the result of some outside influence or damage. There are reports of considerable inheritance of forking both in hardwoods and conifers. For example, Howe (1969) found that stem forking was strongly inherited in sugar maple (*Acer saccharum*). In red oak (*Ouercus rubra*) 41% of the seedlings from one source were multistemmed while two other sources only had 4% multistemmed seedlings (McGee 1968). In *Pinus contorta*, forking was also found to have a relatively strong genetic component (Franklin and Callaham 1970), and Ladrach and Lambeth (1991) found forking of *Pinus patula* when grown in Colombia to have $h^2 = 0.46$, a rather high value. We have observed dramatic differences in forking in clones of *Eucalyptus grandis* in Brazil. In clonal material, such as rooted cuttings, forking can be very strongly inherited and detrimental.

It is difficult to summarize the importance of the genetic control of branch characteristics because of the diverse results reported. Numerous investigators conclude for *Populus*, as did Nelson et al. (1979), that spacing affects branches, but genetic control is also reasonably strong. In addition, there is usually a correlation between limb characteristics, such as branch size and branch angle (Reid 1963) for *Pinus radiata*. Limb variation was reported as being considerable and great differences were found in the inheritance of crown form in Benguet pine (*Pinus kesiya*) according to Shelbourne (1963). In sycamore (*Platanus occidentalis*), Ferguson et al. (1977) found the inheritance of branch characteristics to be large enough to be of practical value. Yet, Campbell (1965) emphasized that only limited gains can be obtained from genetic manipulation of branch characteristics in Douglas-fir.

Summarizing, it would seem that branch characteristics are of enough importance and inherited strongly enough that *they must be included in a tree breeding program or wood quality may suffer.* Unfortunately, this is not generally done because of the belief that branch characteristics are mostly determined by environment and because of the difficulty in measuring branch properties per se as well as in quantifying accurately the effect they have on the value of the final product.

8.3 Juvenile Wood and Genetics

The presence and importance of juvenile wood has been well documented by many researchers; a few examples are Bamber and Burley 1983, Megraw 1985, Senft et al. 1985, Zobel and van Buijtenen 1989, and Clark and Saucier 1991. The presence and importance of juvenile wood is rarely challenged now, although it was previously controversial (Hiley 1955, Pearson et al. 1980, Cown 1992). Juvenile wood may be simply defined as the area nearest the center of the tree where there is a change in all wood properties, as one proceeds toward the bark. Juvenile wood is frequently characterized by changes in specific gravity. It is related to the age of the cambium (i.e., the number of rings from the pith) and varies greatly among species, as well as among individuals within species. It is not related to the age of the tree but to the age of the cambium. Thus, in an old tree, in any given year, mature wood is produced near the stem base but in the same year juvenile wood is also formed near the top of the tree.

For a more complete description of the characteristics of juvenile wood see Zobel and van Buijtenen (1989).

8.3.1 Juvenile to Mature Wood Transition

The effect of juvenile wood can be altered by changing its proportion in a tree through breeding, silvicultural practices, or control of harvesting age. Its quality can be changed by directly manipulating the properties of the juvenile wood per se to make it more desirable. As shown previously, the properties of juvenile wood are fairly closely related to those of mature wood. Thus, even though most breeding is currently done to change the wood in the juvenile zone, such breeding will have a distinct effect on the mature wood of the tree.

The general concept has been that juvenile wood is undesirable, as indeed it is for some products. However, for other products it is useful. As one example, Harris (1983) points out the advantages of juvenile wood for mechanical pulps and states that it is even preferred for newsprint, tissues, and fine papers because of its low fiber coarseness compared to mature wood. Currently, juvenile wood does not have the same adverse connotation as it formerly did for fiber production since TMP, CTMP, and other methods of pulp manufacture have been developed. Juvenile wood is, however, still a major problem for most solid wood products because of its low strength and instability upon drying. The general desire in most cases, therefore, is to either modify or improve juvenile wood or to reduce its amount.

It is generally recognized that the longer-lived species have the longest period of juvenile wood formation. For the southern pines, Megraw (1986) and Clark and Saucier (1991) discussed a number of environmental factors that can alter the extent of juvenile wood formation. They found that tree spacing has little effect on the amount of juvenile wood; this was also reported for *Picea abies* by Kucêra (1994). Geographical location of the plantation can be of importance. For example, loblolly pine in the Piedmont of the southern USA produced 14 rings of juvenile wood, while in the coastal plain only 6 rings were formed. There appears to be little difference in the period of juvenile wood formation among the major species of the southern pines according to Clark and Saucier (1991), who stated that the juvenile wood patterns were not related to species so much as to the length of the growing season and rainfall patterns. Cregg et al. (1988) related the transition from earlywood to latewood to the distribution and amount of rainfall. A support of the "synchronous growth hypothesis" was obtained in 1994 by Kucéra in *Picea abies* when he found that juvenile wood formation clearly coincided with the culmination of the current annual height increment. This was found both for 19- and 29-year-old stands on different sites.

Can anything be done by genetic manipulation to alter the quality and amount of juvenile wood? The very idea was scoffed at originally, but as specific genotypes have been intensively studied, it has become apparent that species, individual trees within a species, and even small populations can have different periods and quality of juvenile wood formation. The usual pattern in the conifers is to have low wood density and short cells near the pith with a relatively rapid increase radially outward followed by a leveling off in the mature wood zone, while some diffuse-porous hardwoods have similar juvenile and mature woods. In some species, however, the pattern is different, with high specific gravity near the pith followed by a lowering for a few years and then a gradual increase, a leveling off or sometimes a continued decrease. As one example, Johansson (1993) reported for Norway spruce that basic density decreased outwards from the pith to the bark. A few other conifers also show a trend toward higher density at the tree center, such as Sitka spruce (*Picea sitchensis*) (Jeffers 1959) or white spruce (*P. glauca*) (Corriveau et al. 1990). The tendency towards having high density at the tree center is particularly evident in some ring-porous hardwoods such as the oaks (*Quercus* spp.) and hickories (*Carya* spp.). Usually, despite the higher density near the center of the tree, the juvenile wood has less strength.

In Douglas-fir, Abdel-Gadir and Krahmer (1992) found that the period of juvenile wood formation of individual trees extended to 11 to 37 rings from the pith. Most of the variation was among trees-within-plots. Significant differences were also found among families-within-provenances. The authors concluded that the period of juvenile wood production for this population of Douglas-fir was under appreciable genetic control, Senft et al. (1986) showed that juvenile wood production in Douglas-fir lasted for about 15 annual rings. They mention that top bolts of 60-year-old trees have juvenile wood properties similar to those of the same age rings at the base of the tree. For radiata pine, Cown (1992) recommends that juvenile wood should be defined in pine as occurring where the density is less than 400 kg/m^3 rather than the standard now often used of ten rings from the pith. He also emphasizes the problem of spiral grain within the juvenile wood zone.

Some Douglas-fir trees approached the transition from juvenile to mature wood sooner than others, and the selection of families with individuals reaching an early plateau would result in a shorter period of juvenile wood formation (Anonymous 1990, Vargas-Hernandez et al. 1994). The family heritability for transition age was $h^2 = 0.30$. There was no relationship between the pattern of juvenile wood formation and growth rate. In loblolly pine, Syzmanski (1991) found that a breeding program selecting for an early transition age to mature wood might be viable for some northeastern and western sources, but he did not recommend any seed source to change the juvenile to mature wood trend. Since a negative correlation was obtained between transition age and juvenile wood specific gravity, Syzmanski (1991) concluded that some improvement for an earlier transition age may be accomplished by selection for high specific gravity. Loo-Dinkins et al. (1985) observed a genetic variation in time of transition from juvenile to mature wood in loblolly pine. Fast growth was found to be negatively correlated with the age of transition of the wood type. The transition age of juvenile to mature wood was heritable ($h^2 = 0.22$) in slash pine as reported by Hodge and Purnell (1993), who stated: "Moderate selection pressure (selection of the top 25%) for transition age would be expected to decrease transition age by approximately one year...."

8.3.2 Changing the Properties of Juvenile Wood

The study of wood properties should be simple, based upon the report by Nepveu and Teissier du Cros (1976) on poplars (*Populus* spp.), who stated: "The expression of the genes concerned takes place at one year of age and does not change up to adult age." Therefore, the choice of individuals with high density wood as early as one year of age is possible since this characteristic is already expressed by the juvenile wood. This simplicity has been doubted by some, especially for characteristics other than wood density.

One example of trees having a different inheritance pattern of juvenile wood was 40 of 1000 loblolly pine trees which had a juvenile wood density similar to that of mature wood (Zobel et al. 1978). It was found that 10-year-old open-pollinated families from the mother trees with a high-density juvenile wood produced progeny with juvenile wood of high density. It was clear that the tendency to have a juvenile wood with a high density was inherited strongly enough to warrant breeding for a change in the density of juvenile wood. The progeny of parent trees with high density juvenile wood had 2.0 lbs/ft^3 (32.0 kg/m^3) dry weight higher density than progeny from average trees. Upon pulping, the high-gravity families produced about 5% greater yields/unit volume or 1% greater yields/unit weight of dry wood. Tear factor of paper made from the progeny of high specific gravity parents was significantly larger than normal. Although most genetic efforts are to increase the density of juvenile wood, especially in the hard pines, sometimes the goal is to produce a juvenile wood with characteristics more similar to those of mature wood, such as having longer tracheids.

It was found that juvenile wood density of Douglas-fir is under strong genetic control and has moderate tree-to-tree variation (King et al. 1988, Anonymous

1990, Vargas-Hernandez and Adams 1991). The studies reported were for 15-year-old trees; since juvenile wood is formed for 10 to 20 rings from the tree center in this species, the wood of the trees assessed was essentially juvenile. The individual tree heritability of juvenile wood production was $h^2 = 0.59$. Selecting families of ponderosa pine with denser juvenile wood, or families that start producing mature wood at younger ages, would improve both wood uniformity and strength (McKimmy and King 1978). Similar conclusions, based on 80-year-old trees, were reached for Douglas-fir when it was stated that selection during the juvenile period would be expected to improve mature wood density as well (Abdel-Gadir et al. 1983a, b).

In clones of radiata pine, Nicholls (1965) found that the wood density near the pith had $H^2 = 0.60$; at the 9th growth ring from the pith, $H^2 = 0.24$; and at the 25th ring, $H^2 = 0.60$. Nicholls hypothesized that different genes could be expressed at different times. Although not shown as heritabilities, the wood density of Douglas-fir at year 7 and later had stronger genetic correlations with the earlier years than those produced at 15 years (Anonymous 1990). It was reported that the specific gravity of 2-year-old wood is a reliable predictor of wood specific gravity at age 25. However, heritability of tracheid length declined with age after 4 years and the coefficient of genetic prediction for juvenile and mature tracheid lengths were close to zero. For radiata pine, Cown (1992) mentioned the possibility of changing juvenile wood through genetic manipulation.

When the heritability of wood density was compared for juvenile and mature wood of a tree, Talbert et al. (1982a) found similar heritabilities for the two woods. The weighted specific gravity of the whole tree also was similar but a little lower than the heritability for 10-year-old trees. Thus, any of these three measurements (genetic correlation of $r = 0.88$) can be used to estimate inheritance of specific gravity of a tree.

Although there was no proof of the inheritance of juvenile wood of Japanese larch (*Larix kaempferi*), Nobori and Fukazawa (1988) recommended lengthening sawtimber rotations over 20 years or locating clones with high density wood at young ages, indicating the potential for inheritance of juvenile wood to improve wood at mature ages.

In semitropical and tropical areas, fast growth produces trees of a merchantable size at young ages. This results in higher proportions of juvenile wood being harvested from fast-grown plantations than from slower-grown forests, making juvenile wood and its manipulation more important in plantation forestry. Since it is suspected that juvenile wood does not always behave genetically in the same manner as mature wood, it is important, especially in the conifers, to learn how to genetically change juvenile wood. There is currently interest in determining whether juvenile wood, which has properties undesirable for some products, can be changed to make the wood suitable for a greater range of products (Fig. 8.4) by choosing parents with the desired wood density.

There has been some progress in using genetics to change juvenile wood production and characteristics. An example is Loo-Dinkins et al. (1985), who reported a genetic difference among trees in the time of transition from juvenile

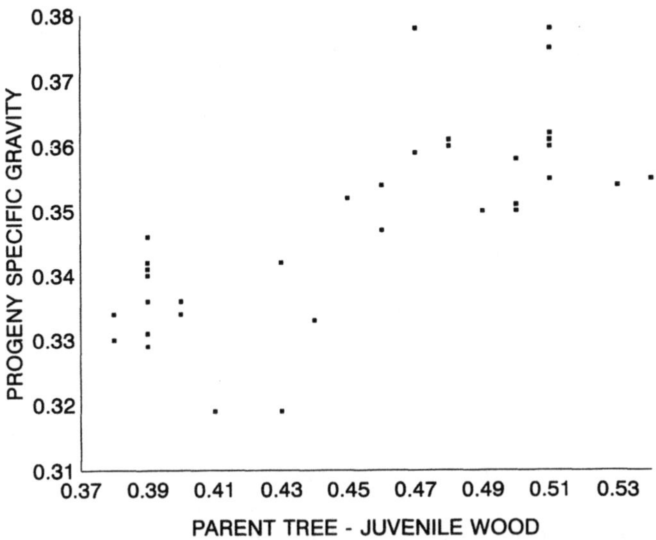

Fig. 8.4. Juvenile wood of the progeny can be altered by choosing parents with the desired juvenile wood properties. This is shown graphically where the loblolly pine parents with high specific gravity juvenile wood produced progeny with high specific gravity juvenile wood

to mature wood and that faster-growing trees tended to have an extended juvenile period when assessed for wood density. Further, they found the age of transition to juvenile wood was not correlated with either specific gravity or tracheid length. For different species, the changeover from juvenile to mature wood was found to be related to the life span of the tree (Dadswell 1958). The pattern of change in wood density was different and earlier than that for cell length although it varied greatly among species.

Although there is an apparent extension of the period of juvenile wood formation when trees are given optimal growing conditions and treatments with nitrogen fertilizers, it is evident that the large differences among trees in the period of juvenile wood formation is still maintained. For example, Zobel et al. (1978) reported that formation of juvenile wood in *Pinus taeda* usually occurs within the first 7 annual rings from the center of the tree but varies from 5 to 11 rings among individual trees. In an attempt to assess the time of change of juvenile to mature wood in the southern pines, Saucier and Clark (1992) found that loblolly and slash pines grown in the same environments have similar transition ages. They concluded that the period of juvenile wood formation, as measured by specific gravity, is more influenced by the environment of a given geographic location than by inherent differences within the tree, a result opposite to that reported by Zobel et al. (1978), who found small geographic differences but a huge individual tree variation within a region. In the coastal plain, loblolly pine produced juvenile wood for the first 6 to 10 rings while in the Piedmont it was 10 to 14 years (Saucier and Clark 1992). Slash pine produced juvenile wood for 6 years in Florida, for 10 to 12 years in the coastal plain of South Carolina, and for 14 years in the Piedmont of South Carolina. The authors offered little discussion of individual tree differences within areas. In contrast, Zobel et al. (1978) found that this is where the bulk of genetic control of juvenile wood occurs. The report by Hodge et al. (1992) agreed with the time change as found by Saucier and Clark (1992); for slash pine the ring density change from juvenile to mature wood was approximately 9.5 years. When assessing changes in earlywood and latewood separately, Hodge et al. (1992) found that the density change in latewood was 2 years earlier, even though the latewood percentage continued to increase, changing at about 10.5 years.

Juvenile wood qualities can vary greatly with the environment, especially in species like *Pinus caribaea* whose wood properties are especially sensitive to environmental differences. For example, Harris (1977) reports for this species that juvenile wood has similar wood density in the center of the stem wherever it is grown, but there is a large increase in density of the whole tree with latitude. In New Zealand, mature wood rarely develops before the 10th ring from the pith, while in Malaysia, Fiji, and Queensland mature wood is fully developed by the 10th ring. Birt and Harris (1970) found that in New Zealand juvenile wood of loblolly pine ended at about 7 years, while for slash pine it was more than 10 years and was more than 14 years for *P. caribaea*. The length of juvenile wood production of species grown in New Zealand differed considerably from that of trees growing in the indigenous ranges of the species.

Density gradients from pith to bark can vary greatly from tree to tree. Bunn (1981) found that the differences in gradients through 25 rings in radiata pine could differ by as much as 150 kg/m^3. On the average, the wood adjacent to the pith had a density of 340 kg/m^3 while that near the 25th ring was 450 kg/m^3.

The formation of juvenile wood has been less well studied in the hardwoods than in the conifers. While assessing the wood of four eucalypt species in Tanzania, Lewark and Hamza (1992) defined juvenile wood as: "... the wood formed near the pith characterized by the progressive increase in fiber length, fiber wall thickness and basic density." They found that 17-year-old trees had not yet reached their maximum fiber length, so all the wood assessed was classified as juvenile. It was of interest to note that among different classes of hybrid poplars, Marton et al. (1967) found stronger genetic control in the mature wood than in the juvenile wood.

Even though juvenile wood can be changed by genetic manipulation, as has been well proven for wood density, present research activity makes it clear that a reduction of juvenile wood by breeding will not be a major goal in the immediate future. There is the genetic potential to use it because of new manufacturing technologies for the use of juvenile wood. The genetics of juvenile wood requires a great deal of additional research.

8.4 Chemistry of Wood

Most emphasis related to the genetics of wood has been on wood and cell morphology and not on wood chemistry. Despite considerable differences within and among individual trees in wood chemistry, the major impact on paper quality resulted from anatomical variability, according to Byrd et al. (1965). However, wood chemistry can play a significant part in the utility of different woods. Some genetic studies have been devoted to wood chemistry and especially to cellulose and lignin, which involve micropulping methods (see Chap. 3.6.4 and Chap. 6.4).

One would assume that the chemistry of wood (cellulose, hemicelluloses, lignin, extractives, etc.) would be closely related to the utilization of the wood. As shown in Chapter 11, pulping characteristics per se can be relatively strongly inherited. The treatment of the wood in pulping depends on the chemical constituents of the wood. The genetic characteristics of a few of the more important chemical constituents of wood will be discussed in the following sections.

8.4.1 Cellulose and Lignin

By far the bulk of genetic studies on wood chemistry have been related to the cellulose and lignin contents. Most assessments have dealt with the importance of these two major wood components to paper manufacture (see Chap. 6.4), and only a few studies have addressed breeding specifically for desired cellulose yield.

In 1961, Dadswell et al. reported that there was little genetic control of cellulose yield in *Pinus radiata*. They commented that this very interesting observation needed further investigation because most workers felt that this conclusion might not be correct. However, an early assessment by Zobel et al. (1966) found that within 48 families of 5-year-old loblolly pine, cellulose yield varied greatly among families as well as among individual trees within families, but that cellulose was inherited in a non-additive manner such that it would not respond to a simple selection program. A later study on the same material when it was 16 years' old also showed that differences existed for both holocellulose and alphacellulose but that the heritability (additive variance) of these traits was so low that it made them unsuitable for a standard selection and breeding program (Jett et al. 1977). The dominance component accounted for about 15% of the phenotypic variance while the additive component was essentially zero. For Douglas-fir and western hemlock (*Tsuga heterophylla*) Wellwood and Smith (1962) found that tree-to-tree variation in percent cellulose was very high. They concluded that cellulose yield has a strong genetic basis but did not hypothesize about the kind of genetic control involved.

If the genetics of cellulose yield are primarily nonadditive, as cited above, then improvement of cellulose yield can only be obtained through a special control-pollination breeding program or by use of vegetative propagation, even when variation in cellulose yields are large. Although control-pollination techniques can be used, they have not been employed for the purpose of improving cellulose yield because of the difficulty and cost of making controlled crosses. When rooted cuttings are feasible, this method is preferred. It has been used very successfully with the eucalypts. By employing vegetative propagation, Campinhos (1980) reports that pulp yields were improved from 47 to 51% using only the high holocellulose producers of *Eucalyptus grandis* and some eucalypt hybrids. Brandáo (1984) showed that the average eucalypt pulp yield could be increased by 23% using clones with the desired specific gravity and high cellulose yields. In poplar, several researchers have reported real differences in cellulose yield among clones and that breeding for cellulose could be done in a selective breeding program (Schönbach 1960a, b). Similarly, based on 75 eastern cottonwood clones (*Populus deltoides*), Olson et al. (1985) found significant differences in alpha-cellulose yield among clones (48.2 to 55.8%). They concluded that alpha-cellulose content was moderately heritable and could be of importance in a breeding program when clonal propagation was used.

Some researchers prefer to work with lignin rather than cellulose because the determination of lignin is more accurate than that of cellulose. Significant differences were found in lignin content among individual loblolly pine trees which were highly correlated with pulp yield (van Buijtenen 1969b). However, the magnitude of the differences in lignin were so small (1 to 2%) that van Buijtenen questioned whether a breeding program would be warranted. Others have also found evidence for inheritance of lignin content, such as Einspahr et al. (1964), who reported that lignin was strongly inherited, having $H^2 = 0.72$ and $h^2 = 0.25$ in slash pine and similar for quaking aspen (*Populus tremuloides*). In a later study, Einspahr et al. (1967) reported a heritability for lignin in the quaking aspen of $H^2 = 0.58$. Significant clonal differences in lignin content were found in poplar by Schönbach (1960a). Also in poplars, a small genetic component relative to lignin content was obtained by Cech et al. (1960). Working with radiata pine in Australia, Dadswell et al. (1961) did not find meaningful heritabilities for lignin. A surprising result was reported by Schütt in 1958, who found significant differences in lignin and cellulose contents among provenances of lodgepole pine when grown in Germany. Most researchers have found that while lignin and cellulose differences occur among individual trees on a given site they usually are not found among stands on different sites or even among provenances.

In the kraft pulping process, about 90% of the wood lignin is eliminated from the pulp. Despite the complexity of the lignin molecule, Eriksson and Dinus (1991) have proposed genetic manipulation of lignin through methods such as somatic embryogenesis or other biotechnological methodologies. They stated: "... we particularly have in mind forest trees with lower lignin content, with lignins of modified methoxyl content, or with a reduced association between lignin and the hemicelluloses". They proposed controlling genetic and environmental factors to develop embryos and seedlings, all from the same donor trees. The studies would include Norway spruce and Eastern cottonwood and if successful, millions of seedlings with an altered lignin content could be produced. The method is not yet operationally used.

An interesting use of genetic manipulation relative to chemical degradation of wood is the development of lignin-degrading *Actinomycetes* in which the degrading organism and not the tree is genetically manipulated. Currently *Streptomyces viridosporus* and *Streptomyces setonii* are used to bioconvert lignin into useful chemicals (Crawford et al. 1982). For this to be really successful they state that: "... genetic manipulation will likely be required to make the bioconversions economic". If successful, this use of genetics would be most beneficial to the forest industry.

8.4.2 Extractives and Gum Yields

A measure of extractives is the amount of resin actually contained in the wood when the resins are extracted or when the wood is pulped. One example of extractive content extremes is heartwood formation (see Sect. 8.4.3). The wood of apparently similar trees contains different amounts of resinous materials. For older trees, percent extractives appears to be an inherited characteristic. However, in 5-year-old loblolly pine, Stonecypher and Zobel (1966) found h^2 to be only 0.05. A very high heritability, near 1.0, was reported for pinene composition in wood oleoresin of *Pinus radiata* by Burdon et al. (1992a). The authors, however, suggested that this would not be a good breeding goal.

The amount of extractives in wood can be of great importance to wood quality and utilization, and there frequently is a strong genetic control of extractives production. In a series of six crosses, all with the same mother trees but different fathers, the wood extractives in loblolly pine varied from 3.9 to 5.0% (Zobel 1973), showing the influence of different fathers on resin production. In 20-year-old radiata pine, Burdon and Young (1991b) obtained $h^2 = 0.51$ for extractives from rings 0 to 5, $h^2 = 0.45$ for rings 6 to 10, and $h^2 = 0.37$ for rings 10 to 20. These relatively high heritabilities indicate the potential for breeding for wood with less resin. However, Rink and Thor (1973) report no significant effects of genetics on extractives content in 6-year-old Virginia pine (*Pinus virginiana*). When assessing for ethanol-benzene extractives and turpentine, Franklin (1974) with loblolly pine, as well as Mergen et al. (1955) with slash pine, found important genetic differences among families for both traits. Yet, van Buijtenen (1967b) reported that the content of extractives varied considerably among southern pine trees but were not inherited very strongly.

Gum yields from standing conifers show strong inheritance patterns (Straus and Critchfield 1982). Gum from the standing tree is obtained when the tree (usually pine) is tapped for naval stores with the gums coming from the surface wood and cambial zone. Gum collection affects wood because the stem areas (faces) from which naval stores are collected usually have a heavy resinification (resin content) of the wood inward and generally upward from the area. It is evident that the severity of resin deposition below the face varies from tree to tree but no genetic assessments have been made. This concentration of resin is so heavy that for some pulps the wood cannot be used and it prevents the most valuable basal log from being used for lumber or plywood. Although dozens of reports like those of Mergen (1954), Squillace and Bengston (1961), and Bernard-Degon et al. (1971) could be cited showing a strong inheritance of naval stores yields, no more will be cited here since naval stores cannot be classified as a wood property. Studies of resins in the wood, such as in *Pinus pinaster*, (Keller 1973) reported the family heritability of the resin content to be $h^2 = 0.57$.

The genetics of extractives from the wood of hardwoods have rarely been assessed. One study on black locust (*Robinia pseudoacacia*) revealed that there were some extractive differences (overall 7.6%) among nine clones (20-year-old trees) when using hexane and ethyl alcohol extractions. Hot water extractives did not vary among clones (Stringer et al. 1982). Heritabilities (H^2) varied from 0.48 for hexane soluble extractives to 0.06 for hot water extractives. A very high broad sense heritability ($H^2 = 0.87$) was reported for extractives in the wood of *Populus tremuloides* by Einspahr et al. (1967).

The percent extractives from healthy older loblolly pine wood was 3.1%, while from diseased trees (fusiform rust) it was 19.4% (Zobel 1973). Breeding for disease tolerance to fusiform rust in pine will greatly alter the resin content of the wood by restricting the amount of disease infection.

8.4.3 Heartwood

Heartwood is a common source of resinous extractives, especially in older trees. There are huge differences in heartwood formation among trees of the same age growing under similar conditions. For example, Harris (1954) found that the heartwood content varied from 1.2 to 24.8% in ten codominant radiata pine at 12 years of age. This led to the hypothesis that there probably was a strong genetic component for heartwood formation. However, few studies have been made to confirm this. Observations by the authors of this book and the patterns found would indicate a reasonably strong genetic control of heartwood formation in mature trees for both pine and some diffuse-porous hardwoods.

Heartwood formation in radiata pine in Australia begins at about 14 years of age (Nicholls 1967b) while Koch (1972) showed that heartwood formation in loblolly pine started at about 26 years. This differs somewhat from the results shown in Fig. 8.5, in which initiation of heartwood formation in loblolly pine varied considerably from tree to tree. In the hybrid poplar *P. nigra* × *P. maximowiczii* little genetic control of heartwood was evident. It seemed to be more affected by environmental conditions (Noh et al. 1987), yet Noh et al. (1986) reported the heritability of heartwood formation for poplars in general to be $h^2 = 0.32$. In larches, heartwood formation begins when the tree is only 5 years old.

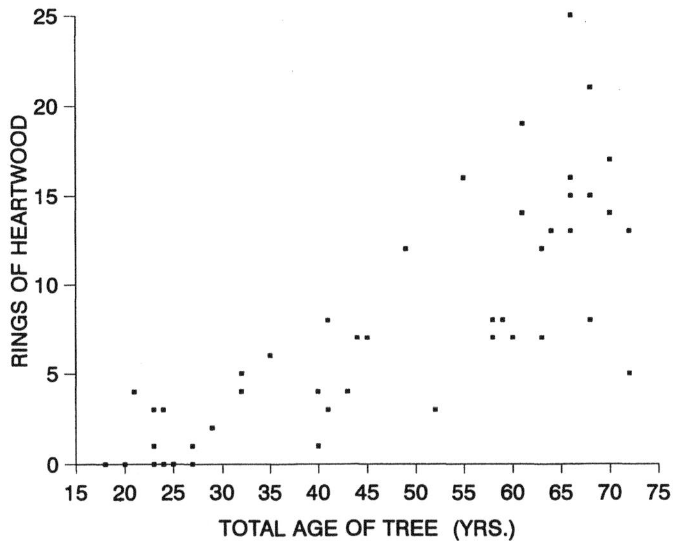

Fig. 8.5. The amount of heartwood produced varies greatly among trees within a species, as shown. Heartwood is commonly believed to be the result of environmental and growth conditions but a fairly strong genetic component of the individual tree is also indicated

Sometimes, it takes many years before heartwood formation occurs and few studies have been continued for that length of time. Some years ago, however, there were reports on tests by Schreiner (1958) to determine the pattern for heartwood formation. He found no conclusive evidence for the inheritance of heartwood formation although Harris (1954) had suggested that the tendency for heartwood formation was genetically controlled. His theory was that genetic constitution, age, and environment are all important to the formation of heartwood. Nicholls (1963) determined the broad sense heritability of heartwood formation of *Pinus radiata* clones to be $H^2 = 0.37$. Also for radiata pine, Cown et al. (1992) reported that the heritability for heartwood formation was $h^2 = 0.49$. For the same species, Burdon and Young (1991a) also found a narrow sense heritability of $h^2 = 0.49$ for this characteristic, indicating the importance of heartwood while Nicholls and Brown (1974) reported $h^2 = 0.20$ and $H^2 = 0.30$ to 0.45 for heartwood formation. These reported inheritance values were moderate to high, indicating that good gains would be obtained by selecting against, or for, early heartwood formation because there also is a large variability in this trait. In giant sequoia (*Sequoiadendron giganteum*) Knigge (1993) found huge differences in heartwood formation and suggested that these may be genetically controlled.

Heartwood usually is highly colored (dark), and for some species the colored heartwood is greatly desired. The actual color of the heartwood may be under strong genetic control as was reported for sugi (*Cryptomeria japonica*) by Honda (1912) and Kurinobu et al. (1990). The latter said proof of inheritance of black heartwood was lacking but recommended that it is better to avoid black-hearted trees in seed collection. (See section 8.5.3 for inheritance of wood color.)

There is little information about the genetics of heartwood formation in the hardwoods, other than that related to color. An exception was for *Quercus robur* and *Q. petraea* in Europe by Savill et al. (1993), who studied the sapwood content (not heartwood) and found that the number of sapwood rings had $h^2 = 0.57$ and $H^2 = 0.83$. This relationship was negatively correlated with growth rate. The authors stated: "... our results ... suggest that including the number of sapwood rings as a selection criterion could be beneficial in both breeding and clonal propagation of *Q. robur* and *Q. petraea*".

8.4.4 Other Chemicals

Among the miscellaneous studies of inheritance of chemical constituents of wood is one made on six trees from each of six control-pollinated *Pinus taeda* families (Zobel 1971). All families had the same mother but each had a different father. The glucose content of the wood varied from 65 to 70% but the differences were not statistically meaningful. The content of other sugars (arabinose, xylose, mannose, galactose) were essentially the same for all crosses. It was concluded that breeding for differences in polysaccharide content should not be done unless larger variation is encountered within a species such as was found in families of *Eucalyptus grandis* in Brazil by Bertolucci et al. (1992). In addition to the large range in wood density (433 to 549 kg/m^3), they noted that the pentosan content varied from 13.9 to 16.3%, extractives (ET/TOL) from 1.09 to 2.91%, and lignin from 27.3 to 30.6%.

Given a reasonable amount of variation in some aspects of wood chemistry, some genetic gain could be obtained because broad sense heritabilities appear to be high. This was shown in a thorough study of the inheritance of chemical properties in the wood of *Eucalyptus grandis* in Brazil by Bertolucci et al. (1992), who reported the following heritabilities:

Pentosans (mostly xylan)	$H^2 = 0.95$
Lignin	$H^2 = 0.79$
Ethanol, toluene extractives	$H^2 = 0.78$

Such high heritability values give great promise of gains using clonal propagation in plantation programs when reasonable variation is present.

Breeding for chemical properties of wood was reported for *Eucalyptus globulus* and *E. nitens* by Orme (1992). Gains of 3 to 11% were predicted for second generation improvements in several chemical properties, in addition to the usual improvement expected in volume and form using standard breeding methods.

Working with clones of hybrid poplars, Marton et al. (1967) found no differences for pentosan, lignin, xylan, or extractive yields, indicating that there was no chance for improving these chemical properties by breeding.

8.5 Miscellaneous Traits

There are a number of additional, unrelated characteristics of a tree that can influence wood properties. Three of the more important are discussed below.

8.5.1 Moisture Content

The inheritance of wood moisture content is not usually considered even though the moisture content of a tree can be accurately assessed from breast height sampling (see Chap. 3, Fig. 3.6). Moisture content in newly cut wood can have a high heritability, as illustrated for *Pinus taeda* by Zobel et al. (1968a). These authors not only described the inheritance pattern of the moisture content of loblolly pine but also reported an inverse correlation between wood density and moisture content within trees of a species growing in similar environments. The higher density trees have lower moisture content in the sapwood portion of the tree, although this relationship can vary greatly after heartwood formation has commenced. Inheritance of moisture content was also reported by Matziris and Zobel (1973).

It is important to understand the relationship of moisture content with tree age. This was clearly shown by Howell et al. (1984) in slash pine. The same tonnage of green wood of young trees has much lower fiber yields than a ton of older trees (Fig. 8.6). Generally, the high wood density trees within a species have a lower moisture content; for example, in loblolly pine with a wood specific gravity of 0.42, the moisture content was 129%, 0.45 specific gravity wood had 108% and 0.47 specific gravity woood had a 93% moisture content, calculated according to the dry weight of wood. The relationship between high density and low moisture content, however, definitely does not always hold between species, as found by numerous investigators.

In sugi (*Cryptomeria japonica*), Kurinobu et al. (1990) reported that moisture content is highly heritable. In his summary of the inheritance of wood properties, Zobel (1973) concluded that moisture content inheritance closely parallels that of specific gravity. In fact, it is often stronger (Zobel et al. 1968a). The reason for the close parallel in inheritance between wood density and moisture content relates to the inverse correlation between wood density and moisture content (Franklin and Squillace 1974). When assessing the moisture content of black spruce, Boyle et al. (1987) found the heritability of this trait to be the same as

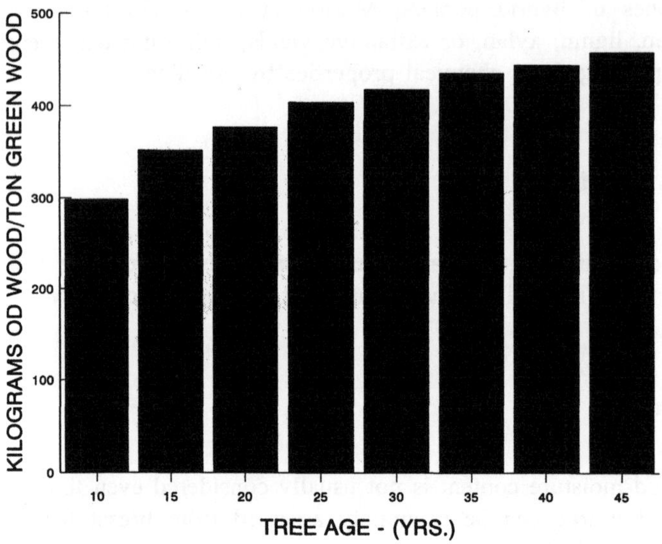

Fig. 8.6. Moisture percent in wood is strongly inherited but it is closely tied to tree age, as shown by the relationship of dry to green weight at different ages. (After Howell et al. 1984)

Table 8.3. Inheritance of moisture content in the wood of some eucalypts (After Dean et al. 1991)

Species	h^2	Kind of inheritance	Age of tree
Eucalyptus	0.60	Family h^2	1.0
grandis	0.28	Indiv. h^2	1.0
E. viminalis	0.52	Family h^2	3.0
	0.35	Indiv. h^2	3.0
E. nitens	0.50	Family h^2	8.0
	0.47	Indiv. h^2	8.0
E. globulus	0.81	Family h^2	8.0
	0.88	Indiv. h^2	8.0

for wood density, with general combining ability accounting for virtually all the genetic variance observed.

Some work has been done on the inheritance of moisture content in the eucalypts, as reported by Dean et al. (1991) (Table 8.3). In *E. Viminalis*, Otegbye and Kellison (1980) also found a strong inheritance of moisture content.

In contrast to the eucalypts, Land et al. (1983) found essentially no narrow sense heritability for stem moisture content in sycamore, with $h^2 = 0.01$. Little breeding has been done to change moisture content of the wood of any forest tree species.

8.5.2 Bark Characteristics

Some might wonder why bark is listed as a wood property when it is normally removed, and either disposed of or used for energy. However, bark is sometimes used along with the wood, as for charcoal production or for pulp with some eucalypts, poplars, and *Gmelina*. Because of its potential value, the amount of bark available and usable is of interest; the volume is huge. For example, in British Columbia, Canada, about 20% of the stem wood volume logged in 1965 (about 300 million cubic feet or 4.4 million oven-dry tons) was bark (Smith and Kozak 1967). Further, bark is often included in the wood volume reported by forest managers, especially in South America. If tree diameter is taken outside bark, actual wood volumes are overestimated. The percentage of bark fluctuates greatly with the species and size of tree; it varies from 8% by volume for some large hardwood species to 40% for small tropical pines. Thus, bark thickness is important in inventory and, if ignored, will have a major effect on the amount of wood which is utilizable. Some species, like *Eucalyptus pellita*, are not favored for pulping by some Companies because of thick, fibrous, dark-colored bark.

Most studies of bark relate to its thickness for commercial trees (Miller 1961 on slash pine) or, as done in British Columbia by Smith and Kozak (1967), for several species. Many foresters and some mill operators know the differences in bark quality, and which barks are usable and which are not. For example, in *Eucalyptus grandis* some genotypes have a smooth whitish-greenish bark very suitable for pulping with fibers similar to those of the wood and of light color. Some trees of this species have a heavier, dark, loose, stringy bark which is not desirable for pulping although this activity has not been publicized, several organizations are selecting and planting those clones with the usable bark and rejecting those with heavy, dark, undesirable bark.

The genetics of actual bark thickness has received considerable attention. In loblolly pine, bark thickness is under strong genetic control in 10-year-old progenies (Pederick 1970) while for young trees of this species Hamrick (1962) reported that bark thickness was not strongly inherited. As expected, Pederick found a positive correlation ($r = 0.5$) between tree size and bark thickness (Fig. 8.7). There were large differences in bark thickness among families and also among individuals of the same size within families. He also reported differences in bark thickness between different provenances when grown in the same environment, with the Piedmont (inland) provenance having thicker bark than the coastal plain sources. The inheritance for bark thickness varied by test location with $h^2 = 0.65$ obtained based on three trials. Pederick concluded that bark thickness is one of the most strongly inherited tree characteristics. In their study of slash pine, Squillace and Bengston (1961) reported a high heritability of $h^2 = 0.52$ for bark thickness. For *Pinus patula* grown in Colombia, bark volume had a heritability of $h^2 = 0.35$ (Ladrach and Lambeth 1991).

In a study of 24-year-old provenances of *Pinus contorta* var. *latifolia* in Sweden by Persson and Downie (1992), the bark thickness varied between the

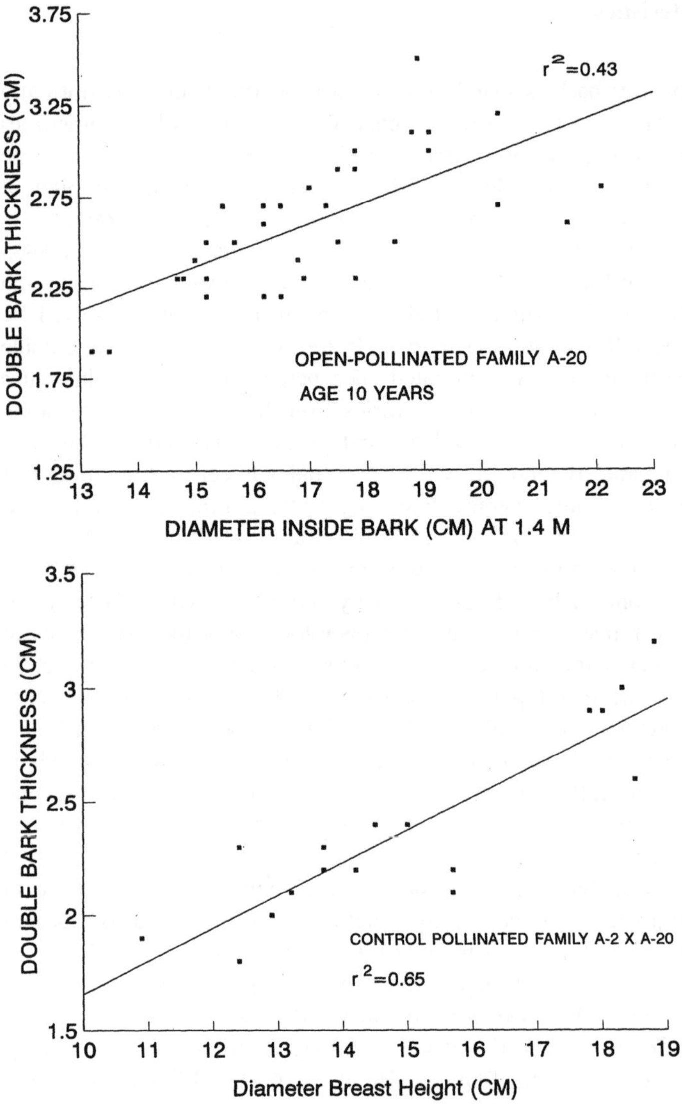

Fig. 8.7. Although bark thickness has a strong genetic component, it is also closely tied to the diameter of the tree stem. Shown are two cases of the relation between bark thickness and tree size (diameter inside bark = d.i.b.) for an open-pollinated (*above*) and a control-pollinated family (*below*) of loblolly pine trees with one common parent. Note the greater spread of values for the open-pollinated family

northern and the southern provenances. There was greater variation in bark thickness to bole size in the north. Among the provenances of British Columbia origin, no differences in bark thickness were found.

In nearly all species studied, specific gravity of the bark has been found to vary greatly among individual trees within a species. Although a genetic study per se was not made, the differences in bark specific gravity reported by Hale (1955) were: balsam fir (*Abies balsamia*) 0.35 to 0.45, black spruce 0.35 to 0.44, white spruce 0.32 to 0.37, jack pine (*P. banksiana*) 0.30 to 0.41, white birch (*Betula papyrifera*) 0.48 to 0.56, and trembling aspen 0.46 to 0.54. Such large tree-to-tree differences within species are a good indication that genetic differences in specific gravity of bark occur. However, care must be taken because both Hale (1959) and Smith and Kozak (1967) found that bark specific gravity increases with tree age as well as varies from tree to tree within species. Only one tree from each of the tree species of the USA was used in the 1967 report by Smith and Kozak. Because of large tree-to-tree variation, bark assessment needs to include at least 30 trees if reliable estimates of bark characteristics are to be obtained. Any assessment for genetic variation in bark density must be based upon tree-to-tree bark variations of a number of trees of the same age. In sycamore, Land et al. (1983) found the heritability of bark specific gravity to be low ($h^2 = 0.14$).

A special problem is related to the bark of red ceiba (*Bombacopsis quinata*) where the profuse, heavy spines that normally occur on the bark make the species unpleasant and difficult to work with. Occasionally, essentially spine-free individuals are found. A half-sib family test in Colombia, which assessed 520 trees, showed that 18% were essentially spineless and the heritability of spinelessness was zero (Kane et al. 1993), indicating that spine production may be primarily controlled by non additive genes. If so, vegetative propagation would be effective in producing spine-free trees. Some families from control pollination are essentially spine-free. An example of inheritance in the quality of bark was given by Kennedy and Wilson (1954) for the smooth and cork-bark alpine fir (*Abies lasiocarpa*). In this instance, trees with different barks had different densities.

8.5.3 Wood Color

Variation in the color of wood can be of major importance. It is often argued whether the color differences are influenced more by genetics or environment. As one example, Phelps et al. (1992) reported for black walnut (*Juglans nigra*) that there was much tree-to-tree, as well as within-tree, variation in color. These authors concluded that color of wood in black walnut is more influenced by the environment than by genetic factors. They found that wood color varied by soil type and geographic location, but they did not discuss the causes for tree-to-tree variability within a site. In sugi, Kurinobu et al. (1990) and Honda (1912) found black heartwood in some trees and surmised that there may be a genetic component for this trait. Much of the color differences in wood of *Quercus petraea* and *Q. robur* could not be explained by the usual factors of age, amount of available water, or soil variability, so Klumpers et al. (1993) hypothesized that wood color is at least partially under genetic control. Heartwood color differences and their possible genetic control in *Sequoiadendron giganteum* were mentioned by Knigge (1993).

Wood brightness is a form of color variation. It is a function of the light scattering properties of the wood and the light absorbance properties of the wood chemicals, and is a key property in the production of paper. Although it is not related to cellulose yield as such, wood brightness is a very important property affecting paper quality. In a special study on brightness of loblolly pine wood, Wilcox (1975) found considerable genetic control in which good brightness was related to low specific gravity and low latewood content. A high brightness was also closely related to a low content of compression wood. Large differences were found in the wood of 12-year-old loblolly pine grafts by Wilcox (1974) ranging from 45 to 52% for brightness. Tests were made to determine if this variation in brightness had a genetic basis. It could be shown that the brightness of unbleached loblolly pine mechanical wood pulps could be improved 1.6 points by selecting clones that had above-average brightness or below-average wood absorption coefficients. These clones also contained less lignin, which makes them desirable for pulping.

8.6 Summary

This chapter includes some tree and wood properties not covered earlier that are genetically controlled and that affect wood properties. Some are of minor while others are of major importance.

1. A topical question relates to *juvenile wood.* Most suggestions are to manipulate it silviculturally but there is also a genetic component (usually individual tree) that is now being considered in some breeding programs. There is a definite difference among trees as to the age of transition from juvenile to mature wood, and this appears to be strongly controlled genetically within a number of conifer species. The properties of juvenile wood vary greatly among species, with some having high (resin-free) wood density near the tree center (some firs, spruces, and ring-porous hardwoods) and others having the lowest (resin free) density there (most hard pines, Douglas-fir, and larch). Diffuse-porous hardwoods show only minor differences in properties between juvenile and mature wood. The quality of juvenile wood also differs among individuals within a species and is under strong genetic control.

2. *Bark* is used in some species, and its amount and quality can be of importance. There is a move to use more bark because of the huge volumes available. Bark thickness is strongly inherited as is the physical quality of the bark. Some breeding is being done to produce a bark more useful to the industry, especially in the eucalypts. The specific gravity of the bark itself varies a great deal and shows some genetic control.

3. One would think that the *chemistry of the wood* would be of major importance in pulping, but all reports are that wood anatomy has the greatest influence on the final product. There have been some genetic studies on wood chemistry,

especially with the eucalypts, that have shown good heritabilities for *cellulose* and *lignin* contents. There has been some emphasis on breeding for more desired cellulose and lignin contents. Heritabilities for lignin are quite high but cellulose appears to have almost all nonadditive genetic variance. Thus a selection program for altering cellulose yield based on additive genetic variation would not be effective. When clones are used, good gains in cellulose percent would be obtained because there is a large amount of tree-to-tree nonadditive variation. Because phenotypic variation in lignin content is low, genetic gains from selection would be small.

4. The inheritance of *extractives* and *gums* in the wood is usually strong, and these resins have a major effect upon utilization of wood for all products. However, there are a few studies showing low inheritance values. Diseases can greatly influence resin production and genetic improvement of disease tolerance is expected to alter wood quality as well. Breeding to change wood extractives is quite feasible biologically but there is some question if it would be economically viable.

5. *Heartwood* formation is frequently associated with a large amount of extractives, both in the conifers and hardwoods. The general belief is that heartwood formation is triggered by environmental influences or that it represents a form of incipient decay. Despite this, several studies report moderate heritabilities for heartwood formation.

6. *Stem form* (taper) is of much economic importance, and it varies greatly from species to species and site to site. Proof of genetic control strong enough to allow a change in taper by breeding is currently very scarce.

7. *Stem form* or *sinuosity* is important. A few examples of inheritance are shown in Chapter 11, Table 11.1, but many more are available from the literature. It is clear that stem straightness is moderately to strongly inherited, and there is a lot of variability for this trait. Improvement of stem straightness through genetic manipulation is usually considered to be of prime importance in breeding programs. It is a fact that by improving straightness, the amount of reaction wood can be reduced, which is important for both solid wood and pulp products. In fact, producing straight boles of both conifers and hardwoods is one of the best and quickest ways to improve wood properties by limiting reaction wood formation.

8. Although they are more difficult to change genetically, *limb characteristics* have an important influence on wood quality. Some limb properties, like branch angle, are strongly inherited, but limb size is more influenced by the environment in which the tree is growing, although some moderate heritabilities for this trait have been reported. Limb size is usually strongly related to tree size, but exceptions are known. There is interest in producing uninodal trees in some species like radiata pine. Several studies on branch numbers and locations indicate that these traits have moderate to high heritabilities. Despite the complexity, branch characteristics are inherited strongly enough so that they should be included in a tree improvement program if the best wood is to be obtained.

9. *Wood color* is generally thought to be influenced by the environment, but a reasonably strong genetic component has been found for some species.

10. *Moisture content* has high heritability, very similar to that of wood density. Trees are found with differing wood moisture contents even when they have the same wood density.

Chapter 9

Wood Genetics Related to Provenance and Seed Source

9.1 The Meaning of Provenance and Seed Source

The wood produced by trees grown from different seed sources or provenances can differ significantly. If the effects of provenance or seed source on wood are to be understood, the terms need to be briefly defined. The definitions by Zobel et al. (1987) provide a useful understanding, as given in Chapter 9.1.1.

9.1.1 Provenance, Geographic Source, or Geographic Race

Provenance is defined as: "... the original geographic area from which seed or other propagules were obtained .... If, for example, seed of *Eucalyptus grandis* were obtained from Coffs Harbour, New South Wales, Australia, and grown in Zimbabwe, they would be classified as the Coffs Harbour provenance ... ." Seed source refers to the area from which the seed were obtained. "If seed from the trees grown in Zimbabwe were harvested and planted in Brazil, they would be referred to as the Zimbabwe seed source and the Coffs Harbour provenance." *Seed source is where the seed are collected, provenance denotes the indigenous origin of the trees.*

9.1.2 Confusion and Complexity of Terms

Although the definition of provenance (and seed source) is relatively simple and clearcut, a great deal of confusion can occur. This results from equating the wood properties of trees where they are growing naturally with the wood from the same provenance/seed source when grown in a different environment. Sometimes, provenances growing in a new area produce wood quite different from that produced in their indigenous location. Until definitive tests are made, it is impossible to determine if these differences are primarily caused by the new environment or are the result of genetic differences attributed to the provenance or seed source. However, when relative differences in wood among several provenances (seed sources) are maintained in new environments, then it is possible to hypothesize that genetic control is related to provenance (source) where the seed were obtained. The average wood of a provenance in the new environment may differ considerably from that in the indigenous range because of the environment (Zobel 1981); but if provenances retain their relative differences in the new environment similar to the differences in the original environments, then genetic control by provenance is indicated (Fig. 9.1).

The above descriptions can be illustrated by the wood density of loblolly pine (*P. taeda*). In their indigenous environments, wood density of a Florida provenance will be much higher than that of a Virginia provenance. This difference may be due to either the environment or to genetics – one cannot tell without a test. However, when the Florida provenance is grown in South Carolina alongside the Virginia provenance, both usually have similar wood densities. This result indicates that the provenance differences in the indigenous range are primarily environmentally caused and little would be gained in trying to improve wood density by using seed from the Florida provenance and growing it further north.

Fig. 9.1. A major task in introducing species and provenances is to make sure they have suitable wood when grown in the new environment. Shown is a magnificent Mexican pine with good wood properties being collected by the CAMCORE Cooperative and tested in a number of different environments. The decision concerning using this provenance is based upon the kind of wood produced by its progeny in the new environment. (After Zobel and Talbert 1984)

It is essential, therefore, to look very carefully at reported results of differences in wood of differing provenances or species (Fig. 9.2). Do the reported values truly indicate the possibility of genetic differences or are they caused by an unknown relationship between genetic and environmental control? For example, Caribbean pine (*Pinus caribaea*) grown in New Zealand, Queensland, Fiji, and West Malaysia showed great diversity in wood density, which varied from 400 to 700 kg/m^3 (Harris 1977). Increases in overall density, latewood density, and latewood ratio were found with decreasing latitudes. However, Harris (1977) warned that such test differences can be due to chance selection of sites within a region in addition to those directly related to latitude. All these factors must be taken into account when assessing provenance differences. Genotype × environment interactions, which sometimes occur, can cause great confusion. This happens when the genotypes of one provenance respond differently to a given environment relative to the genotypes of another provenance. Although some genotype × environment interaction seems to be generally evident, it usually is so small among provenances that it can be ignored.

Fig. 9.2. Frequently, species are rejected because the provenance is satisfactory for growth but the wood is thought to be undesirable. The correct provenance was used, but it was claimed that wood of Douglas-fir (*Pseudotsuga menziesii*), when grown as an exotic, had too much resin in the wood. In fact, the resin content of the wood of Douglas-fir is similar as an exotic to that in the indigenous range of the species. (After Zobel et al. 1987)

Even though little information can be obtained directly about genetic control by assessing wood of provenances in their indigenous sites, it is of value to study trees there because it gives one an estimate of the amount of wood variation present in a species. For example, there was a great deal of interest in the potential for use of *Pinus tecunumanii*, but no one knew anything about the variability in its wood. To learn something about the wood of this species, a study of indigenous stands in different provenances was made by Eguiluz-Piedra and Zobel (1986), which indicated that there was only a slight variation in the specific gravity of 35- to 60-year-old trees (0.51 to 0.54), and the tracheids were long, averaging 4.1 mm. The results showed that the wood of this species is the best of those known for the Central American pines for use in solid wood products and paper manufacture. Such information is helpful even though it is not possible to designate genetic differences among provenances from a study of indigenous stands.

Because of the complexity and potential confusion, it is essential to understand the reported differences in wood properties between various provenances. There are hundreds of references reporting differences in the wood properties of provenances of species in their indigenous environments. Since it is not possible to determine whether these differences are environmental or genetical in nature they *will not be emphasized in this book.* **The discussion will primarily concern wood of provenances in trees grown in a new environment where one can make a reasonable assessment whether genetic control is involved in producing the differences** (Fig. 9.3).

Related to this are differences in the wood related to where the provenances are tested. Frequently, average wood properties differ under different test conditions, but the inheritance patterns usually remain the same. For example, Lima (1987) reported on provenance tests of *Pinus oocarpa* made in three different environments, two in Brazil and one in Colombia. The average specific gravities were quite different in the three areas but the heritabilities for wood density were $h^2 = 0.68$, $h^2 = 0.78$, $h^2 = 0.58$. While variable, all indicate strong genetic control of wood density.

9.1.3 Assessment of the Wood of Provenances

There are several ways in which provenance effects on wood can be evaluated (see Chap. 3). The most common is to use a radial section near breast height, either weighted or nonweighted for juvenile wood, that is, to use the increment core or wedge samples unadjusted. Sometimes, whole tree values, properly weighted for all heights in the tree, are used. There have been a few studies that assessed only the juvenile wood in fast-growing species such as the eucalypts and *Gmelina*. Very few studies on provenance differences have been reported for mature wood only.

Fig. 9.3. When the wrong provenances are used, such as this provenance of *Pinus oocarpa* in Colombia (*top*), the wood will be nearly useless because of the very high proportion of compression wood produced. If the wrong species is used, such as *Pinus elliottii* planted in Colombia (*bottom*), a similar result will be obtained. (After Zobel and Talbert 1984)

It is essential to understand whether provenance and seed source results are based on whole tree values, or only on a sample from the tree. Frequently, trends and rankings will be the same but especially for juvenile wood of the hard pines, Douglas-fir (*Pseudotsuga menziesii*), spruce (*Picea* spp.), fir (*Abies* spp.), and larch (*Larix* spp.), the patterns may vary considerably. Results related to juvenile wood differ because of the varied definitions of juvenile wood and upon what criteria are used to estimate it. For example, Jett et al. (1991), using trees with 9 rings from tree center, reported on the provenance variation of specific gravity of juvenile wood. The juvenile wood zone of loblolly pine, as estimated by specific gravity, is considered to be 7 to 11 rings from the pith, but normally 10 annual rings are used for testing; this includes the transition wood.

Two methods are used to test for provenance differences in wood properties. One is to use bulk collections of a number of trees and assess wood properties of the planting as a unit. This method gives a satisfactory differentiation of provenance but not the variation in wood properties within the stand. If bulk testing is done, about 30 trees must be used for each provenance to obtain a reliable estimate of wood properties. A bulk test with only a few trees, or where a large proportion of the seed of the provenance tested comes from a few trees, may give incorrect results. Most researchers now use the system of mother trees within provenance for testing the wood properties of provenances. This is standard with the CAMCORE (Central America and Mexican Coniferous Resources) Cooperative. Its value was outlined by Ladrach and Zobel (1986) as: "Many of the provenance trials (in the tropics) are done by parent tree, which makes possible the selection of outstanding, unrelated genotypes. A good land race will often produce 30–50% more usable wood than the original stand." Much of the added yield comes from improved wood properties. There have been questions as to the value of selecting for provenance wood density and whether the differences would be maintained in the provenance tests. Dvorak and Wright (1993) reported: "... wood density of selected trees of *P. tecunumanni* in Central America were significantly correlated with wood density of open-pollinated progeny grown in Colombia."

9.2 The Overall Effect of Provenance

It is amazing how little sound information exists relative to the genetics of wood as it is affected by provenance or seed source. There is information about the site effects on wood of trees from different provenances but less is known related to where the seed came from. The effect of provenance on the timber and pulp properties of *Pinus caribaea* when grown in South Africa was studied by Falconhagen (1979). When assessments of provenance effects are made, there frequently is little effect due to provenance. Usually, the bulk of the variation in wood properties is related to the test site (Zobel et al. 1972) along with differences among families or individual trees within the provenance being tested.

Numerous researchers state that the lack of significance in wood properties among provenances when grown in plantations is due to the overall effect of growing conditions and large within-provenance variation.

Often when provenance studies are made, they are not intensive enough to give reliable results. Frequently provenance differences are assessed at too young an age. Extreme provenance differences in wood can appear immediately, but tests must be followed to a more advanced age before reliable results can be obtained. This is necessary because of the different extent of juvenile wood production and different adaptability of the trees to the new environment.

Usually the provenance differences that affect wood are related to extremes of the environment, not to small changes in soil pH, nutrient, or other minor environmental variations. Thus, sampling for provenance assessment must be carefully planned. It is essential to know the main environmental patterns, to prevent either too wide sampling or taking excessive numbers of samples from a given area. An example of the former was cited by Langlet (1959) for Scots pine (*P. sylvestris*), in which too wide sampling led to erroneous conclusions.

Differences in wood among provenances is often the result of the effect of the environment on the physiology of the tree. The physiological reasons for wood variability are many and complex. One simplified example was cited by Kennedy (1969), where differential cessation of cambial activity among provenances was related to climate severity and affected the wood quality of jack pine (*Pinus banksiana*). The differences in the earlywood-latewood ratio, as affected by environment and/or genetics, are often cited as a major reason for wood differences among provenances grown in a different environment. Based upon the findings of Rees and Brown (1954) with *Pinus resinosa*, Saucier and Taras (1967) with *P. taeda*, Kennedy (1971) for *P. banksiana*, and Worrall (1975) with *Picea abies* (Norway spruce), it appears that the environment where the trees are grown is the major controller of latewood content. All four studies showed that the earlywood-latewood ratio differences among seed sources were slight when trees were planted in the same environment and there was no systematic trend related to the environment where the provenance or seed source originated. In loblolly pine the percent latewood appeared to be greater from the more northerly sources but differences were very small.

9.2.1 Genetic Differences in Wood Properties Among Provenances of the Hard Pines

Many of the reported provenance differences in wood properties are among the hard pines. Some initial observations and studies on the wood density related to seed source, especially among the southern pines in the southeastern part of the United States, show only limited effects on wood, while those from the southwestern part exhibit some real differences by provenance. Additionally, red pine in the northeastern United States showed very little wood differences by provenance (Peterson 1968). All but one of the red pine seed sources were similar and the provenance that differed had a specific gravity that was only 0.02 higher than that of the others. The large wood density differences in the original stands as well as provenance plantings, were primarily the result of a response to the environment.

Because of these early findings, the idea became generally accepted that provenance differences in the wood of hard pines were too small to be of any concern to the forest tree breeder. However, as more complete and more recent studies have shown, in some instances genetic control of wood density in provenances is strong enough to be a tool that should be used by the tree breeder. A few studies on wood properties as affected by provenance in the hard pines are listed in Table 9.1. A few studies are not included in Table 9.1 but are instead discussed in somewhat greater detail in the following.

In an extensive study of provenances of loblolly pine 10 years of age, Byram and Lowe (1988) found provenance differences in 24 of 28 plantings. The more northern sources had the highest specific gravity (0.42) in all plantings, while the North Carolina source was lowest (0.38) in the five plantings that contained that provenance. Of the western sources, the southern Louisiana provenance was low at 0.40. The North Carolina provenance produced 8% less fiber per unit volume of wood than the local source when planted in Arkansas. As reported by Byram and Lowe (1988), in the west: "The difference between south Arkansas and south Louisiana represented approximately 26 kilograms of dry matter per cubic meter (162 pounds per cunit). The south Louisiana family with the highest specific gravity had a lower specific gravity than the lowest from south Arkansas." The low specific gravity of the south Louisiana source was evident in all of the plantings. A point of special interest is that families generally ranked the same in all tests, indicating strong genetic control and little genotype × environment interaction. The authors concluded that studies of the specific gravities of indigenous stands should not be used for making seed transfer decisions. This can only be done accurately based on test plantings in the new environment.

In a study of the juvenile wood of 12-year-old loblolly pine, 18 open-pollinated families from widely divergent parts of the range were sampled in seven widely spread plantations by Jett et al. (1991). Four of the families were unstable and thus contributed to a large genotype × environment interaction. Trees grown from parents from the most southern part of the species range consistently displayed low specific gravity wherever they were grown; families from Florida and Livingston Parish (Louisiana) showed low specific gravity at all test locations.

In New Zealand, tests of *Pinus taeda* by Birt and Harris (1970) indicated only minor differences in wood density among the various provenances. Several provenances were found to produce useful general-purpose softwood timber, but individual tree selection would be more effective than provenance selection in improving wood density.

When a number of provenances of *Pinus caribaea* var. *hondurensis* were planted in several countries on 11 different sites, Wright (1990a) found that

Table 9.1. Some examples of genetic variation in wood properties, resins, and chemicals of provenances of hard pines

Species	Reference	Comments
Pinus banksiana	Schoenike 1962	When grown together, the slow-growing sources from northern Canada and Nova Scotia had a much higher specific gravity than the fast-growing sources from Wisconsin, Minnesota and Michigan. All wood traits studied showed about 30% difference among them, depending on provenance
	King 1967	The northerly sources produced the highest specific gravity and shortest tracheids. The between-stand component of genetic variation was much greater than the within-stand variation
P. caribaea	Wright et al. 1986a	When grown in Malaysia, considerable differences were found among provenances. Those from Guanaja, Mountain Pine Ridge and Santa Clara had the highest wood density as well as the fastest growth
	Wright et al. 1990	The use of the best seed from the best provenances would considerably increase the productivity of Fijian plantations compared to the commonly used source
	Barnes et al. 1977	Wood specific gravity of 6-year-old trees grown in Rhodesia varied from 0.34 to 0.36 among provenances
var. *hondurensis*	Wright 1990a	There was a negative correlation between wood density and site. There were significant differences in density among provenances with the coastal provenances lower in density than the inland ones
P. contorta	Schütt 1958	Differences in both cellulose and lignin were related to the provenance or the seed source
	Schütt 1962	Three different sources produced trees with different wood densities and tracheid lengths. Variations related to origin were greater than those among individual trees. No criteria could be found for using seed of specific provenances
	Polge 1963b	The lodgepole pine originating from the highest elevations produced the highest wood density
	Henderson and Petty 1972	The southern coastal provenances produced tracheids 20% shorter than the northern sources
P. echinata	Posey et al. 1970	In Oklahoma, the specific gravity of shortleaf pine grown in two plantations was related to seed source. The same source had the highest wood density in both plantations. Specific gravity and tracheid length differences were greater among provenances from north to south than those east to west
P. elliottii	Derr and Enghardt 1960	In a 22-year-old trial grown in Louisiana, no significant differences were found among seven sources for wood specific gravity or tracheid length

Table 9.1. (*contd.*)

Species	Reference	Comments
	Barrett and Bengston 1964	Oleoresin yields were similar when five different sources of slash pine were grown together
P. kesiya	Zavarin et al. 1968	There were differences in beta-pinene, longifoline, and beta-phellandrine among four provenances when grown in Zambia
P. maximinoi	Wright and Osorio 1993	There were significant differences in wood density of 14-year-old provenances when grown in Colombia
P. nigra	Lee 1979	There were significant differences in tracheid length among the 27 different seed sources tested
P. oocarpa *P. patula* ssp. *tecunumanii*	Corea 1989	Considerable differences on internode length were evident in different provenances in *P. tecunumanii* and provenance selection can affect limb numbers and dispersion. All of these affect wood quality
	Wright et al. 1989a	There were no statistically significant differences between provenances in wood density for these species when grown in Ecuador
	Wright 1990b	The provenance differences were small and the trees produced no variation in wood properties when grown in Brazil, Ecuador, Kenya, Puerto Rico, South Africa, and Zambia. The wood at all six sites exceed the 0.45 specific gravity considered minimum for kraft linerboard
P. palustris	Saucier and Taras 1966b	When grown in Virginia, five sources of longleaf pine had similar wood specific gravity. The sixth source, from Florida, had significantly lower specific gravity. The authors question whether this might result from frost susceptibility of the Florida source
P. pinaster	Nicholls 1967a	Four provenances of *Pinus pinaster* grown in Australia showed small differences in tracheid length. The basic density of one (Corsican source) was markedly lower than those from the Landes and Leira provenances. One provenance combined high wood density and long tracheids with good growth and form
P. ponderosa	Echols and Conkle 1971	Seed from lower elevations produced higher specific gravity wood when planted at high elevations; higher elevation sources produced lower density wood wherever planted
	Echols 1973	There was no relationship in tracheid length among the provenances (elevation zones) of the seed parents
P. radiata	Bannister et al. 1962	When grown in New Zealand, there were significant differences in alpha-pinene content among the three major California sources
	Burdon et al. 1972	There were differences in wood density among progeny from the three mainland populations (Monterey and Año Nuevo were highest and Cambria the lowest) but they were so small as to be nonsignificant

Table 9.1. (*contd.*)

Species	Reference	Comments
	Burdon and Young 1991b	Of the four provenances studied, the Guadalupe source had the highest basic density
	Burdon et al. 1992a	There were large differences in beta-pinene content between populations, with Año Nuevo averaging 80% and Monterey 71%
	Burdon and Low 1992	Wood density was about 10% higher from the slower-growing island populations than from the mainland ones. Compression wood was also more frequent from the island populations
P. sylvestris	Petrini 1959	In Sweden, the best timber was obtained from the inland source of Scots pine. The German and southern Swedish sources produced trees with poor wood quality, resulting from large branches and poor stem form
	Dorn 1969	The more northern sources produced shorter tracheids
	Remröd 1976	Seed from northern provenances of Scots pine produced trees with somewhat higher wood density, but the differences were considered to be of little importance operationally
	Miler et al. 1979	The Netherlands sources had long tracheids when compared with those from Finland
	Stahl et al. 1986	The northern sources had lower basic density when grown in Sweden. The high-yielding provenances also had low basic density
P. taeda	Strickland 1960	Differences were evident among sources from Maryland to Texas when grown in Georgia. There was a relationship between specific gravity and latitude as well as longitude ($r = 0.98$) and ($r = 0.95$). Highest gravity was from northern sources, longest tracheids from the southern ones
	Crow 1962	Sources from around Louisiana were grown together. Seed from the northern part of the state produced trees with higher specific gravity wood than did that from the south
	Gilmore et al. 1966	Neither percent latewood nor specific gravity was related to seed source but the wood was different when grown in differing environments in the test areas
	LaBorde 1966	Specific gravity was affected by geographic seed source. Those from the northern part of Louisiana produced wood with higher specific gravity than the sources from the southern part of the state
	Saucier and Taras 1967	Of the nine sources of 14-year-old loblolly pine tested, only the Maryland provenance produced high density wood

Table 9.1. (*contd.*)

Species	Reference	Comments
	Lantz and Hofmann 1969	Trees from the Piedmont, Sandhills, and Coastal Plains seed sources had similar specific gravity within tests. Differences in specific gravity among plantations (sites) were much greater than among seed sources, showing the great effect of environment
	Harris and Birt 1972	When grown in New Zealand, only minor differences were found in wood density by provenance. Wood from young trees grown in North Carolina and New Zealand differed considerably, showing the effects of environment
	van Buijtenen 1978	In a test in Texas, seed sources from the "Lost Pine" area and from widely distributed provenances differed significantly in wood specific gravity
	Talbert and Jett 1981	Small regional differences in specific gravity were found in juvenile wood and larger differences in mature wood. Differences among regions should not be interpreted as being genetic in origin
	Byram et al. 1984	The Arkansas source of loblolly pine had the highest specific gravity, while the southern Louisiana source had the lowest from 13 plantations in the Western Gulf Region
	Tauer and Loo-Dinkins 1990	When grown in Arkansas, definite provenance effects were found. Trees from the western and northeastern areas produced the highest specific gravities
	McAlister and Clark 1992	Seed source location was not significantly related to bending properties of either juvenile or mature pine wood whether grown in Coastal Plain, Piedmont, or Upper Coastal Plain
	Syzmanski 1991	The environment of the test site was most important in influencing specific gravity. The western and northeastern sources tended to have an early transition age from juvenile to mature wood
P. tecunumanii	Wright et al. 1986a	Wood density was different among provenances of 6-year-old trees grown in Zambia. Provenances specific gravity values were 0.49, 0.47, 0.45, and 0.45. These differences are small, but do show some genetic control by provenance
	Wright et al. 1986b	For 7-year-old trees, three provenances in Brazil had wood of 0.51, 0.49, and 0.49 specific gravities
	Osorio and Dvorak 1993	Wood density was higher for the Central American provenances than for those from Mexico (358 kg/m^3 vs. 330 kg/m^3) for 8-year-old trees
P. virginiana	Rink and Thor 1973	Wood variation among plantations was large but independent of seed source. The authors concluded that specific gravity of Virginia pine is less affected by seed source than by the environment. They estimated a gain of 14% in wood density if the 4 best of the 12 origins tested were used

the sites were more important than the provenances, all of which were from Honduras, in determining wood density. However, the coastal provenances generally produced lower density wood with less variation than the inland provenances but differences were small. If care is taken in choosing sites, provenance choice is not too critical relative to wood properties since there was no provenance × site interaction for either density or within-sample variation. However, if planted on the wrong site, wood properties can be greatly changed.

Several other studies related to provenance variation in wood have been made with *Pinus caribaea*. Among these is one by Garcia de Leon (1982), who assessed seven provenances grown at two sites in Australia and one in Brazil. He found essentially no wood differences related to provenance; however, large differences were associated with site. In 1977 Barnes et al. studied eight different provenances of *P. caribaea* at different sites in Zimbabwe and did not find any wood differences associated with provenance. In another study of five provenances, Barnes et al. (1983) again reported no major difference in wood related to provenance; however, some small provenance differences were found within sites. With the same species, Gibson (1982) also observed the same pattern for several traits. Finally, in Fiji, only small differences in wood density were found among seed sources of *Pinus caribaea* compared to the large variability among trees within seed sources (Cown et al. 1983).

Wright (1990a) reports that the wood density of temperate pines such as *P. sylvestris, P. strobus, P. resinosa, P. banksiana* and *P. echinata* have a clinal trend in wood density. This sometimes enables the production of wood of the desired type when provenances are planted in new environments; this concept was based upon studies by Echols (1958), Gilmore (1968), Posey et al. (1970), and Kennedy (1971).

Although most studies related to wood and provenance source deal with wood density, other wood properties, including resin content, may be related to provenance. An example was given by Burdon and Young (1991b) for 20-year-old *Pinus radiata* where resin contents differed markedly among provenances, with the Guadalupe source having the most resin. However, no differences in heartwood formation were found among sources. Two sources had almost double the compression wood when compared to the other two. The authors also reported that there were spiral grain differences among provenances. Very small provenance and large site differences in wood and pulping characteristics were found by Wright (1987) in *Pinus tecunumanii*. Wood density differences related to provenance were small, in both test areas, although the Mountain Pine Ridge provenance had higher density than that from Camelia.

Although it is commonly considered to be a varietal difference, the varieties *Pinus elliottii* var. *elliottii* and *P. elliottii* var. *densa* could be considered provenances. Despite the variety name *densa*, 11-year-old var. *elliottii* had considerably higher wood density than var. *densa* when grown on the same site. (sp. gr. 0.57 compared to 0.53) (White and Saucier 1965).

9.2.2 Genetic Differences in Wood Properties Among Provenances in Conifers Other Than the Hard Pines

Some reports about genetic control of wood properties among provenances in conifers are rather clearcut. An example is the study done on 46-year-old Douglas-fir by McKimmy (1966). He assumed that differences among plantations were environmentally caused, while differences due to seed sources or provenances within plantations were genetical in nature. The report reads: "Variation in specific gravity and summerwood percent from plantation to plantation was many times more than the variables from seed source to seed source." This study confirmed the frequent observation that major differences in specific gravity are related to the sites on which the trees were grown rather than to the provenance or seed source from which the seed had been obtained. However, certain sources had somewhat higher average specific gravity than did others. Also, McKimmey found the interaction of seed source and plantation for specific gravity to be small but highly significant.

In Douglas-fir, McKimmy and Nicholas (1971) observed that progeny of individual parents from different seed sources showed heritable differences in both tracheid length and specific gravity at age 47. Families with long tracheids or high specific gravity were consistent in these traits over all sites. Both specific gravity and tracheid length varied among plantations, indicating environmental control (specific gravity varied from 0.42 to 0.48 and tracheid length from 3.0 to 3.7 mm). For nursery-grown Douglas-fir, Haigh (1961) showed that if the effects of provenances were not considered, specific gravity was inversely related to growth rate; when the effect of provenance was considered, it was provenance and not growth rate that exerted the dominant influence on specific gravity. When tests of Douglas-fir were made at three sites in Coastal British Columbia, the fast-growing and most southern source was the best at all sites and maintained this superiority from ages 7 to 16 although the fastest-growing provenance had the lowest density (Loo-Dinkins and Hamm 1990). A study of 55-year-old Douglas-fir grown from an unreported seed source in Victoria, Australia, had a wood similar to that of the same aged material grown in New Zealand, except for the wood property commonly referred to as toughness. In New Zealand, five provenances of Douglas-fir grown at five locations failed to show provenance effects on wood, according to Cown and Parker (1979).

In a study of one of the oldest known provenance tests, Abdel-Gadir et al. (1993a) reported for Douglas-fir that there was some genetic variation in both earlywood and overall density resulting from an 80-year-old provenance study. They summarized that selection for families within populations could influence latewood proportion and wood density, but selection of provenances had less effect.

In 3-year-old seedlings of 25 provenances of *Abies grandis* (grand fir) grown in Europe, latewood percentage was strongly correlated with provenance. A few provenances exhibited both high wood density and fast growth (Polge 1968).

There is much discussion relative to growth and wood properties of Norway spruce (*Picea abies*) of German and Norwegian origins. In 1965, Klein found that the growth of the German origin continued longer in the autumn. Higher specific gravity and latewood percent was found in the spruce of German origin compared with the Norwegian origin. Klein also reported that there was no difference between German spruce grown in Norway and in Denmark. In several studies with *Picea abies*, Parrot (1960) found highly significant variation in wood density among 12 provenances. However, there was no correlation between mean density and the latitude of the provenances. He concluded that genotypic differences existed among certain populations in respect to wood density. In a study of Norway spruce in Poland (Anonymous 1968b), it was found that lowland spruce sources produced higher specific gravity than mountain sources. Variation in wood density among individual trees in both the lowland and mountain spruce sources was larger than the average between lowland and mountain sources. Exactly the same results were obtained by Ericson (1960b) and Worrall (1975) with Norway spruce when they reported that northern seed sources and those from high altitudes produced less dense wood. In contrast, Knudsen (1958) and Parrot (1960) did not find differences in specific gravity related to seed origin. In trials of Norway spruce in the northern United States, Stairs (1969) found only small differences in specific gravity related to seed origin, but he expressed the opinion that these were large enough to have an effect in an operational program. However, in Michigan, Lee (1979) could not detect any differences among geographic sources of Norway spruce.

Growth termination conditioned by provenance is one explanation of wood differences for different sources of Norway spruce. In 1963, Dietrichson reported this phenomenon in the 1938 provenance tests in Sweden and Norway. His results were confirmed in 1964 when he observed a greatly delayed latewood formation in the west-central European provenances grown in Norway. Dietrichson frequently found incomplete lignification of the outer latewood zone, which resulted in low wood density.

When Sitka spruce (*Picea sitchensis*) was grown in Scotland, the Queen Charlotte Island origin had higher wood density than the other three provenances included in the test (Jeffers 1959). However, Delaporte (1983) found few provenance differences in tests of this species. In Great Britain, Scott and MacGregor (1952), studying both growth rate and wood density for Norway and Sitka spruce, found that the wood differed considerably according to the source used. They concluded that source differences merited study as much as growth rate because of the vital importance of source for wood quality. Definite differences among provenances of white spruce (*P. glauca*) were reported by Beaulieu and Corriveau (1985) from tests in Canada.

Based on 300 stands of *Pinus longifolia* grown in South Africa, Paterson (1967a) reported that different seed sources had different amounts of spiral grain. No differences in specific gravity were evident among several seed sources of white pine (Lee 1974).

In Japanese larch (*Larix leptolepis*), some source differences in wood properties were found in plantations grown in the Lake States of the United States, but no geographic trends with seed sources were evident (Lee 1975). When 20 sources of Japanese larch were grown in Michigan and other Lake States, Loo-Dinkins et al. (1982) found highly differing specific gravities (0.39 to 0.42) related to seed source with 17-year-old trees. Although not related to seed source, but to species, some, such as *Cupressus lusitanica* (Mexican cypress), seem to have very similar wood despite the differing areas where the species is grown.

For conifers other than the hard pines it is evident that the provenance source usually has a small to moderate effect on wood properties. Environmental effects are usually moderate to large, and individual tree values within provenance are large, overshadowing the smaller provenance effect.

9.2.3 The Importance of Provenance in Determining the Wood Properties of Hardwoods

There are only a few reports of wood property differences related to provenance or seed source in some conifers, and there are even fewer reported for the hardwoods. Most show that hardwoods have a closer relationship between the wood from the new site and the site of origin than do the conifers. However, as shown for 36-year-old teak (*Tectona grandis*), variation in wood exists when seeds are obtained from different origins (Purkayastha et al. 1973). When genetically unimproved material was compared with two sources, large differences were found in the eucalypts (Camphinos and Claudio-da-Silva 1990) as shown in Table 9.2.

Table 9.2. Wood Properties from different sources of *Eucalyptus grandis* grown at Aracruz, Espirito Santo State, Brazil. (After Campinhos and Claudio-da-Silva 1990, Table 5)

Source of seed	Basic density	Pulp yield (kg m^3)	Bark content (%)
Commercial Brazil	480	47	18
Zimbabwe and South Africa commercial	490	47	15
First generation rooted cuttings (from select trees)	490	49	12
Second generation rooted cuttings (more intensive selection)	520	52	10

In provenance tests of a number of species of eucalypts growing in China, Harding et al. (1989) found significant differences in wood of 5-year-old provenances of *Eucalyptus camaldulensis*, *E. tereticornis*, and *E. grandis*. However, the largest variation was among trees within a provenance, and the provenance effect was relatively small. In Australia, provenance tests were made with *Eucalyptus globulus* from Victoria and from Tasmania (Dean et al. 1990). Pulp yield differences were huge, varying from 40 to 56%, which is economically very significant. In another study by Dean et al. (1991) with *E. obliqua*, differences in yield among provenances were nearly as large, 40 to 53%. In Brazil, wood was quite different from different provenances of *Eucalyptus saligna* and *E. grandis* (Vital and Lucia 1980, De Souza et al. 1986).

Recent intensive studies on the properties of *Eucalyptus grandis* provenances and pulp made from them when grown in South Africa have been reported by SAPPI (South African Pulp and Paper Industry). Some of these are summarized below (SAPPI 1993).

Region of origin	Wood density kg/m^3	Yield %	Viscosity (CPS)	Brightness (% ISO)
Atherton	402	48.1	55.1	54.8
Gympie	401	47.7	51.1	53.8
New South Wales	396	48.4	50.3	53.2
South of 31°S	396	48.1	47.2	51.9

Despite the widely differing sources, the wood and pulp property differences are minor and not sufficient for choosing sources for more productive commercial use.

In the Amazon region, Woessner (1983) reported useful differences in the wood of *Gmelina arborea* related to seed origin. In 17-month-old trees, he found wood density differences among provenances to be from 0.29 to 0.38 g/ml. He concluded that the wood differences related to provenance are large enough to significantly improve future yields from Jari plantations. No differences in wood density were evident among *Leucaena* provenances planted in Colombia (Ladrach 1984).

Considerable differences in fiber length of *Populus trichocarpa* and *P. deltoides* were reported by Gabriel (1956), depending upon their geographic source. One clone of *P. trichocarpa* from Washington had considerably longer fibers than the average of three clones from Alaska. Unlike many others, Gabriel found that different provenances had fiber length differences when grown in the same environment. This can be contrasted with the results obtained by Meyer-Uhlenreid (1959), who assessed ten species of poplars and noted that the Tacamhaca sources all had similar fiber lengths when grown under uniform conditions. Large differences in wood were obtained from different provenances of *Populus*, and the assumption was that some of the differences were genetic (Scaramuzzi 1960). However, Thorbjorensen (1961) reported that the fiber length differences among test sites with *Liriodendron tulipifera* (yellow poplar) in Tennessee were due to site differences and not to genetic differences related to origin of seed.

In *Platanus occidentalis* (sycamore), Jourdain and Olson (1982) reported that both specific gravity and fiber length showed significant differences with seed source. No relationship between specific gravity and seed source could be found (Land and Lee 1981, Land et al. 1983). Nebgen and Lowe (1982) also reported that family variation was more important (accounting for 79 to 100% of the total variation) than regional differences. Strong seed source differences existed for *Fagus sylvatica* (beech) in Germany, where provenance was the principal source of variation directly influencing the technological value of the wood (Koltzenburg 1966). For sweetgum (*Liquidambar styraciflua*), Johnson and McElwee (1967) found no provenance-related variation in specific gravity or fiber length but in white ash, (*Fraxinus americana*), diploid trees from the south had somewhat longer vessel elements and fibers than diploid trees from northern seed sources (Armstrong and Funk 1979). Provenance is important in green ash (*Fraxinus pennsylvanica*). Lowe and Greene (1990) found for 42 10-year-old open-pollinated families representing seven east Texas provenances and planted in three locations that specific gravity was significantly affected by plantation, provenance, family within provenance and the plantation × family within provenance interaction. No consistent pattern of variation was evident for fiber length.

9.3 Summary

Wood properties as related to provenance or seed source are a complex and somewhat confusing matter. Strong relationships are usually not found, but occasionally seed source plays an important role in determining wood properties. In some species and under some circumstances, wood differences related to provenances are large enough to be important in an operational forestry program (Table 9.1). When a trend is present, specific gravity frequently decreases with increasing latitude and elevation of the parental source. In the southern pines of the United States, specific gravity of wood from natural stands increases from north to south, but the trend when using seed of different provenances is for the northern sources to have the highest density. Usually, differences related to seed source are relatively small compared to individual tree or family variation. This was summarized by Zobel and van Buijtenen as "... tree-to-tree variation in wood properties is greater than that resulting from provenance or species differences." The complicating factor of genotype × environment interaction in wood proper ties among provenances may occur, but it usually is of limited or no importance. Even though one cannot use definite guidelines, Zobel and van Buijtenen (1989) summarize three general trends related to wood properties and provenance. These are briefly mentioned below:

1. Wood from trees grown in severe environments has a closer affinity to trees from different provenances than do those grown in milder environments.
2. Strong to modest provenance control appears in *Pinus contorta*, *P. banksiana*, *Abies* spp. *Picea* spp., the tropical pines, and to some extent in Douglas-fir.

Most hardwoods show wood with only moderate to no relationship with seed source or provenance with the exception of some eucalypts.

3. Genotype × environment interaction plus the strong genetic control of wood by individuals and families make it difficult to predict the response of wood to origin of seed with any degree of accuracy. Although foresters often choose provenances with the desired wood properties for their planting programs, only occasionally has such action been shown to be highly successful in supplying the desired wood.
4. **The only sure way to tell what kind of wood trees from a given provenance will produce in a different environment is to grow them there.** Before untested provenances are planted wholesale, one must be certain, through testing, that the wood produced will be acceptable. At least 30 unrelated trees should be used to represent a provenance. Ideally, the trees should be grown for a whole rotation or two but, practically, assessment after one-half rotation will usually give reasonably accurate results.

Chapter 10

Correlations Among Wood Properties and with Growth Rate

10.1 General Concepts

If any breeding program for wood properties is to be successful, it is essential to understand how the different characteristics relate to one another. The relationships are usually reported as correlations. There are both phenotypic and genetic correlations reported in the literature. One would think that the two should be similar and they sometimes are. Frequently, however, they are different and, can be very different, as shown for white spruce (*Picea glauca*) by Yanchuk and Kiss (1992), who found a phenotypic correlation of $r = -0.40$ between wood density and height growth but observed essentially no genetic correlation. It would be ideal to report only genetic correlations but nearly all values available in the literature are phenotypic correlations. Even more confusing, many references do not state the type of correlation being reported. Thus, in this chapter we are forced to report correlations as found in the literature, assuming that when unstated most are phenotypic. However, when specifically indicated, genetic correlations will be mentioned.

If improving one desired characteristic results in a reduced value for a second desirable property a very careful consideration must be made as to the most effective breeding plan. It is always important to avoid problems related to adverse correlations; to do so, one needs to know both the extent of genetic relationships among characteristics as well as their economic value. This was stated plainly by Zobel et al. (1962b): "... that if really meaningful results regarding genetic control of wood are to be achieved, it is necessary to determine whether characteristics are wholly independent, or whether several are so closely interrelated that if some are changed, the others will change also." Effective progress through the application of genetic principles is unquestionably dependent to a very large degree on the nature and extent of interrelationships existing between different characteristics. In eucalypt breeding, Dean et al. (1991) took an extreme position and stated that it is not appropriate to breed for several criteria at the same time if there are adverse correlations among traits. Whether this is true depends on the strength of the correlations and the economic value of the traits assessed; a number of researchers have indicated how this can be done.

If the relationship is positive, then improvement in one trait will bring benefit from the correlated trait; if correlations are antagonistic then attempting to improve both traits at once may cost time and money while achieving very little.

The correlations among wood properties have generally not been well determined, although some are reported, such as moisture content, heartwood color, and wood density in sugi (*Cryptomeria japonica*) (Kurinobu 1990). From what is now known, it appears that most correlations among wood properties are generally weak or lacking entirely (Fig. 10.1). Usually, only those wood properties that would seem to be logically related, such as wood density and cell wall thickness, have strong correlations. Generally, wood properties such as cell length and wood density, are not, or are only weakly correlated. In their summary of loblolly pine (*Pinus taeda*), Zobel et al. (1962b) found that trees with almost all possible combinations of tracheid characteristics were present and were not correlated. This indicated that there is considerable hope of developing strains of trees with various combinations of wood characteristics through selective breeding. In their work with 50-year-old Douglas-fir (*Pseudotsuga menziessii*) families, McKimmy and Nicholas (1971) found no consistent pattern of relationship among specific gravity, tracheid length, and growth. Sometimes the correlation coefficient even changed signs from one plantation to another.

As mentioned in Chapter 3, Franklin and Squillace (1973) have pointed out a common source of error that will result in fallaciously low estimates of correlations. They suggested that density is determined on a unit volume basis but usually moisture content, extractive yields and pulp yields are estimated on a dry weight basis. When correlations are made between them it is equivalent to dividing each unit on the volume basis by density which will automatically induce

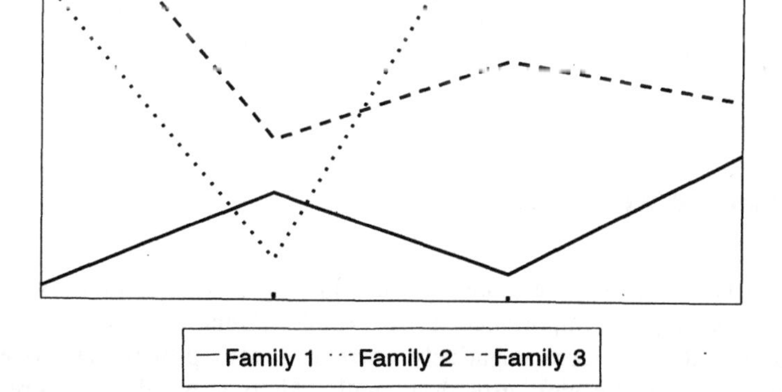

Fig. 10.1. Wood properties are usually not closely correlated genetically. Shown is the comparative variation in characteristics for three loblolly pine trees of the same age growing on the same site

an inverse relationship between wood density and the other characteristics. This may magnify the low or negative correlations found between wood density and volume growth.

10.2 Growth Rate and Wood Properties

There have been many estimations of the correlation of growth rate, tree form, and limb characteristics with wood properties. This subject has been covered in literally hundreds of papers which were summarized by Zobel and van Buijtenen (1989) as well as by several other researchers. Overall, many papers report that there is little relationship between growth rate and wood properties; some show a negative relationship and a few indicate a positive relationship. An example was given by Pereira (1992) for fast-grown 6-year-old *Eucalyptus globulus* trees, where she reported that no relationship existed between specific gravity or fiber length with growth rate. In their summary, Eisemann et al. (1990) state for hoop pine (*Araucaria cunninghamia*) that substantial genetic gains for growth are possible without adversely affecting wood quality. The most discussed is the relationship between growth rate and wood density.

There is, however, one question: What is used to measure growth rate?. Diameter growth is sometimes used and height growth is even more popular. In older trees, merchantable volume is commonly employed and, currently, tons of dry wood production are sometimes used to make correlations. It is not unusual to find a correlation of wood properties with height growth, while there is no similar relationship with diameter growth (Squillace et al. 1962, Shelbourne et al. 1972, Sohn and Goddard 1975, Bridgwater et al. 1983). This is confusing since the relationship between height and diameter growth within a species is usually very strong.

As emphasized earlier, growth rate has a direct and strong influence on the weight of wood produced (Fig. 10.2), but this can be influenced by the density of the wood grown. The relationship of growth rate to wood properties is thus very important.

10.2.1 Growth Rate and Wood Density

The relationship between growth rate and specific gravity has been intensively examined and can be of great importance. For example, if one selects for high wood density, and if density were negatively correlated with growth rate, one would unintentionally select for slow growth as well. This type of adverse correlation must be avoided or adjusted for (see Chap. 13.3). In Table 5.2 of Zobel and van Buijtenen (1989) there are 59 references covering the interrelationships of growth rate and wood density for the *hard pines alone*. Of the studies, 35 showed

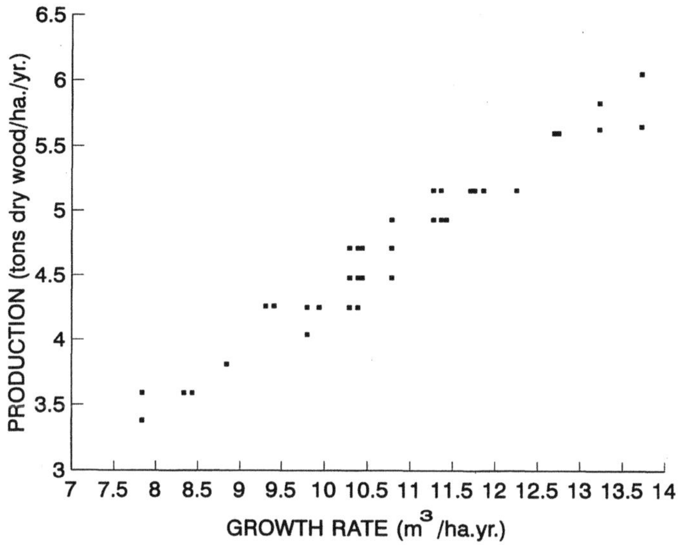

Fig. 10.2. Growth rate alone is very important in affecting the amount of wood produced per hectare per year. However, the production shown can be considerably altered by the density of the wood grown. Wood density is important in determining both yields and quality

no relationship between growth rate and specific gravity, 9 exhibited a small correlation, while 11 showed a significant reduction in specific gravity with faster growth rate, and only 4 showed a higher specific gravity for the fastest growing trees (Fig. 10.3). Yet there is a common acceptance among many foresters that if one grows hard pine trees rapidly, low density wood will result (Fig. 10.4). In Zobel and van Buijtenen's Table 5.3, which showed growth rate and density for other than the hard pines, all combinations were found. Some genera, like *Picea*, usually showed a negative correlation while there was no pattern for the others. Some studies of growth rate and wood density, many of which were not included in the earlier tables, are listed in Table 10.1.

Although it is not the subject of this book and will only be partially covered, it is of value to make a few summarizations relative to growth rate and wood density:

1. Most of the conifers with dense wood, especially the hard pines, show little or no meaningful relationship between growth rate and specific gravity (Fig. 10.4). There are, however, numerous exceptions; for example, King (1986) states for coastal Douglas-fir that there is a negative correlation between wood density and volume, and King et al. (1988) found a strong negative correlation of $r = -0.53$ between diameter growth and wood density (see also Chap. 13.3). Similarly, van Buijtenen (1969b), reports a negative correlation for *Pinus taeda* in Texas, although there are a number of papers showing no relationship between

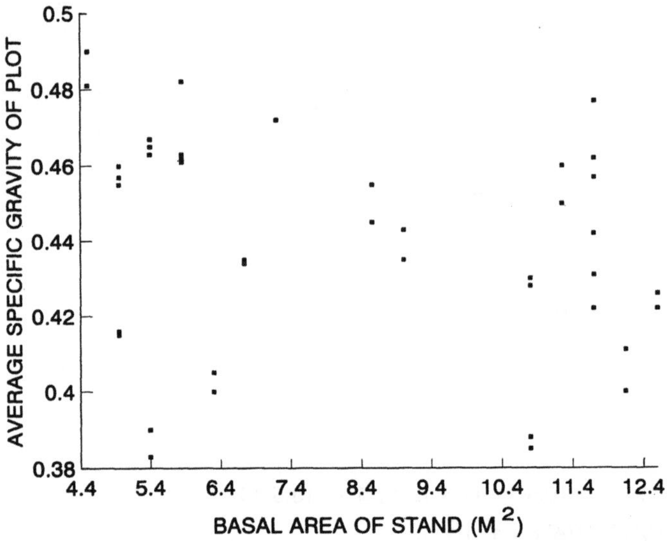

Fig. 10.3. The relationship of growth rate to wood density is frequently debated. For some species, like the loblolly pine shown here with the same-aged trees on adjacent but varied sites, wood density is the same regardless of the growth rate, when expressed as basal area. For some species and provenances there is a negative relationship between growth rate and wood density

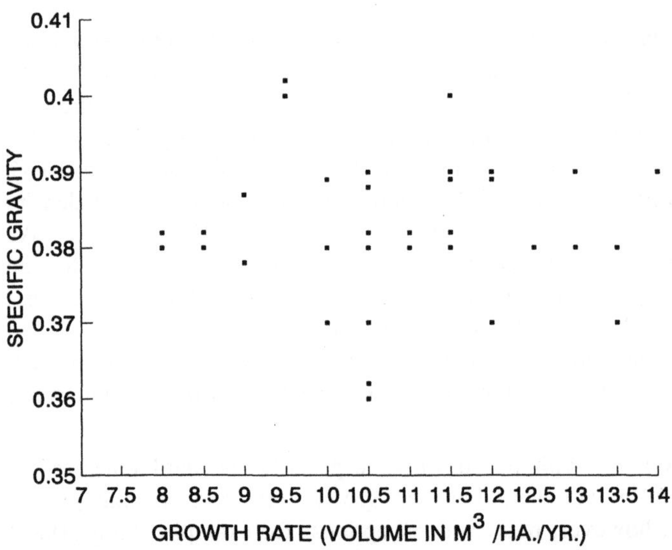

Fig. 10.4. Much work has been done on the correlation between volume growth rate and wood properties, especially with wood density of the hard pines. A knowledge of any correlations is absolutely essential for an efficient breeding program. Correlations differ with species (shown here is the essential absence of any correlation in *Pinus taeda*) but also with individuals within species.

Table 10.1. Relationship of wood density to growth and volume in certain conifers – a random sample of some species[a]

Species	Reference	Comments
Abies spp. *Tsuga* spp.	Hawthorn 1961	The relationship between growth rate and specific gravity was negligible for both genera
Cunninghamia lanceolata	Yitai et al. 1992	There was a significant negative genetic correlation between wood density and volume and tree height
Picea abies	Stairs 1969	There was a strong negative correlation between growth rate and wood density ($r = -0.87$)
	Lacaze and Polge 1970	Most spruce showed a negative relationship between vigor and wood density; correlations were −0.21 for height, −0.32 for diameter
	Mothe 1983	At all levels of measurement (genetics and environment) vigor of the tree and wood density were negatively correlated
P. glauca	Corriveau et al. 1990	There was a slight negative correlation of wood density with the width of the annual growth rings, both at the individual and population level. However, there were deviations from this trend
	Yanchuk and Kiss 1992	Phenotypic correlations were $r = -0.40$ with height growth and $r = -0.46$ with diameter. The genetic correlations were zero
P. rubens	Barbour et al. 1992	The faster-growing trees from thinned plots had the same wood density after 15 years compared to the unthinned plots
P. sitchensis	Brazier 1967	The fastest-growing trees usually had lower wood density
Pinus caribaea var. *hondurensis*	Kanowski 1986	There was no correlation between growth rate and wood density
	Harding et al. 1991	There was an adverse correlation between growth rate and wood density
P. elliottii	Squillace et al. 1962	Specific gravity was inversely phenotypically related to diameter but positively related to height. Modest gains were obtained by breeding
	Hodge and Purnell 1993	Both positive and negative correlations were found. Any negative correlations were so small that it will be possible to improve both wood density and growth rate at the same time

Table 10.1. (*contd.*)

Species	Reference	Comments
P. nigra	Rendle and Phillips 1958	The higher density of older wood was due to an age effect rather than to diminishing ring width. Rapid growth can produce high density wood
P. ponderosa	McKimmy and King 1978	More often than not it was found that the faster-growing families of the same age produced denser and stronger wood
P. radiata	Nicholls et al. 1964	Basic density was independent of ring width
	Nicholls et al. 1980	There was a small, nonsignificant genetic correlation between ring width and average density. The small negative correlation may disappear in older growth rings
	Bannister and Vine 1981	There was a weak negative phenotypic correlation between growth and wood density. Multiple regression relating density to height and diameter growth accounted for only 2.2% of the total variation
	Cown et al. 1991a	There was no clear correlation between growth rate and wood density. Tree age, not tree growth rate, was the key determining factor for wood density at all sites
	Burdon and Young 1991a	There was a strong negative correlation between wood density and growth rate in rings 6 to 10, weaker in rings 10 to 20 and absent in rings 0–5
	Burdon and Low 1992	Wood density had negative genetic and phenotypic correlations with diameter growth but a positive correlation with height, despite the strong correlation between height and diameter
P. sylvestris	Velling 1974	The relationship between wood density and cubic volume growth showed a predominantly negative correlation
P. taeda	Zobel 1956	Although a weak negative relationship occurred between growth rate and wood density, it was so small as to be meaningless
	Geyer and Gilmore 1965	There was a great deal of variation in specific gravity but only a small amount could be related to growth rate

Table 10.1. (*contd.*)

Species	Reference	Comments
	Zobel et al. 1969	Based on data from 3000 trees, and from both open- and control-pollinated tests, growth rate, rated either by tree diameter or volume per unit area per year, was not closely related to specific gravity
	Pearson and Gilmore 1980	Fast-growing trees had similar mechanical properties to trees of slower growth of the same age from similar environments
	McKinley et al. 1982	Selection for growth alone would result in a negative response for wood specific gravity on trees grown in Texas
	Bridgwater et al. 1983	There was a weak (−0.07) genetic correlation between volume and specific gravity at age 10
	Loo-Dinkins et al. 1984	Genetic correlations between specific gravity and height and diameter at age 20 years had a strong negative relationship
	Byram and Lowe 1988	At the provenance level, fast growth was strongly negatively related to wood density
	Syzmanski 1991	No correlation was found between basal area and specific gravity
Pseudotsuga menziesii	Littleford 1961	At fast growth rates, one tree is more competent than another to produce wood of high strength. A moderate percent of latewood more than compensates for any detrimental effect that fast growth might have on strength
	Anonymous 1990	It will be impossible to improve wood density without limiting gains in bole volume
	Loo-Dinkins and Hamm 1990	There was no consistent relationship between volume and density of trees within the provenances, but the fastest-growing of the three provenances had the lowest wood density
	Vargas-Hernandez and Adams 1991	It is possible to obtain a substantial gain in bole volume without loss in wood density
	Abdel-Gadir et al. 1993a	No correlation existed between ring density and ring width in an 80-year-old plantation. The potential exists for improving wood density by selection with only a minor impact on radial growth

Table 10.1. (*contd.*)

Species	Reference	Comments
	Vargas-Hernandez and Adams 1994	Selection for increased wood density will cause an earlier transition to latewood formation, negatively affecting growth rate
Sequoiadendron giganteum	Knigge 1993	Specific gravity was not correlated with ring width or height of the tree. There is no fear of growing trees fast at wide spacings

[a] These are mostly additional to the 59 references of growth to wood density on the hard pines cited in Table 5.2 and the 44 references in Table 5.3 in Zobel and van Buijtenen (1989).

growth and wood density for this species when grown in other areas (Zobel 1970, Megraw 1985). Lack of a relationship was specifically indicated for loblolly pine by Zobel et al. (1962b) when they summarized: "It would seem to be a fair summary from our data to say, then, that on the basis of analyses of over 670 trees ... growth rate plays a very minor part in influencing specific gravity values." A similar result was found for *Pinus occarpa* when this species was grown in widely differing environments (Lima 1987).

The greatly increased growth rate of genetically improved progenies had no effect on their wood densities, as Cown et al. (1992) summarized for radiata pine (*Pinus radiata*) and stated: "... the average for the progenies overall was almost identical to that of the control trees." They further commented that this was true despite the well-publicized negative genetic correlation between growth rate and wood density for this species. In New Zealand, the average density of all the selected (fastest-growing trees) was essentially identical to the average for all trees sampled on similar sites in radiata pine (Carson 1993, pers. comm.[4]) also showing a minor effect of growth rate. Exactly the same result was observed by the authors of this book for loblolly pine. In western hemlock (*Tsuga heterophylla*), the fast-grown plus trees had essentially identical wood density compared to the rest of the stand (Wellwood and Smith 1962). Megraw (1985) found for loblolly pine that there was no difference in specific gravity between the fastest and slowest growing trees on a plot.

There has been controversy concerning the effect of growth rate on the wood density of *Pinus radiata*. A recent paper by Cown et al. (1991a) summarized their own and various other studies as follows: "There is no clear correlation between growth rate and density, although a weak negative relationship is not uncommon."

[4] Carson, M. Forest Research Institute, Rotorua, New Zealand.

This was also reported for different growth rates within clones (Burdon and Harris 1973) when they reported inconclusive relationships between growth and wood density.

2. There are a number of reports of a negative relationship between growth rate and wood density in several genera such as spruce (*Picea* spp.) and fir (*Abies* spp). For black spruce (*Picea mariana*), Boyle et al. (1987) obtained a negative correlation of wood density with both height and diameter growth of $r = -0.35$.

3. There are contradictory reports on the effect of growth rate in Douglas-fir. In one study by McKimmy (1966) on 46-year-old trees, it was found that growth rate was not related to a significant additional proportion of specific gravity variation. A similar result was shown in two papers cited in Table 10.1 but others indicate a negative relationship between wood density and growth rate.

4. There are contradictory reports about the relationship between growth rate and wood properties in the hardwoods. Usually, the diffuse-porous hardwoods show little or no relationship between growth rate and wood density. For example, in quaking aspen (*Populus tremuloides*) Einspahr et al. (1967) found specific gravity to be only weakly, if at all, influenced by rate of growth. Similarly, Yanchuk et al. (1983a) reported a slight negative phenotypic correlation. However, Olsen et al. (1985) did report a negative relationship and suggested use of a selection index to improve volume, wood density, and alpha cellulose yield at the same time. In the Euramerican poplars, Nepveu and Teissier du Cros (1976) found no correlation between growth rate and wood density. They suggested selection of rapid growth individuals from among those with high wood density. The same was reported for eastern cottonwood (*Populus deltoides*) by Farmer (1970) In European white birch (*Betula pendula*), Velling (1983) found no unfavorable relationship between growth rate and wood density when 46 full-sib families were assessed in southern Finland.

Growth rate is usually independent of specific gravity for the eucalypts; an example was given for several species by Harding et al. (1989). Many other papers could be cited relative to the low, or lack of, correlation between growth rate and specific gravity of the eucalypts, such as the one by Borralho et al. (1992), which showed a correlation of only 0.08 for 8-year-old *E. globulus* in Portugal. They reported that selection for fast growth would lead to strong improvement in dry weight but only a slight deterioration in wood density. Indices combining height with dry weight gave the greatest improvement in weight of wood produced, while allowing wood density to improve slightly. Clarke (1990) in a study of *E globulus* in South Africa, found no statistically significant genetic correlation between growth rate and wood density for a 9-year-old open-pollinated test. He reported a phenotypic correlation of $r = 0.33$ between wood density and volume growth. Several authors have reported that faster-growing eucalypts produced somewhat denser wood.

In a study of 460 clones of *Eucalyptus grandis* in Colombia, 4 years of age, Wright and Endo (1993) concluded that there was no correlation between growth rate and wood density. This was true for both volume and height growth.

In *E. viminalis*, Otegbeye and Kellison (1980) found a very small negative correlation between growth rate and wood density. Not much has been done with the tropical hardwoods relative to growth rate and wood properties. For *Triplochiton scleroxylon* (abachi) from West Africa, growth rate showed no relationship to fiber length, but wide rings contained more parenchyma and thus there was a decrease in specific gravity (Oteng-Amoako et al. 1983).

5. Frequently, but not always, the ring-porous hardwoods such as the oaks (*Quercus* spp.) and hickories (*Carya* spp.) show a positive correlation between growth rate and wood density. This means that the faster-growing trees often have higher wood density than the slower-growing individuals (Zhang et al. 1993). They found that for the durmast and English oaks (*Q. petrea* and *Q. robur*) grown in France that wood density increased with increasing ring widths until the ring width approaches 2 mm, after which wood density levels off. The concept is that only a given amount of vessels are produced early in the growth period and any extra growth later in the year results in the production of a greater amount of thicker-walled fibers resulting in a higher density wood (Nepveu 1984b). However, for green ash (*Fraxinus pennsylvanica*), Lowe and Greene (1990) report that there was no relationship between growth rate and specific gravity. This result is surprising because those who use ash (when strength is needed) usually utilize the younger, fast-growing stems. In the ring-porous East Liaoning oak (*Quercus liatungensis*), Zhang and Zhong (1991) found that growth rate is not important as a factor in controlling specific gravity. However, they did find for the slower-growing trees that specific gravity increased rapidly with increasing growth rate but for trees with wider rings, specific gravity was constant. Also, there was essentially no effect of growth rate on specific gravity in the juvenile wood while sometimes there was such an effect in mature wood.

10.2.2 Growth Rate and Other Wood Properties

There is only limited information relating growth rate to other wood properties. One example is shown for several species in East Africa in Tables 10.2 of the kind of information badly needed but rarely available.

For tracheid length, Bissett et al. (1951), along with many others, showed that an increase in growth rate *within* a tree results in shorter tracheids. Yet, results with many species have clearly shown that genetically faster-growing trees do not necessarily have shorter tracheids. A correlation coefficient of only $r = 0.11$ was found by Zobel et al. (1969) with 30 year-old *Pinus taeda*. This lack of relationship of growth rate to tracheid length was also reported by Nicholls et al. (1964) in radiata pine. Thus, trees that grow faster from environmental manipulation have shorter tracheids. Because fast growth and tracheid length are not strongly related genetically, fast-growing trees can have either long or short cells. An example has been reported by Stairs et al. (1966), who found that cell length, diame-

Table 10.2. Percent change in wood properties and tree form resulting from the selection and breeding for faster growth rate of three conifer species grown in East Africa. (After Paterson 1967b)

Property	Percent change *Cupressus lusitanica*	*Pinus patula*	*Pinus radiata*	Importance
Basic density	−2.5	−4.3	−4.6	Important
Fiber length	−1.0	−1.0	−1.5	Important
Grain angle	−6.0	−13.5	+5.0	Important
Cellulose prod/acre	+3.9	+11.3	+8.0	Very important
Reduction of compression wood	−28.2	−18.5	−33.8	Very important
Taper	0.0	0.0	0.0	None
Knottiness	+12.1	+1.3	+15.8	Very important
Branch angles	+11.6	+8.9	+12.9	Little importance

ter, and wall thickness in Norway spruce were of equal size in both fast- and slow-grown trees. There are, however, always exceptions.

In 12-year-old slash pine (*Pinus elliottii*) selection for greater height and diameter produced increases in spiral grain, microfibrillar angle, and latewood content according to Allen 1977, who believed that reduction of latewood content with faster growth would be difficult because of its low heritability and large adverse genotypic correlation. For microfibrillar angle and spiral grain, containment is more likely because of their higher heritabilities and lower adverse correlations with height and diameter. Also working with slash pine, Hiller (1964) made an in-depth study of nine factors that affect fibril angle. He found that the strongest correlations with fibril angle were the age from the pith, $R^2 = 0.58$ and percentage latewood, $R^2 = 0.70$. Age of ring from pith is the easiest way to estimate the angles.

In ponderosa pine (*Pinus ponderosa*), McKimmy and King (1978) found that accelerated growth does not adversely affect the modulus of elasticity. In Japanese larch (*Larix leptolepis*), no relationship between growth rate and the angle of spiral grain was evident Mikami (1973). In 20-year-old half-sib families of *Pinus radiata*, Cown et al. (1992) found that improved growth rate resulted in somewhat increased compression wood and heartwood resin content, while tracheid length decreased.

Published results are scarce relative to growth rate and miscellaneous properties in hardwoods. Usually only a slight relationship has been found; this was reported for *Populus* × *euramericana* by Scaramuzzi (1958), who detected no correlation between fiber values and vessel volume and growth rate. Similarly, Farmer (1970) did not find a relationship between mean clone diameter and fiber length. However, fiber length was found to be strongly correlated with height and

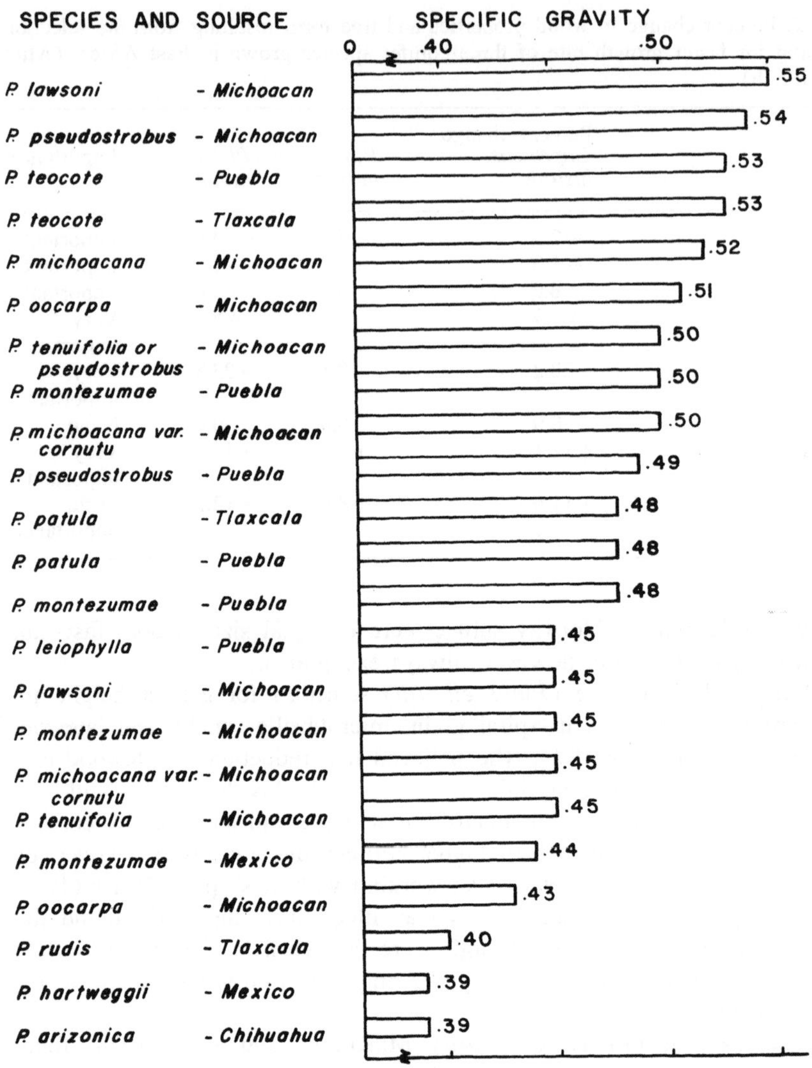

Fig. 10.5. Many wood properties are genetically independent traits. This is indicated for a number of Mexican pines, arranged in descending order of specific gravity (*left*). With the species in the same order, tracheid lengths have been plotted (*right*). Note how wood density and tracheid length are essentially unrelated. (After Zobel and Talbert 1984)

diameter in trembling aspen, but specific gravity only weakly so (Einspahr et al. 1967). For *Populus* in general, Göhre (1960) concluded that density was not related to ring width. In the tropical hardwood *Triplochiton scleroxylon*, specific gravity was related to cell wall volume, fiber volume, and double wall thickness. Growth rate was not related to fiber length but was associated with parenchyma

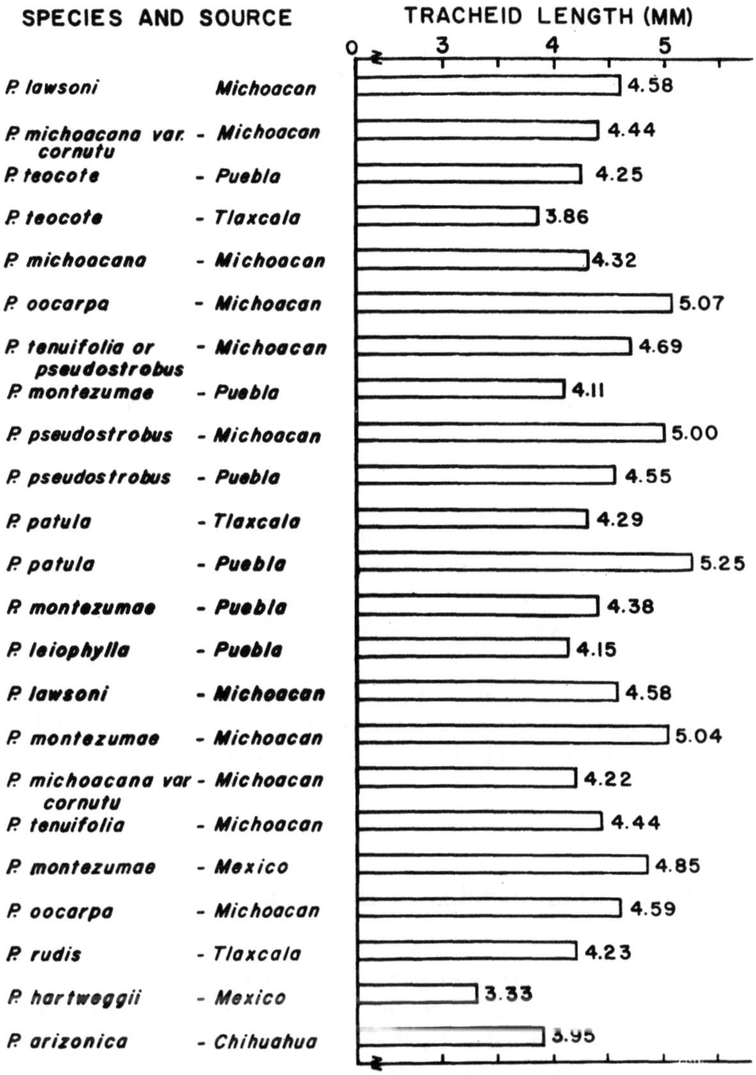

volume and a reduction in specific gravity because of a decrease of volume and wall thickness of the fibers (Oteng-Amoako et al. 1983).

Although rarely studied as a genetic characteristic of wood, volumetric shrinkage was inherited fairly strongly in birch (Velling 1983). There was a favorable relationship between growth rate and volumetric shrinkage (i.e., the faster-growing trees showed less shrinkage) in *Betula pendula*.

A very small, negative correlation ($r = -0.11$) was detected between growth rate and fiber length in *E. viminalis* by Otegbeye and Kellison (1980). They found that the faster-growing trees had slightly wider fibers ($r = 0.13$). A very

Table 10.3. Phenotypic correlation coefficients (r) between wood specific gravity and some other traits for earlywood and latewood (Goggans 1964)

	Specific gravity of the tree	Specific gravity of latewood	Specific gravity of earlywood
Earlywood traits			
Earlywood sp. gr.	0.37	0.04	–.–
Double wall thickness	0.22	0.13	0.30
Radial lumen diameter	−0.12	0.18	−0.34
Radial tracheid width	−0.09	0.19	−0.30
Tangential tracheid width	−0.05	0.01	−0.06
Latewood traits			
Percent latewood	0.65	0.30	0.14
Latewood sp. gr.	0.47	–.–	0.04
Double wall thickness	0.21	0.40	−0.05
Radial lumen diameter	−0.04	−0.29	0.04
Radial tracheid width	0.06	−0.04	0.01
Tangential tracheid width	0.03	0.00	−0.09

small correlation ($r = -0.04$) was reported by Malan (1989) for *E. grandis* and by Bhat and Bhat (1983) for *E. tereticornis* ($r = -0.05$) for growth rate and fiber length. However, Yanchuk et al. (1983b) found a positive correlation of growth rate and fiber length with the faster-growing aspen trees (*P. tremuloides*) having the longest fibers. Genetic correlations showed that this relationship was under moderate genetic control.

10.3 Wood Property Relationships in the Conifers

It is possible to relate wood properties to one another as well as to certain tree characteristics and to paper properties. There are many of these, but comparisons will only be discussed here for those that are the most important. In general, wood properties are generally not closely genetically correlated except when they measure similar characteristics like wall thickness and specific gravity (Fig. 10.5). One detailed study by Goggans (1964) with *Pinus taeda* in an open-pollinated test in Georgia related four cell properties with the specific gravity of latewood and earlywood. A glance at Table 10.3 shows that there are essentially no meaningful correlations, except the earlywood and latewood specific gravity, percent latewood, and wall thickness, with the specific gravity of the whole tree. It is evident that almost all characteristics studied were phenotypically independent of one another.

There may be a relationship between tree form and wood properties but this is usually not the case except for compression wood. An example is *Pinus oocarpa*, where essentially no correlation was found between specific gravity and tree straightness (Lima 1987). This is the usual situation since the genetic control for specific gravity is considered to be independent of that for straightness, and they are inherited separately. Surprisingly, for slash pine in Australia, Allen (1985) found no correlation between stem straightness and compression wood. Yet he cites the possible effects of straightness on compression wood and warns that his findings may not be typical. A correlation between tree form and specific gravity was reported by van Buijtenen (1965) in which the loblolly pine progenies with high specific gravity exhibited the better form. Allen (1977) concluded that improving stem straightness in slash pine will have no adverse effect on any wood properties.

For loblolly pine, Franklin (1974) reported no correlation between tree straightness, extractive content of the wood, and turpentine yield. However, he did find that trees with small limbs yielded less extractives and turpentine than those with large limbs. As expected, he also found that wood from trees infected with fusiform rust (*Cronartium quercum* f. sp. *fusiforme*) produced more turpentine. The percent extractives from healthy, noninfected older loblolly pine wood was 3.1% while from trees diseased with fusiform rust it was 19.4% (Zobel 1973). Breeding for resistance to fusiform rust in loblolly pine will result in better wood because of a reduced resin content. For *Picea abies*, Klem (1942) reported that lower density trees have more knots, associated with an increase in resin content.

By far the most studies correlating wood properties are those associated with wood density and tracheid length. This is primarily because of the importance of these two properties but also because they can be simply and accurately measured. However, numerous other types of correlations have also been studied. For example, based on a study of a 15-year-old progeny trial of *Araucaria cunninghamii*, Eisemann et al. (1990) showed generally favorable genetic interrelationships among the wood properties studied. These would enable small or moderately favorable predicted indirect responses to selection compared to direct responses when individual traits are selected. Relationships among a number of wood properties are shown in Table 10.4.

The predominance of correlations involving specific gravity and tracheid length is evident from Table 10.4. For specific gravity, the correlations with percent latewood are strong. (This subject is covered in Chap. 4.1.1). Wall thickness was strongly related to wood density and the cells with larger lumens had lower density, as would be expected.

The correlation between cell length and specific gravity has been most studied. All of the five references listed in Table 10.4 indicate essentially no relationship between the two. Although a positive correlation is often assumed between latewood percent and cell length, this often is not the case. Both positive and negative relationships have been reported (see Jackson and Greene 1958, Jackson 1959, Ericson 1960a, Goggans 1962b, Zobel et al. 1962). One still commonly hears the concept that latewood cells are longer than earlywood cells and this is the usual situation. Tracheid length and other cell dimensions were positively related but the correlation coefficients were unexpectedly low. The longer cells were negatively related to spiral grain but had a greater micellar angle.

Table 10.4. Miscellaneous relationships among wood properties

Relationships	Species	Reference	Comments
Specific gravity and percent latewood	*Pinus elliottii*	Dadswell and Nicholls 1959	This strong relationship is expected and generally found
Specific gravity and wall thickness	*P. taeda*	Goggans 1964	Wall thickness was positively correlated with wood density, as expected
Specific gravity and cell width		Goggans 1964	The wider cells had lower specific gravities
Specific gravity and lumen diameter		Goggans 1964	There was a negative correlation between the two, as also shown by Larson (1973)
Specific gravity and tangential wall thickness	Conifers in general	Larson 1973	Wall thickness was highly correlated with wood density
Specific gravity and radial lumen diameter	Conifers in general	Larson 1973	An inverse relationship was evident with higher gravity with smaller diameter lumens
Specific gravity and compression wood	*Pinus elliottii*	Allen 1985	The trees with denser wood had more compression wood; $r = 0.80$
Specific gravity and compression wood	General conifers	Timell 1986	There was a positive relationship between specific gravity and compression wood
Specific gravity and spiral grain	*Pinus elliottii*	Allen 1985	A correlation of $r = -0.21$ was found
Tracheid length and specific gravity	*P. taeda*	Zobel et al. 1960	The study spanned seven states and included thousands of trees. Correlation was low with $r = 0.10$
Tracheid length and specific gravity	*P. elliottii*	Zobel et al. 1962a	In grafts, $r = 0.03$ indicating no relationship between cell length and specific gravity

Tracheid length and specific gravity	*P. radiata*	Harris 1961	No relationship was found between tracheid length and specific gravity
Tracheid length and specific gravity	*P. taeda*	Jackson and Strickland 1962	Tracheid length was not correlated with specific gravity
Tracheid length and specific gravity		Goggans 1964	No relationship was found
Tracheid length and specific gravity	*P. elliottii*	Strickland and Goddard 1966	Tracheid length was negatively correlated with specific gravity
Tracheid length and specific gravity	*P. taeda*	Matziris and Zobel 1973	A small negative correlation was found between tracheid length and specific gravity
Tracheid length and specific gravity	*P. elliottii*	Allen 1985	There was essentially no relationship between tree cell length and wood density. However, latewood cells were longer than earlywood cells with $r = 0.97$
Tracheid length and compression wood	*P. taeda*	Zobel et al. 1960	Tracheid length was negatively correlated with compression wood
Tracheid length and compression wood		Allen 1977	The correlation was $r = 0.50$ between tracheid length and compression wood, a surprising finding since compression wood tracheids are only half as long as those in corresponding normal wood
Tracheid length and tracheid width		Zobel et al. 1962b	The correlation coefficient was $r = 0.44$, lower than would be expected
Tracheid length and cell diameter		Matziris and Zobel 1973	Longer tracheids had larger cell diameters

Table 10.4. (*contd.*)

Relationships	Species	Reference	Comments
Tracheid length and wall thickness		Zobel et al. 1962b	An $r = 0.49$ was found
Tracheid length and latewood percent	General	van Buijtenen and Zobel 1994	In general, latewood cells are slightly longer than those of the earlywood
Tracheid length and latewood percent	*Pinus taeda*	Goggans 1962	A positive relationship was found with longer cells in the latewood
Tracheid length and spiral grain	East African conifers	Paterson 1967b	There was a negative correlation between the two, with trees with longer cells having less spiral grain
Tracheid length and spiral grain	*Pinus elliottii*	Allen 1977	The $r = 0.17$ in the 14-year-old trees is so small as to be quite meaningless
Tracheid width and wall thickness	*P. taeda*	Zobel et al. 1962b	A surprisingly low $r = 0.32$ was found between cell width and wall thickness
Wall thickness and cell diameter	*P. resinosa*	Larson 1960	As wall thickness increased, cell diameter decreased
Wall thickness and fibril angle	*P. elliottii*	Hiller 1964	Fibril angle was greatest in thicker walled cells and positively related to wall thickness
Tracheid length and microfibril angle	*P. radiata*	Donaldson 1993	Microfibril angle was strongly correlated with tracheid length
Tracheid length and micellar angle	*P. elliottii*	Dadswell and Nicholls 1959	There was a high positive correlation between cell length and micellar angle
Tracheid length and fibril angle	*P. sylvestris*	Echols 1958	The two were highly correlated
Tracheid length and fibril angle	General	Megraw 1985	Cell length and microfibril angle were closely related

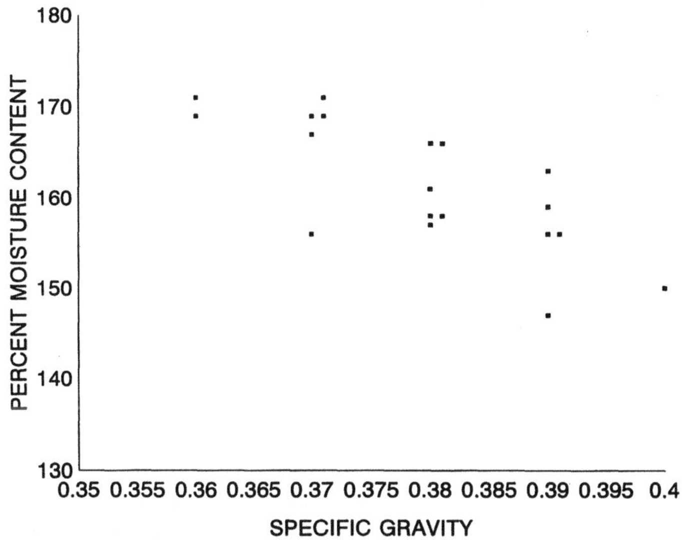

Fig. 10.6. Within a species the higher density trees have a lower moisture content. Therefore, when one breeds for high specific gravity, a lower moisture content will result. This does not always hold among different species, but usually the wood of high density trees contains less moisture

Another rather common type of correlation is to compare wood properties with pulp or paper properties. There is much literature dealing with such relationships but a detailed discussion of it is not possible here. Our interest is to determine how strongly a wood characteristic is inherited and from this to determine how a change in the characteristic will affect the final product. This is illustrated for slash pine by Einspahr et al. (1964). A correlation between tracheid length and percent lignin was found and this affected pulp yield. Also, high handsheet density was related to short cells. As a result, handsheets with a similar amount of cellulosic material often showed considerable variation in density that was negatively correlated with cell length. In loblolly pine, van Buijtenen et al. (1961) reported that zero-span tensile strength was correlated with tracheid length and sheet formation, but was independent of lignin content and yield. In a later paper, van Buijtenen (1967b) concluded that pine trees with small tracheid diameters produced pulps with higher tear. Increased latewood content was related to specific gravity, which in turn produced decreased tensile, bursting strength, and apparent density in paper. Moisture content was strongly related to wood density (Fig. 10.6).

10.4 Relationships Among Wood Properties in Hardwoods

It is difficult to locate good studies on relationships of wood properties in hardwoods. There are some reports, such as that by Einspahr et al. (1963) on triploid *Populus tremuloides* in which the correlation coefficient between specific gravity and fiber length was small (about $r = -0.39$). This tendency for shorter fibers to be associated with higher wood specific gravity is not usual. More common is the result obtained with a totally different species (*Gmelina arborea*) in which specific gravity and fiber length were not correlated (Akachuku 1984). The effect of wood properties on the final product were shown for *Populus tremuloides* where 62% of the pulp yield could be accounted for by the percent lignin, the specific gravity, and the fiber length (Einspahr et al. 1967).

Working with yellow poplar (*Liriodendron tulipifera*), Lowerts and Kellison (1981) found essentially no correlation between sensitivity to rot and discoloration of the wood with specific gravity, wood moisture content, or growth rate at breast height. None could be used as an indicator of susceptibility to rot or discoloration (see Chap. 11.2.1.1). A few reports relating wood properties among hardwoods are listed in Table 10.5.

Table 10.5. Relationships among wood properties in hardwoods

Species	Reference	Summary
Betula pendula	Velling 1983	There was an unfavorable positive correlation between wood density and volumetric shrinkage
Eucalyptus globulus	Clarke 1990	There was no relationship between tear and wood density or between burst and wood density
Fraxinus americana	Hiller 1968	A positive relationship was found between wall thickness and spiral grain
Platanus occidentalis	Huber and Bongarten 1981	Fiber length was unchanged by changes in specific gravity
Populus spp.	Göhre 1960	Density was correlated with the number of fibers per unit area of cross section and with reaction wood. Surprisingly, it was not related to thickness of cell walls
P. tremuloides (*triploids*)	Einspahr et al. 1963	Specific gravity and fiber length had small correlation coefficients, about −0.39, indicating a small tendency for shorter fibers to be associated with a high specific gravity
Triplochiton scleroxylon	Oteng-Amoako et al. 1983	Specific gravity was directly related to cell wall volume, fiber volume, and double wall thickness but negatively related to parenchyma volume

10.5 Relationship of the Wood Properties of Coppice, Rooted Cuttings, and Grafts to Donor Trees

In several species, regeneration of stands through sprouts (coppice) rather than by seedlings and by rooted cuttings is common. There has been a question as to whether the wood of the coppice sprout would be the same as that of the tree which was coppiced or which supplied branches for the rooted cuttings.

The assumption is usually made that since the original tree and the coppice tree are both grown on the same root system and have exactly the same genotypes, the wood ought to be the same if the same number of annual rings are involved. This is, in fact, the usual result when the age of the wood of the coppice tree is the same as that of the original tree although the amount of reaction wood in coppice is often greater, especially if more than one shoot is left to grow per stump and the trees lean away from each other with limbs only on one side. (Zobel and van Buijtenen 1989). Although differences were found in wood density for sprouts from different stumps of 1-year coppice of red maple (*Acer rubrum*), all sprouts from the same stump were essentially the same in red maple (Saucier and Taras 1966a). In a study of 3-year-old *Eucalyptus saligna* in Hawaii, King (1980) found that the individual shoots from a stump did not all have the same wood properties, with the largest shoot having the highest specific gravity. He reported that of the total variation in specific gravity, 35% occurred within the coppice clones. Of most importance, King found: "The specific gravity of coppice wood was similar in value to similarly-aged tree wood of this species."

Exceptions to this relationship have been reported. One example is in *Eucalyptus bicostata, E. globulus*, and *E. viminalis*. Ferrari (1992) sampled two 10-year-old shoots from each of the approximately six on each stump (the sprouts had not been reduced to one or two per stump as is done in managed coppice stands). The basic density of the coppice shoots was found always to be less than that of the original tree. Differences were considerable, for example in *E. bicostata* 0.55 to 0.50, *E. globulus* 0.53 to 0.48, and *E. viminalis* 0.49 to 0.45. Ferrari also reported very slight differences in the alpha-cellulose and lignin contents. This result, which differs from most reports, may be because six shoots were left per stump for coppice compared to the single stem of the tree grown from seed. The six shoots would quickly utilize the stored nutrients in the stump and roots.

In *Eucalyptus camaldulensis*, the wood properties of coppice were different from those of the original trees in having lower wood density but longer fibers (Sesbou 1981). Yet, in 1991, Sesbou and Nepveu, working with the same species grown in a number of countries, concluded that for basic density and fiber length, there is a positive correlation between the original trees and the coppice. The large tree-to-tree variability usually found is also present in the coppice; thus, coppicing does not result in greater wood uniformity than is present in the original plantation. They did find a small negative relationship between growth rate and basic density.

Oaks are frequently coppiced. The wood specific gravity of the original forest and the coppiced trees of *Quercus sessilifolia* were similar (Todorovski 1969).

Yet in *Populus*, Phelps et al. (1983) reported that the specific gravity of the original 3-year-old trees was greater than that of their coppiced sprouts of the same age.

There has been an understanding that cuttings will have the same wood properties as the donor plant at the same age. To test this, Nicholls and Brown (1971) set up a trial of radiata pine. Although quite similar, they found that ramets had somewhat greater cell length, larger spiral grain and somewhat less dense wood than did the donor tree (ortet). They pointed out that these differences probably were genetic in origin, but that the rooted cuttings were grown on different sites under differing cultural conditions, which might account for some of the wood property differences. It is also assumed that grafts will have wood similar to that of the donor tree at the same age. One example of this was in *Pinus sylvestris*, where Ericson (1960a) found a strong relationship between the basic density of the donor tree and the grafts when grown in different environments. In contrast, Zobel et al. (1962b) obtained a correlation of only $r = 0.28$ for specific gravity of donor tree to graft in loblolly pine and a correlation of $r = -0.10$ between tracheid length of graft and donor tree. However, the two were not of the same age. When the wood of clones and progeny of *Picea abies* which had identical rates of growth were compared, no evidence of an unfavorable effect of vegative propagation was found (Persson 1972). Wood specific gravity of cuttings made with seedlings was the same at one location and slightly lower at another in 37-year-old rooted cuttings of white pine (*Pinus strobus*) (Struve et al. 1984).

10.6 Wood Property Relationships Between Chemical Composition and Pulp Properties

It is evident that as wood anatomy changes, there will probably also be some variation in the chemistry of the wood. Only a few studies have been made of the relationship between morphology and chemistry, although more has been done relative to tree form and its effects on wood, as described in Chapter 8. The effects can be large, as illustrated here for yield and tear factor.

Cellulose characteristics were related to tracheid characteristics and specific gravity in a study with loblolly pine (Zobel et al. 1962b). The results showed that cellulose content was only correlated with wall thickness and wood density; all other cell characteristics were poorly related to wood chemistry (Fig. 10.7). However, because of the difficulty of analysis and the gross method of cellulose assessment the authors stated that a cellulose to wall-thickness relationship should be viewed with caution because it may simply reflect ease of penetration of the sodium chlorite used in the analysis for cellulose. Yet it would appear that the value of $r = 0.56$ indicates that a closer study is necessary. It their studies with *Pinus taeda* and *P. patula* in South Africa, Wright and Sluis-Cremer (1992) found that correlations between tracheid characteristics and pulp and paper

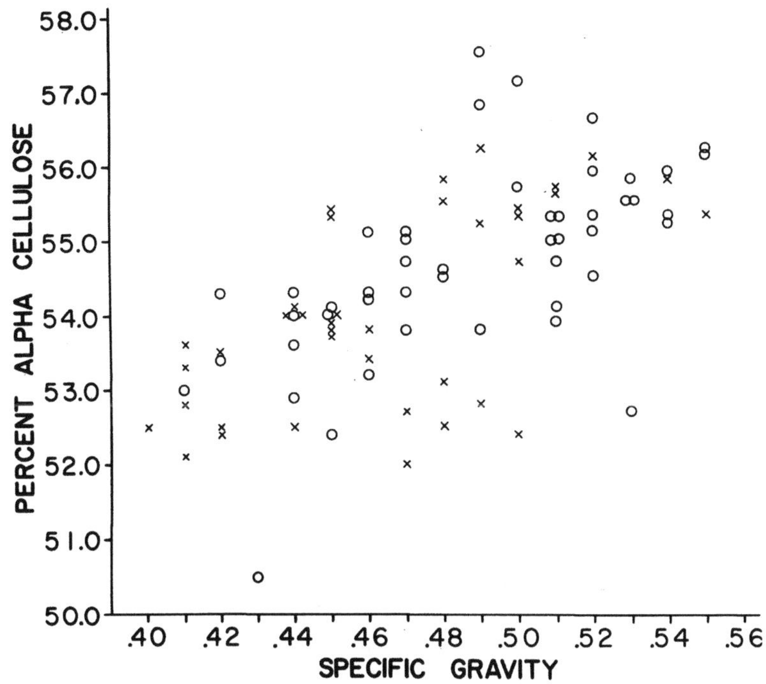

Fig. 10.7. Fiber properties and chemical composition are sometimes correlated, as is shown for specific gravity and alpha cellulose. The trees with the higher wood densities have the greatest alpha cellulose content

strength properties indicated that no simple tracheid measure could account for more than 50% of the observed variations in the pulp. They concluded that direct testing of wood by the use of micropulping is preferable.

In Norway, there was little or no connection between the percentage of lignin and the true specific gravity in spruce (Klem 1942). With faster growth rate resulting from greater planting spacing, both the lignin and resin contents of the wood may increase. For loblolly pine, Zobel (1973) reported that differences in polysaccaride content were not closely related to specific gravity. For many hundreds of loblolly pine trees growing in seven states, Zobel et al. (1960) found that compression wood percent and cellulose yields were rather strongly negatively correlated, as would be expected.

Differences in paper strength using loblolly pine were found to be associated with the additional lignin in compression wood by van Buijtenen et al. (1968), who reported that an increase in lignin decreased yield and zero-span tensile strength. The cross sectional dimensions of the tracheids had a strong effect on paper properties, with the diameter of earlywood cells affecting zero-span tensile strength, apparent density, and perhaps sheet formation. Percent latewood primarily affected zero span tensile strength. As reported by many, Dadswell and Watson

(1961) found the most important relationship to tearing strength to be fiber length, but wall thickness was also important. Thick-walled fibers adversely influenced bursting strength, tensile strength, and folding resistance. Dadswell and Watson stated that the ratio of cell length to diameter had only a slight influence on paper properties. This is contrary to others, who consider that such ratios are important.

Sheet bulk was affected positively by cell wall thickness and cell length, and negatively by cell diameter, as reported in a summary paper by Ifju and McLain (1982). Mean fiber diameter accounted for 89% of the variation in Canadian Standard Freeness. Tear was most sensitive to cell length and average surface density of the fibers, but lower cell volume, cell diameter, and cell wall thickness were all important.

In his study of wood brightness in 12-year-old grafts of loblolly pine, Wilcox (1974) found that high brightness was associated with low compression wood content as well as with a low density and a low latewood content. Also for *Pinus taeda*, Watson and Hodder (1954) found that the microfibrillar angle had no influence on sheet properties and cell length was of limited importance over the range of 1.5 to 3.5 mm. Tear and bulk were only slightly influenced by tracheid length and microfibril angle although tensile strength was related to tracheid length, according to a general report by Watson et al. (1952).

Campinhos and Claudio-da-Silva (1990) have discussed the importance of wood chemistry in pulping *Eucalyptus* and its relation to other wood properties. They recommend that more variation and genetic studies are needed in this area. For example, preliminary results indicate that wood pentosans decrease with wood density. If confirmed, this can be an important fact because of the strong inheritance of wood density.

Studies have been reported from South Africa by SAPPI Forests in relating certain pulping characteristics to wood density in *Eucalyptus grandis* (Anonymous 1988). It was found, surprisingly, that pulp yield was slightly negatively correlated with wood density. The Kappa number (lignin content) and tear factor were not related to density, but the burst index had a negative correlation ($r = -0.40$) with tear.

10.7 Summary

It is essential to know how different wood and growth characteristics relate to one another. Many traits are genetically independent; but if a change in one characteristic results in a change in another, then the relationships and magnitude of the changes need to be known. If a relationship is positive such that a desired change in one characteristic results in a desired change in the other, all is well; an example would be cell wall thickness and wood density; but if there should be a negative relationship, such as for growth rate and wood density, then the geneticist who wishes to improve both faces a real problem.

Generally the relationships, with the exception of that between growth rate and wood density, (as well as tracheid length) are not well known genetically. Even then, apparent relationships can be confusing. For example, in pines, any silvicultural manipulation, such as thinning or fertilization, which increases growth rate of a tree, will result in shorter tracheids. Yet the genetically fastest-growing individuals in a species may have either long or short tracheids. An increased growth rate due to genetics does not affect tracheid length in a manner similar to an environmental effect.

In general, for the hard pines, there is little or no relationship between growth rate and wood density of individual trees, as Megraw (1985) summarized for loblolly pine: "Taken together these studies provide overwhelming evidence that specific gravity and growth rate cannot be inversely related genetically." However, there are occasions where fast growth is negatively related to growth rate. It is not unusual to find that the faster growing provenances have on the average a lower wood density.

For the firs and spruces, most reports show a negative correlation of wood density with growth rate but even then, such as with Norway spruce, reports in the literature are not unanimous. There are similar confusing reports for Douglas-fir. Some of the confusion is related to whether one refers to phenotypic or genotypic correlations, which sometimes are opposite in direction. The relationships for 32 studies of various softwood species and genera are shown in Table 10.1.

Growth rate is usually found to be independent of wood density for the diffuse-porous hardwoods, but there are exceptions. Generally, there is a positive correlation between growth rate and wood density in the ring-porous hardwoods, but this relationships has not been well studied genetically.

The relationship of growth rate to other wood properties cannot be summarized because they are so variable. Some are obviously closely related genetically, while others are independent from one another. A number of examples are cited in this chapter.

The genetic relationships among wood properties themselves (and tree form) are of importance. Usually, there is only a close genetic correlation where one property directly affects another, such as cell wall thickness and wood density; these can be high. Yet, the most studied relationship, that between cell length and specific gravity, shows essentially no correlation. All the miscellaneous relationships that could be found in the literature have been listed in Table 10.4. These include wood as well as pulp and paper properties. No generalized conclusion can be drawn from the table, although some negative relationships will be of key importance in a tree improvement program.

Generally, the wood of coppice, rooted cuttings, and grafts is very similar to that of the donor tree, as would be expected. There are a few exceptions, however. Usually, the anatomical wood properties are not too closely correlated with the chemical properties of the wood, although the pulp and paper characteristics are usually closely related to anatomical wood properties.

Chapter 11

The Genetics of Miscellaneous Factors That Affect Wood

11.1 What Are Miscellaneous Factors?

This book has attempted to generally classify genetic factors that affect wood properties. However, several important items do not fit any general categorization and are reported in this chapter. Disease and insect attacks can have a major effect on wood and hybridization is rapidly becoming more important since the hybrids can be used with vegetative propagation techniques. Currently, biotechnology is being used in an attempt to improve wood quality.

11.2 Diseases and Insects

Normally, diseases and insects (generally referred to as pests) are not equated with wood quality unless they directly attack the wood, as indicated in Chapter 11.2.1.1. However, several types of pests, in addition to the organisms that cause wood decay and discoloration, can significantly affect wood quality. Genetic control of the pest, or the genetic manipulation of the response of the tree to the pest, can improve wood quality. When dealing with pests and wood properties, two kinds of genetic control are possible. The first is to genetically control the pest itself; this is rarely attempted. The second is to alter genetically the tree host so it is less attractive to the pest. Several examples will be given to illustrate the value of genetics to change the host to reduce the impact of pests upon wood quality.

11.2.1 Diseases

Fungal diseases and other organisms destroy wood as rots in one of two ways; the cell walls are attacked (white or brown rots) or as sap stains that attack stored carbohydrates in the wood. The latter do little damage to wood structure but often discolor it (Rydholm 1965). Large amounts of low quality timber result because of fungal rot or stain. It is known that there are species differences, and even provenance differences, related to such fungal damages as blue stain (Levi 1978, 1981, Sinclair and Ifju 1979). However, no intensive genetic efforts have been made to breed strains of trees with resistance to these fungi, which are of major economic importance.

Although many specific examples of direct disease effects on wood could be given, it is clear that any disease attack has some effect on wood quality because the disease impacts the physiology of the tree which in turn affects wood properties. When diseases are controlled genetically, better quality wood can result. Good silvicultural practices are also necessary to retain high wood quality since tree wounding often results in formation of wetwood, shake or other defects resulting from disease infection (Bauch 1980). Even foliage damage can affect wood properties. In New Zealand, radiata pine (*P. radiata*) is attacked by a needle disease called *Dothistroma*. In their studies, Carson and Carson (1991) found a weak negative correlation between *Dothistroma* resistance and increased wood density. This resulted from the correlations between *Dothistroma* resistance and increased growth rate and between growth rate and wood density.

11.2.1.1 Wood Decay and Discoloration

It has long been recognized that the wood of some species decays more easily than others. Less well known is the fact that some trees within a species are more decay-resistant than others. This is evident even when the trees are the same age, grown in the same environment, and tested under similar conditions. Whether or not this obvious difference is due to genetics is not generally known.

Zabel (1956) found that in some hardwoods, such as white oak (*Quercus alba*), there was more decay in test blocks from susceptible trees than in those from resistant trees. The same was also observed in the generally decay-resistant black locust (*Robinia pseudoacacia*). In his discussion, Zabel quotes others as stating that wood durability is under rigid genetic control. In trembling aspen (*Populus tremuloides*), individual clones grown in Manitoba, Canada, showed distinct differences in the amount and type of rot present. Growth rate was not related to the amount of infection in this study (Wall 1971).

Discoloration of wood, often the result of disease infection, is a relatively common cause of degrade, and frequently represents the first stage of decay. This is especially important in high quality woods. Infection, whether through wounds or branch stubs, is random, but they spread at different rates in individual trees and there seems to be a relatively strong genetic component involved (Shigo et al. 1977). As stated by Garrett et al. (1976) for hybrid poplar (*Poplus* spp.) that was inoculated by *Septoria musiva*: "There is some evidence to suggest that (wood) infection by some diseases may be under strong genetic control..." The growth of hybrid poplar was found to be independent of the amount of discoloration, and the individual clone was more important than the parents involved. Two clones with the same parents were at opposite ends of the discoloration scale. Working with the same pathogen and 18-year-old *Liquidambar styraciflua* progeny tests and *Populus deltoides* clones in Mississippi, Garrett et al. (1979) found that the rate of closure and size of the discolored columns associated with the wounds were inheritable traits that would respond to a selection program (Garrett et al. 1979). It was observed by Waterman (1954) that different clones of hybrid poplar also reacted differently when infected by *Septoria musiva*. She concluded that the best method to control this pathogen was to use resistant clones.

The spread of discoloration and decay from an infection site is only limited by the ability of a tree to respond and wall-off or compartmentalize the infected tissue from healthy tissue. The compartmentalization process is described in detail by Shigo and Marx (1977). It is also treated in depth in the recent monograph edited by Blanchette and Biggs (1992). Discoloration and decay resistance has been shown to be under genetic control in several hardwood species. For yellow-poplar (*Liriodendron tulipifera*), a valuable hardwood, Lowerts and Kellison (1981) reported individual heritability as $h^2 = 0.49$ and family heritability as $h^2 = 0.57$ for resistance to discoloration and decay resulting from stem wounds. The large amount of tree-to-tree and family variation and the moderately high heritabilities would allow large genetic gains from a program of within-family selection for higher wood quality. Similarly, Noh et al. (1986, 1987) found resistance to discoloration in poplar clones (*Poplus nigra* × *P. maximowiczii*) grown in Korea to be under moderate genetic control ($h^2 = 0.41$). Selection to increase resistance to discoloration in poplar clones would be effective. Two clones responded so favorably that they were recommended for use. In California, heart-stain in tan oak (*Lithocarpus densiflora*) showed no genetic control (Prestemon 1967), but Santamour (1979) found that there was a genetic component in Silver maple (*Acer saccharinum*) for compartmentalization of discolored wood (Fig. 11.1).

Although not directly involving genetic studies, the report by Rudman et al. (1967) stresses the necessity of selecting plus trees of teak (*Tectona grandis*) in New Guinea for resistance to heart rot. There was a close similarity between decay and termite tolerance. Four of the 19, 13-year-old trees selected were completely unacceptable and four trees were found to be very superior for resistance to heart rot. Selecting for tolerance will result in a moderate improvement. Decay susceptibility in loblolly pine (*Pinus taeda*) is related to low wood density according to Schmidtling and Amburgey (1982). Therefore, breeding for high density wood should reduce decay susceptibility if rotations are long enough to make a difference.

11.2.1.2 Other Effects of Disease

One prime example of the effect of disease on wood is the fungus *Cronartium quercum* f. sp. *fusiforme*, the fusiform rust of the southern pines. Literally hundreds of references could be cited on the cause and genetic control of the disease; some were summarized in Zobel and Talbert (1984). Less clear has been the effect of the disease on wood quality but it can be of major importance; a few examples are listed in Table 11.1. It is evident how greatly rust infection affects some wood and product quality (Fig. 11.2). Tolerance to the disease is very

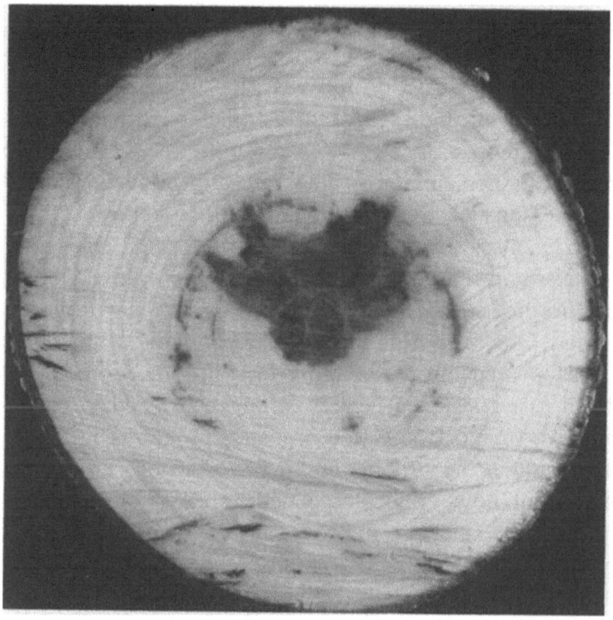

Fig. 11.1. There is a reasonably strong genetic control relative to the severity of decay in wood. Shown is a *Liquidambar styraciflua* tree which has produced a "protective wall" of cells which effectively stops the spread of the decay

Table 11.1. Wood and pulp properties from fusiform-infected and nondiseased trees obtained from approximately the same level[a]

Characteristic	Healthy wood	Diseased wood
Unextracted specific gravity[b]	0.44	0.51
Extracted specific gravity[b]	0.42	0.42
Percent resins[ab]	3.1	19.4
Pulp yields (%, based on dry wt. wood)[b]	52	26
Fold (paper)[b]	209	78
Tear[b]	157	150
Tensile[b]	4300	3600
Burst[b]	32	26
Brightness[b]	15	21
Total carbohydrates[c]	74	50
Alpha-cellulose[c]	41	26
Hemicellulose[c]	29	19
Lignin[c]	22	22

[a] Within each loblolly pine tree.
[b] Zobel (1973).
[c] Rowan (1970).

pronounced and good gains are obtainable by breeding (Powers and Mathews 1987).

In their study of the effect of fusiform rust on loblolly pine wood, Veal et al. (1974) reported that bolts of wood from light, medium, and heavy rust infection classes yielded 4, 10, and 23% less pulp, respectively, and required 4, 8, and 20% more alkali to obtain comparable kappa numbers. The infected wood produced paper with 10% lower burst and tensile strengths. In an internal report at NC State University, Kinloch (1963) observed that shorter and deformed tracheids were more pronounced in fusiform diseased wood compared to healthy wood. Differences in specific gravity are not so obvious but it appears that lower wood density occurs in infected wood, after the high resinous depositions within the wood have been removed. Rust is even more damaging to solid wood products. The genetic control of fusiform rust resistance is very strong and a reduction in infection rate is a major objective of most pine tree breeding programs in areas where rust is a problem; the result is the production of better wood. From the examples cited above, it is evident how important genetic control of fusiform rust is to wood and paper properties, since an improvement in wood properties is possible by genetic improvement of rust tolerance. Many other examples could be given of the effect of diseases on wood and the value to wood properties that can result from a tree breeding program (Fig. 11.3).

Fig. 11.2. Although not often considered in suggestions about methods to genetically improve wood, one of the best is to breed for resistance to diseases of the tree stem. Shown are cross sections from below the fusiform rust gall, through the gall, and above the gall of a loblolly pine. Note the low quality of wood in the diseased area. Pulpwood yields and quality as well as solid wood products are greatly reduced because of the disease

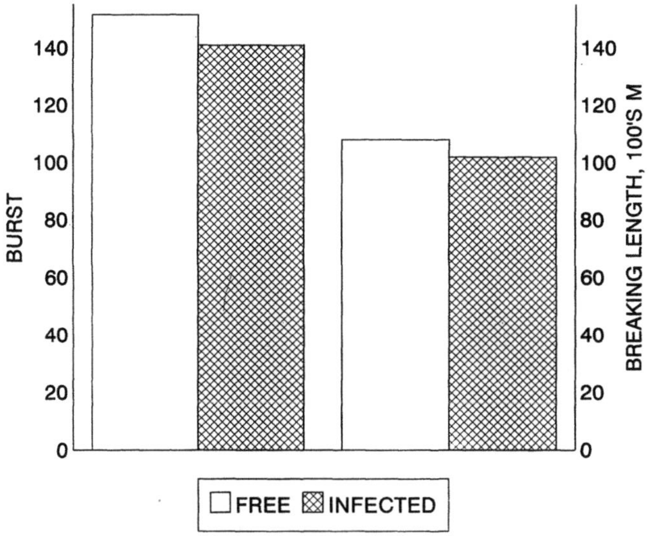

Fig. 11.3. Fusiform infected trees not only give poor yields of solid wood products and pulp but they also produce paper with somewhat inferior properties, as shown

Numerous diseases of forest trees have responded to breeding for tolerance; as stated earlier, some effects on wood always are present. Sometimes the disease directly affects the wood. One example is the canker disease of sycamore (*Platanus occidentalis*), which seriously damages the tree bole. Both family and geographic source differences of the seed were related to the susceptibility to the disease (Coggeshall et al. 1981).

11.2.2 Insects

Insects have a great effect on wood quality, both directly and indirectly as a source of entry for diseases and secondary insects (Levi 1981). Despite the huge losses from insects, very little effort has been applied to produce genetically tolerant trees. A couple of examples are the spruce budworm (*Choristoneura fumiferana*) in the eastern United States and Canada and the defoliator gypsy moth (*Porthetria dispar*). Both insects have a major effect on wood quality (Govette et al. 1983, Garges et al. 1984) (Fig. 11.4). Although there is some indication that tolerant individuals exist, foresters have generally assumed that genetic control would not be feasible and little breeding for resistance has been attempted.

Another group of insects that has had a devastating effect on wood quality are the bark beetles (*Dendroctonus* spp.) (McNab 1983, Plank 1984). Some genetic

testing has been done, especially with the southern pine beetle (*Dendroctonus frontalis*), that indicates strong individual tree resistance to the insect in some species (Powers 1992). Little use of this information has been made because of the mobility of the insects, which usually means that it is not possible to predict where beetle attacks will occur. In contrast, for most diseases, it is possible to designate disease centers in which tolerant trees should be planted. However, despite much well-funded research on the control of bark beetles, the destruction of wood quality by bark beetles has been only slightly modified by breeding programs.

One direct effect of insects on wood properties is the formation of a compression-type wood which results from aphid attack. There are differing tree-to-tree responses to aphid attack but the only breeding program attempted to overcome this problem has been with loblolly pine in South Africa (van der Sijde et al. 1985). Apparently, aphid-tolerant families of loblolly pine are available, making possible the use of this species, which had been rejected by the industry because of undesirable wood from infected trees. Some controversy exists about aphids being the major cause of the wood quality problem but there is no doubt that there are aphid-tolerant families. An abnormal type of wood is also formed on aphid attack in other species, such as grand fir (*Abies grandis*) which has wood with shorter tracheids, greater fibril angle, a greater percentage of latewood, a higher specific gravity, reduced module of elasticity, and greater longitudinal shrinkage when attacked by the aphid *Adelges piceae* (Foulger 1968). This wood is in almost all respects identical with compression wood (Timell 1986). The relationship of wood defect to tipmoth (*Rhyacionia frustrana*) attack in pine was covered by Hedden (1991).

Despite the widespread damage to wood properties from primary and secondary effects of insect attack, little work has been done on the genetics of resistance to insects because of the difficulty of breeding for resistance, the age needed for testing, and the uncertainty where and when attack will occur because of the mobility of the insects.

11.3 Wood Uniformity

Uniformity is one of the greatest needs and goals to be expected from the application of genetics to wood properties. Reducing variability is usually a major objective of a breeding program involving wood properties, and results in higher quality products and increased efficiency of manufacture (van Buijtenen 1967a, Zobel et al. 1983, Blair and Olson 1984). The necessity for and the value of better wood uniformity obtained by breeding has been emphasized through-

Fig. 11.4. Although they are defoliators, the gypsy moth (defoliated trees shown *above* in midsummer) and the spruce budworm (shown *below*) have a major effect on wood quality

out this book. Uniformity can be obtained by improvement of tree form (which reduces reaction wood), by breeding for less within-tree variation (such as reducing juvenile wood), and by breeding for more uniform chemical components (Ladrach 1986). It is, of course, axiomatic that wood uniformity can be greatly improved by breeding and using trees with similar wood properties. Clonal forestry results in trees with the most uniform wood properties among trees if not within the tree (Zobel et al. 1983). There are many references emphasizing that uniformity can be enhanced by using trees with genetically desired wood combined with proper fertilization (Zobel and van Buijtenen 1989).

Although juvenile wood was discussed in Chapter 8.3, it will be further discussed here in relation to wood uniformity. A number of investigators have been interested in improving wood uniformity by manipulating the process of juvenile wood formation. The results have been somewhat contradictory. For example, it was reported for Douglas-fir (Anonymous 1990) that it is impossible to increase uniformity of wood density across growth rings by selecting trees that have more gradual age trends of wood density from the center outwards. In radiata pine, Burdon and Harris (1973) found the opposite and reported that the objectionable pith-to-bark density gradient could be reduced by clonal selection. For slash pine (*Pinus elliottii*), Hodge et al. (1992) and Hodge and Purnell (1993) hypothesized that selection for a decrease in transition age of juvenile wood production will tend to increase juvenile wood density and decrease that of mature wood. This will result in a more uniform density profile, which will improve mill efficiency. We have found that selection for increased mature wood and juvenile wood density will increase whole tree density by 0.02 g/cm^3.

We have observed that pine trees with high density mature wood frequently have high density juvenile wood. Also related to juvenile wood, the change of density with age is important relative to uniformity within a tree. The density to age gradient for 15-year-old *Pinus radiata* was 7.9 kg/m^3/yr; heritability of the gradient was estimated at $h^2 = 0.27$ (Bannister and Vine 1981). Although the heritability is small, this gives some encouragement for developing more uniform wood by the use of genetic manipulation.

Uniformity can be related to time of flushing, which is genetically controlled. Late-flushing Norway spruce (*Picea abies*) had lower wood density than those that flushed earlier; this caused a reduction in the latewood density, resulting in more uniform wood in the annual ring (Dietrichson 1964).

For ponderosa pine (*Pinus ponderosa*), McKimmy and King (1978) reported that a careful selection of trees with rapid growth and high wood density would reduce variation in wood quality among families. Additionally, use of families with low within-family variation would increase wood uniformity even more. Nicholls (1967c) emphasized that the within-ring variation in *Pinus radiata* is not as important as is the uniformity within the species. The wood users in New Zealand made a strong demand for uniformity of *Pinus radiata* wood in a symposium on the kind of forests that are needed in that country (Burdon and Thulin 1965).

In a study comparing the wood of four sources of *Pinus tecunumanii* with *P. patula*, *P. elliottii*, and *P. taeda* grown in South Africa, Malan (1993) found

Pinus tecunumanii to be much more uniform with a latewood percentage about half of that of *P. patula* and one-third that of *P. taeda* and *P. elliottii.* The earlywood of the *P. tecunumanii* was more dense than in the other species even though the latewood densities were similar. As a result, *P. tecunumanii* had a low degree of variability (more uniformity) in density both in the radial and in the axial directions on the stem compared to the other three species. As reported for Japanese larch (*Larix leptolepis*), the within-ring uniformity of wood density was considerable (Isebrands and Hunt 1975). Fertilization made the uniformity even greater. The authors stated: "Uniform wood properties within rings such as reported here improve the efficiency of processing the raw material."

One method to increase uniformity is to make greater use of vegetative propagation. The greater wood uniformity obtainable by using vegetative propagation is widely recognized by many foresters, since all trees of a clone have very similar wood. A comparison of wood properties of 5-year-old loblolly pine seedlings and tissue culture plantlets grown together on three different test sites indicated that the within-site variation in specific gravity for a single clone was only 29% of that from the open-pollinated family from which it was derived, Frampton and Jett (1989) concluded: "When tissue culture techniques become practical, operational clonal plantations of loblolly pine should offer substantial improvement in the uniformity of wood produced relative to the heterogeneous seedling-origin plantations currently being established."

The increased uniformity that results from using vegetative propagation has been exploited on an operational basis by several organizations (Zobel et al. 1983, Brandão 1984). In Brazil, Aracruz has been very successful in obtaining a higher uniformity with vegetatively propagated eucalypts. The specific gravity of the original *E. grandis* trees ranged from 0.35 to 0.85 but after intensive selection, clones were chosen that varied from 0.52 to 0.63. The clones were also selected for high cellulose yields. This uniform wood has been most valuable in the production of both high quality and uniform pulp, which is one of the reasons that paper manufacturers around the world prefer *Eucalyptus* wood.

11.4 Hybridization to Change Wood Properties

Generally, hybrid progeny have characteristics that are intermediate to those of the parents (Pryor et al. 1956). This is particularly true for wood proper ties and was shown for hybrid larch (*Larix eurolepis*), where both lumen diameter and wall thickness of the tracheids were intermediate to those of the parents (Chowdhury 1931). In *Eucalyptus*, Pryor (1950) found the wood density of *E. rossii* to be 41.0 lbs/ft^3 (621 kg/m^3) and that of *E. robertsoni* to be 35.6 lbs/ft^3 (534 kg/m^3). The hybrid had the intermediate value of 37.1 lbs/ft^3 (556 kg/m^3). Pryor concluded: "The results obtained thus far with the two groups of parents and their hybrids are compatible with the hypothesis that the mechanical and physical properties of the hybrids are intermediate between the parents." The pine hybrid *P. caribaea* × *P. elliottii* had wood properties intermediate between those of the two parents, as could be expected (Harding 1990). This use of mid-parent values enables assessment of the progeny without actually doing wood analyses which saves a huge amount of work and time. The intermediate value concept was found in crosses within species. It has enabled using mid-parent values for specific gravity to assess progeny values for loblolly pine with about 70% accuracy (Talbert et al. 1982a).

However, the reported intermediacy does not always occur. According to Harding (1990), the wood specific gravity of 9-year-old hybrid trees between *Pinus caribaea* var. *hondurensis* and *P. tecunumanii* was significantly higher (sp. gr. of 0.43) than those of both parental species (sp. gr. of 0.40 and 0.40). The *P. caribaea* parent and hybrid had about the same amount of grain spirality but the *P. tecunumanii* had higher spirality (3.0 compared to 4.3 degrees). The average wall thickness of the hybrid between *Pinus thunbergii* and *P. densiflora* was greater than that of *P. thunbergii* but quite similar to that of *P. densiflora* (Mergen and Furnival 1960). *Pinus rigida* × *P. taeda* hybrids had longer tracheids than those of both parents (Koo and Hong 1967).

Hybrids are common in hardwoods, especially in the oaks and eucalypts. When Pryor (1951) and Pryor et al. (1956) developed eucalypt hybrids, the wood produced was intermediate between that of the parental species, a common

Fig. 11.5. Hybridization can produce the desired type of wood since the hybrid is usually (but not always) intermediate between the two parental species. Shown is a *E. grandis* × *E. camaldulensis* hybrid in South Africa that has wood intermediate between the parental species

result in the genus *Eucalyptus*. In our operational work with hybrid eucalypts in the tropics, the wood of the hybrid is usually intermediate between that of the parent species. This enables the development of hybrids between species that will have desired wood properties. As an example, when *E. grandis* is crossed with *E. camaldulensis*, the high wood density of the latter species is reduced enough so the hybrid wood has general utility (Fig. 11.5).

In poplar, Son and Chung (1970) produced a hybrid (*Populus alba* × *P. glandulosa*) that had longer fibers and higher wood density than had the commonly used *P. euramericana* hybrids. Sometimes hybridization has no potential for changing wood properties; for example, Marton et al. (1967) found that pentosans, lignin, xylan, or extractive yields were not affected in hybrid poplars because there were no initial differences among the clones. Similarly, in the hybrid *Populus nigra* × *P. maximowiczii*, essentially no genetic control of heartwood formation was found by Noh et al. (1987) and hybridization therefore had no effect on this property.

In addition to hybrids from crossing species, the term hybrid is sometimes used when provenances are crossed. This was done for *Picea abies* by Nilsson (1963) where the Swedish × Swedish crosses were compared with Swedish × "continental" (German) provenance crosses. It was found that the wood between these two different provenance crosses was identical.

11.5 Effects of Polyploidy on Wood

Although rare, polyploid trees[5] are sometimes recommended for use in forestry because polyploids might have larger cells than the normal diploid trees. In the genus *Populus*, which has very short fibers, the hope has been to lengthen the fibers (and increase growth rate) by polyploidy. Aspen (*Populus tremuloides*) triploids had somewhat higher specific gravity than diploids although the cellulose and lignin contents were the same according to Einspahr et al. (1963). However, the fibers of the triploids were 26% longer and 10% wider than those of the diploids. These differences are large enough to have some effect on paper properties. Einspahr et al. (1963) obtained a broad-sense heritability of $H^2 = 0.86$ for the 30th ring and $H^2 = 0.50$ for average fiber length in the triploids. Similar results were obtained by van Buijtenen et al. (1959) for polypoid aspen with a fiber length increase from 21 to 26% over diploid trees. It is reported that tetraploid poplars have longer cells than diploids (Bakulin 1980). In white ash (*Fraxinus americana*), the longest-celled trees were those with polyploid genomes, according to Armstrong and Funk (1979). Therefore, the limited data currently available indicate that polyploid trees may have longer cells than the

[5] Polyploids are trees that have more than the normal diploid number of chromosomes. The most common are triploids and tetraploids (three and four sets) but trees like redwood (*Sequoia sempervirens*) have six sets and thus are hexaploids.

diploid trees, but there is disagreement about the operational importance of these differences.

11.6 The Effect of Tissue Culture and Biotechnology

Major concerns relative to the development of trees using new methods like tissue culture, or other biotechnological methods, are whether the wood will be different from that of trees now being established, or whether biotechnological methods can change the wood. As of now, there are no definitive answers, but information will be available in the next few years. At present, the consensus is that wood resulting from biotechnological manipulation will not be greatly different from the wood produced from seed or rooted cuttings (Talbert et al. 1982a), but that it may be possible to make some improvement, as well as to predict wood properties.

The only information currently available regarding the effect of tissue culture on wood relates to the juvenile wood of *Pinus taeda*, where seedlings were compared to tissue culture plantlets by Frampton and Jett (1989). They found that wood properties from 13 families, grown at three different sites and sampled at ages 2, 3 and 6 years, was only different for the 2-year-material but the same in the older trees. Apparently the wood was not changed by the use of tissue culture.

A suggestion for improvement of wood is to change the structure of the lignin by using biotechnology (Eriksson and Dinus 1991) and several organizations are taking this approach. They are trying to adjust the chemical structure and amount of lignin in trees to facilitate bleaching and to reduce damaging effluent. The objective of the research is to alter the biosynthesis of lignin in wood by biotechnological methods, including the use of somatic embryogenesis. Results are not yet available although considerable research is now underway.

It is too soon to predict what part biotechnology will play in changing wood. Most biotechnological methods are merely ways to multiply desired genetic material; but it now appears possible to predict such things as wood density following gene mapping procedures. Will it be possible to transfer genes for a desired wood property into otherwise desirable genotypes? No one knows, but there is a lot of enthusiastic activity in this field.

11.7 Wood for Energy

Wood is widely used for energy, for both domestic and industrial purposes. According to Goldstein (1980), the largest volume use of wood on a worldwide basis is for energy. A lot of wood is used for industrial purposes, such as for production of charcoal for energy in the steel mills in Minas Gerais, Brazil.

In his somewhat philosophical treatment relative to the use of wood for energy, Boulding (1977) emphasized the importance of genetics in increasing wood production, part of which will be used for energy. He stated that a more efficient capture of the energy by trees (he hypothesized the increase from the one percent now being obtained to three percent by genetically improved trees) makes: "The concept of the energy plantation is... by no means absurd." The eventual tradeoff between the use of wood for energy or other products is important and the part that genetics will play in this competition is important.

Effort is underway to breed fast-growing trees with the high wood densities desired for charcoal. This is relatively simple because of the high heritability of wood density. The problem faced by the breeder desiring wood best for charcoal is whether to use the high density species, most of which have slow growth, or to breed for higher density wood in the faster-growing species. The latter is preferred because the genetics of wood density allow for rapid wood improvement of the fast-growing individuals while it is genetically more difficult to breed for fast growth in the slower-growing, but higher wood density species.

There are some tropical species that exhibit both fast growth and dense wood, such as *Leucaena leucocephala*, for which a great deal of tree improvement effort has been expended (Brewbaker and Hutton 1979). Only a limited amount of the genetic effort has been used to improve wood for the desired high density, high BTU content, low moisture content, and burning with little smoke and sparking. The efforts and methodologies used to produce wood genetically for better energy

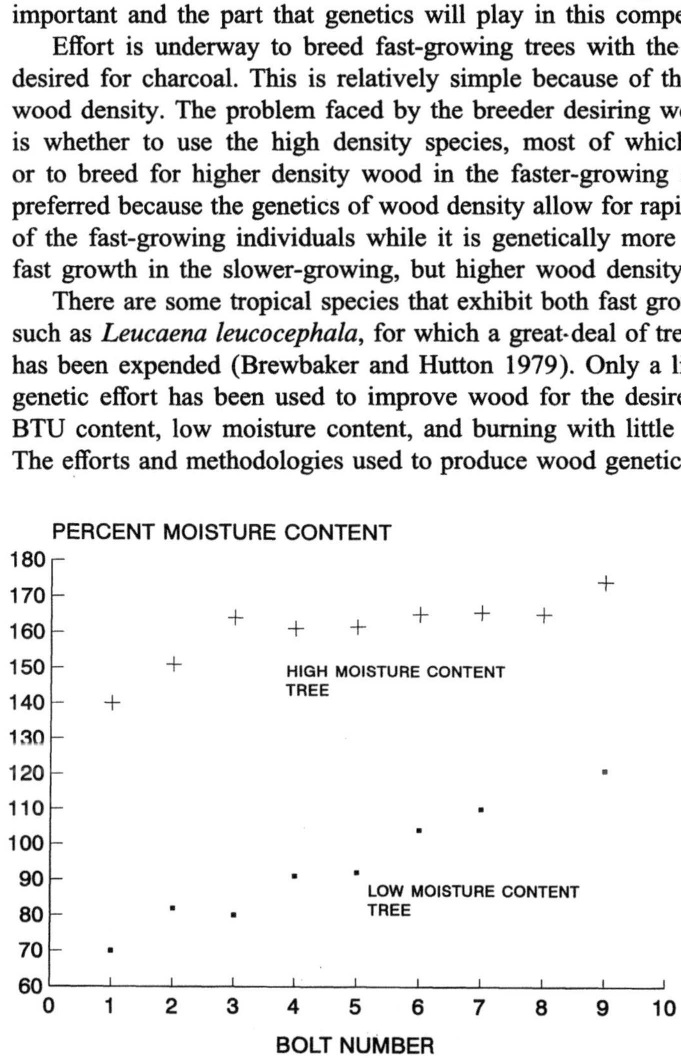

Fig. 11.6. There is some interest in breeding trees especially suitable for energy production. The amount of water in the wood is very important because it has to be evaporated to obtain net calories. Shown are the differences in moisture content between two 25-year-old trees and the variation in moisture content with stem height. These must be assessed in the determination of the genetics of net caloric content

production were summarized by Zobel (1980). Here it was made clear that as the demand for wood for energy expands, it will be very feasible to develop wood especially useful for energy through genetic manipulation (Fig. 11.6).

Many studies have been made on the caloric content of wood but few on the genetics of caloric content. Most of the information related to the energy produced by different woods is proprietary. The differences in caloric content between woods of different species are large. Within the southern pines alone, the caloric value averaged 8600 BTU/ovendry pound of stemwood (Howard 1973). The differences are also large among trees within a species. In sycamore, Ezell et al. (1983) obtained large differences in BTU production even from trees with the same specific gravity. Although not proven, this pattern indicates considerable genetic control of caloric content of wood independent of specific gravity.

The few studies available on the genetics of caloric content show a usable inheritance pattern. In fact, Stringer et al. (1982), working with black locust clones, found that the caloric contents of nine 20-year-old clones varied from 4646 to 4821 cal/g. Caloric content had the highest heritability of all the wood properties they studied ($H^2 = 0.68$). When serious breeding for maximum efficiency in energy production commences, it will be possible to choose the high density individuals of the best species that are also the best caloric producers.

11.8 Summary

This chapter includes a number of miscellaneous factors that are genetically controlled and affect wood properties. Some are of minor while others are of major importance.

1. *Diseases* and *insects* have a major effect on wood quality. These pests can sometimes be controlled genetically and when this is done, wood improvement follows. Little breeding has been done for control of insects but considerable efforts have been committed to disease control. Wood decay and discoloration have been shown to have a genetic component, which at times can be quite strong. Resistance to rot in living trees appears to have a strong genetic background. Clones are available in several species that are relatively resistant to rot, which is usually caused by fungi. Wood colorization, often mistakenly called heartwood, is the first stage in degradation by fungi.

2. All manufacturers desire *uniform wood*, no matter what the final product may be. Both genetic and silvicultural methods can improve wood uniformity among trees. The nonuniformity within a tree resulting from juvenile wood can definitely be improved by breeding and clonal selection, although the heritability for such a change is moderate. The best way to improve between-tree uniformity is by cloning.

3. *Hybridization* can produce "new" types of wood because usually the wood of the hybrid has properties intermediate to those of the parents. With the strong inheritance of wood properties, the wood of a hybrid can be quite accurately

Fig. 11.7. The wood of hybrids is usually intermediate to that of the parents. Shown is a *Eucalyptus grandis* parent surrounded by *E. urophylla* to produce the *urograndis* hybrid. Its wood is intermediate between that of the two species

predicted if the woods of the parents of the hybrid are known. Production of hybrids must take into account the wood of the parental species (Fig. 11.7).

4. The use of *polyploidy* to change wood is of only minor importance but there has been much interest in increasing cell size in the poplars, where the fibers in polyploid trees can be nearly 25% longer than those in diploid trees.

5. There is much interest in the potentials of *biotechnology* to improve wood but no positive results are yet available that can be used in an operational program.

6. *Energy* and *caloric content* of wood can be changed by breeding. Specific gravity may not be very closely related to caloric content. The dilemma for the breeder is whether to work for higher wood density in fast-growing species or to increase growth in the slower-growing species with dense wood. The former is usually chosen because of the inheritance patterns involved.

Chapter 12

Determination of Wood Properties to Be Used in a Tree Improvement Program

12.1 Using Genetic Information

Discussions as to which wood properties should be included in a breeding program occur in a number of places in this book. Some are repeated and summarized in this chapter; this is especially true of Chapter 13 (for example see Chap. 13.2.1). The summary, and thus the repetition, has been done purposely because we wanted to bring together in one place ideas and examples of how to choose wood properties for a genetics program, which characteristics that should be included, and, briefly, why they are chosen, although this aspect has been covered in the chapters dealing with specific wood properties. Once a determination is made as to what wood will be used, it is not easy to change. Choosing wood properties for a breeding program must therefore be done very carefully, using all information available (Fig. 12.1).

The requirements for selecting wood properties were expressed early by Paul (1953), who emphasized the importance of producing wood suitable for the desired product, an aspect also stressed by Mitchell (1960). They both emphasized that a knowledge of the environmental effects on wood must be taken into account. Further, the emphasis in any breeding program should include wood and tree quality as well as volume production.

In their chapter on genetics and breeding of wood, van Buijtenen and Zobel (1989) listed the stages of development in a tree breeding program and pointed to the need to follow these stages in the genetic improvement of wood. Good gains are possible because of the general strong inheritance and wide variability present for most wood properties. Tests of the wood produced in a breeding program are needed as soon as possible so that adjustments can be made in order to continually improve wood quality.

It is difficult to predict what wood properties will be needed in the future. The one certainty is that wood uniformity is of key importance (Blair and Olson 1984). Along with this, the value of wood density is frequently noted, but other wood properties can also be useful. For example, Nicholls (1967b) states: "Fiber length is of major importance for tearing strength ... but basic density provides the most important guide to paper properties."

Information on the genetics of wood production will be only academic unless it can be used to improve trees through a tree improvement program. Genetic information must be determined and reported in a manner that will enable the forest geneticist to readily incorporate it into the breeding program. This requires an understanding of inheritance patterns as well as development of methods to

Fig. 12.1. When growing exotics, such as this beautiful stand of *Eucalyptus* in South Africa, an integral part of the tree improvement program must include wood properties. In addition to species, and possible provenance variations, one nearly always finds very large genetic differences from tree-to-tree within a species. (Courtesy R.J. Poynton, S.A. Forestry Research Institute)

assess wood properties rapidly and efficiently. For example, mid-parent wood density values can be used to assess progeny rather than having to sample wood of the progenies themselves (Talbert et al. 1982a). Such a technique, although only 70% effective, provides a good estimate of the wood properties of the progeny and their potential use with a minimum of effort and expense.

The North Carolina State-Industry Tree Improvement Program is an example of how wood properties were incorporated into a tree improvement program.

When started in 1956, most industrial members of the Cooperative were using loblolly pine (*Pinus taeda*) to make kraft pulp, primarily for linerboard and boxes. The best raw material for this product was high density wood with thick tracheid walls. In more recent times, however, tissues, fine papers, and newsprint have become of major importance. The best wood for these products is low density wood with thin tracheid walls. Although the earlier breeding and seed orchard establishment within the Cooperative was mostly for high density wood, when outstanding low density trees were found, they were not discarded but were put into clone banks (reserves) for potential future use if trees with that kind of wood should ever be needed. When a company product line changed to tissues and fine papers best made from low density trees, the geneticists were able to rapidly use trees from the clone banks to establish new seed orchards to produce lower density wood.

As markets and manufacturing technologies change, the desired type of wood may change. A prime example is the current usage of TMP (thermomechanical pulp) and CTMP (chemi-thermomechanical pulp) pulping methodologies which make the once-scorned low density juvenile wood of value for some products. Once a seed orchard is established and large acreages are planted with its seed, there is no option but to use the type of wood produced. Thus, there is a need to try to visualize the future wood properties that will be preferred by each organization. The difficulty in and fear of making such a decision has resulted in the establishment of many seed orchards with little or no emphasis on wood quality.

Organizations that are making a specific product can determine which wood properties are important and then can incorporate them into the tree improvement program. The result is a uniform, desired wood. Government organizations, however, or those supplying seed to a number of organizations that manufacture greatly differing products, have a more difficult problem. Better wood uniformity is possible by concentrating on average wood and deleting the extreme types. This will improve the wood for solid products, as will longer cells for many fiber products. It is certain that no manufacturer will desire compression or tension wood from crooked trees, or large, steep-angled knots with their accompanying wood defects.

Realistically, company control of wood properties is not as simple as implied above. In addition to using wood from their own plantations, many companies obtain considerable wood, sometimes a majority of its needs, from outside wood purchases. Under these conditions, improvement of wood quality through genetics can be only partially achieved.

Among those who have discussed planning for wood properties in a genetics program, May (1958) suggested avoiding, as an objective, the improvement for a specific paper product or property. (This differs from our previous statement that an organization that makes a certain product should concentrate on wood properties best suited for that product.) May suggested that those wood or fiber properties should be selected for improvement that appear most likely to benefit several paper properties, although he recommended avoiding objectives covering too broad an area of quality improvement. May further stated that a breeding program should assess probable future product changes that would require changing the wood raw material.

It appears to us that *breeding for maximum wood property uniformity* should be added to May's suggestions, since, as previously mentioned, the most important improvement in wood is increased wood uniformity. This objective can be partially achieved by breeding. It can also be obtained by the use of vegetative propagation in which the wood of all trees from a clone will be very uniform. Clones with similar wood can be combined to give uniform, desired products from a forestry operation and still maintain the genetic diversity supplied by using a number of clones.

The shorter the rotation age, such as 6 years in *Eucalyptus*, the easier it becomes to delineate the kind of wood that should be developed. The problem becomes more difficult when very long rotation ages are involved because new wood product manufacturing techniques are constantly being developed. When this occurs, the best wood for today's products and methodologies may not be the best for future needs.

12.2 Selection of Trees for a Genetics Program

If wood properties are to be included in a tree improvement program, there must be a definite plan as how to proceed. In some ways the decision is easy because values for most wood properties are quantitative and do not depend upon the opinion of the selector; but the decision can also be difficult because often the economic value of wood property improvement and product quality is not known. For example, how much improvement in product quality and manufacturing cost will be derived by changing the average specific gravity within a species from 0.45 to 0.47? Or how much will be gained by increasing uniformity so that 90% of the trees fall within the range of 0.46 to 0.54 specific gravity compared to the former 50% that fell within that range for a population without improved wood? Such questions are not only difficult to answer but there will be key differences in the value of changes, depending on the manufacturing methodologies used and the products to be produced.

Gain from a wood genetics program must be estimated by knowing the selection differential and heritability along with the value of unit changes in the wood property involved (Fig. 12.2). Normally, wood is not the primary objective for selection in a tree improvement program. Usually growth rate, tree form, adaptability, and pest resistance come first; then trees that meet these selection criteria are assessed for their wood properties. Such an approach was recommended for the tropical species *Gmelina arborea* by Akachuku (1984), who suggested: "... tree selection should be based on tree form and volume growth rate, followed by a second selection on the basis of desirable wood properties." In their work with *Pinus caribaea* var. *hondurensis,* Burley et al. (1983) found the large range

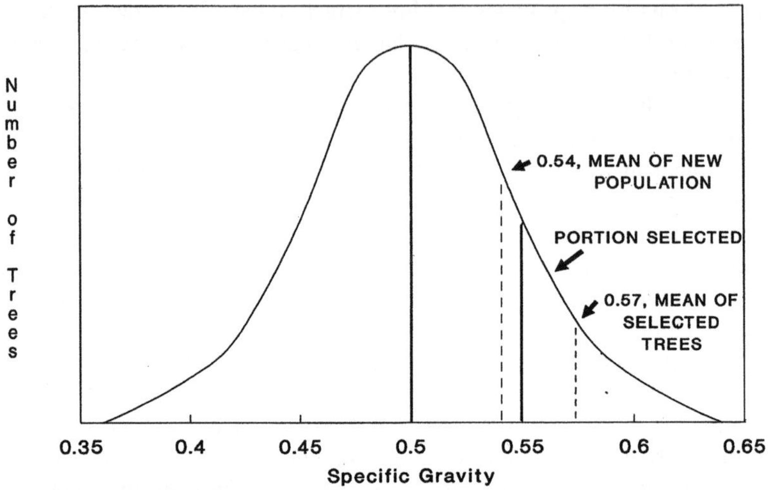

Fig. 12.2. A schematic representation of the gain in average wood specific gravity of a plantation if trees only 0.55 specific gravity and greater were chosen where the average was 0.50 and h^2 was 0.60 (Gain = $h^2 \times$ selection differential). Note the mean of the new population would be 0.54 rather than the present mean of 0.50

of family means in wood density (to the overall mean) of 13 to 25% and the considerable individual tree variation suggested that there are good potential gains from selective breeding. The inclusion of wood properties in a selection program will result in a reduction in the numbers of trees originally chosen that will be acceptable for use. Thus the selection differential is reduced and the gain for primary items such as growth and form might be less. However, the inheritance patterns for most wood properties are so strong that despite a somewhat reduced selection differential, useful gains can still be obtained. This was emphasized for black pine (*Pinus nigra*) by Nanson et al. (1975).

There are, however, restrictions to the flexibility in choosing wood properties. Some products have limits to the range of wood properties that will make it acceptable. For example, for the manufacture of linerboard from pine, van Buijtenen et al. (1974) determined that wood below a given specific gravity was not satisfactory. Thus, in choosing trees for a selection program, including wood, not only are there limits on the genetic control, and on the selection differential, but also on what constitutes usable wood. A good example is in the eucalypts used to make tissues where the specific gravity of some species, as well as individuals in other species, have a wood density too high to make a product with a satisfactory absorbency.

Numerous references have been made in this book to the potential use of tree improvement programs for changing wood properties in conifers in general. Assessments and suggestions are worded in many ways. One study of white spruce (*Picea glauca*) by Corriveau et al. (1990) had a summary ending as follows: "This study confirmed the desirability of using relative density as the basis for making mass selections within fast-growing white spruce populations to genetically improve the quality of wood." Other statements are general, like the summary made by Burdon and Thulin (1965) for radiata pine (*Pinus radiata*): "Available information indicates that many wood properties are highly heritable and that responses can be expected from a mass selection program through seed orchards." They go on to emphasize the necessity of technological improvement in pulp and paper manufacture but point out that these would not be successful without developing improved wood properties. It is statements such as these that give encouragement to improvement of wood by genetic manipulation.

Improvement of some wood properties can result from improvement in tree form and from silvicultural manipulation of growth. Large gains have been obtained by developing genotypes that produce straight trees. Another large gain in wood properties is sometimes possible through development of strains of trees tolerant to pests. For example, huge gains in both volume yield and wood quality have been obtained from breeding pine trees tolerant to fusiform rust (*Cronartium quercum* f. sp. *fusiforme*). Such gains usually can be obtained rapidly and are an easy way to secure better wood.

Most of the previous discussion on improving wood through genetics deals with what might be termed conventional methods of selection. Much more advanced and complicated methods have been proposed, most of which require a selection index. One highly theoretical method, called marker-aided selection, has been proposed by Williams and Neale (1992) for wood density. Marker-aided selection is dependent on genetic mapping and complex genetic assessments. It is clear that as better technologies and understanding are developed, selection methods will become more precise and enable more accurate selection at an earlier age based on parentage.

12.2.1 Considerations for Selection

Several scientists have emphasized the necessity of including wood in tree improvement programs. For example, Zobel (1973) stated: "A well rounded tree improvement program must include wood qualities." In the discussion of what would be needed to do this, Zobel summarized as follows:

1. Develop straighter trees to reduce reaction wood. The inheritance of straightness is so strong that one generation of intensive selection can produce trees straight enough so that this characteristic no longer needs to be emphasized for wood improvement (Fig. 12.3).
2. Although branch and limb characteristics are only moderately strongly inherited, they are so important to the final product that they should be included.
3. Use disease-tolerant trees to improve wood quality. The inheritance of most diseases is relatively strong and disease can have a major effect on wood quality.

Fig. 12.3. A tree improvement program will not be successful unless better wood is obtained. The first step is to obtain good breeding stock, such as the selected *Pinus taeda* shown here. This tree not only has good straightness and small limbs, which results in good wood, but it also has good fiber properties. These characteristics are all reasonably strongly inherited and are independent of each other

For example, logs infested with fusiform rust give lower pulp yields and lower burst and tensile strength. The major increase when the diseased logs are pulped is in resin yields.

4. In most species, first select for the best-formed trees with outstanding growth rate, then make the selection for specific wood properties within that group of trees. Because of the high heritability of wood properties and their relative independence from each other and other tree characteristics, it is possible to tailor-make trees through genetic combinations that encompass a combination of the best adaptability, form, growth rate, and desired wood.

5. Juvenile wood is always present and has properties very different from mature wood, particularly in the conifers. The percentage of juvenile wood can be reduced by longer rotations, although this approach has usually proven to be non-economic. Juvenile wood density can be improved by breeding (Zobel et al. 1978).

6. Generally, anatomical wood characteristics are more important than chemical characteristics in determining product quality.

7. When using exotic species, the wood must be tested and local land races need to be developed. Moving the exotic to grossly differing environments can sometimes drastically change wood (Fig. 12.4).

8. Sometimes unusual wood properties appear to respond well to selection. As just one of many examples, Garrett et al. (1979), working with families of sweetgum (*Liquidambar styraciflua*) and clones of cottonwood (*Populus deltoides*), stated that: "... selection for compartmentalization of discoloration and decay resulting from wounding would be valuable ... and could be applied immediately to existing improvement programs ...".

Compared to most plant species, trees are unique because of the independence of characteristics, the tendency toward strong additive variance, and the lack of genotype × environmental interaction when environmental differences are moderate. (Wood can be dramatically changed by extremes of environment). Inheritance of wood properties is strong enough to make meaningful changes possible in both yield and quality of pulp and paper, solid wood products, and other products, such as charcoal for energy by normal selection and breeding. One exception is percentage of cellulose yield which is under non-additive genetic control; it therefore does not respond to a regular breeding program but it can be exploited using vegetative propagation.

In Queensland, Australia, Harding and Woolaston (1991) determined what the result would be for species where straightness and wood density have an effect on compression wood if wood is not included in a selection program. For hoop pine (*Araucaria cunninghamia*) they found that selection procedures that emphasized growth and straightness but ignored wood quality would increase basic density, decrease spiral grain and lead to a small increase in compression wood.

In developing a tree improvement program involving wood, there is always concern as to which should be the primary wood properties. Wood density (cell wall thickness) is overall the most important of the wood properties and it

Fig. 12.4. It is particularly important to have a breeding program for the development of proper wood when exotics are used. Shown are 2.5-year-old *Gmelina arborea* grown in Colombia. Wood variation from tree to tree is large, so it is necessary to develop a local land race with desired wood properties. (After Zobel et al. 1987)

responds very well genetically despite its complexity. Usually, several different properties are suggested that also have a strong genetic control. However, from a practical breeding standpoint, inclusion of too many different wood properties is impractical because of the dilution effect (reduced gain for the important properties) caused by each additional characteristic. Thus the decision usually must be made as to which one or two wood characteristics will be used in the genetics program. As is usual for all kinds of characteristics in a breeding program, one must include those wood properties with the most economic worth and those that are most easily manipulated genetically. The characteristic usually included is wood density, as has already been mentioned repeatedly. This was mentioned by Wilcox (1977), who concluded for tropical forest trees that: "Combined among- and within-family multiple trait index selection had been used in New Zealand to select generation plus trees with high wood density in the juvenile core, fast growth rate and good stem and branch characteristics."

Wood properties to consider in a breeding program depend upon the species and desired products, and Wilcox (1977) stated that selection for wood properties such as spiral grain, latewood percent, and cell length will nearly always be in conjunction with selection to improve volume growth, stem form, and disease resistance. For radiata pine, increased tracheid length, increased basic density, and minimum longitudinal shrinkage should be the objectives of a breeding program for wood improvement (Dadswell and Nicholls 1959). Even for spiral growth, Harding (1990) recommended that emphasis should be given to breed against this problem in *Pinus tecunumanii.*

Less well documented than for the conifers is the choice of wood properties to include in a hardwood tree improvement program. In *Gmelina arborea*, Akachuku (1984) found sufficient variation in wood density, fiber length, and fiber proportion to include in a selection and breeding program. The wood criteria that should be improved in the eucalypts were indicated by Dean et al. (1991). Their list is similar to the conifers, including basic density, fiber length, and pulp strength properties. They closed with the following statement: "But it is not enough to list traits of interest... . Does the trait have a level which is critical to success? Or is the trait of economic importance? And what is its importance relative to other traits?" They then made the suggestion that all traits must be assessed by their relative economic importance if they are to be judged for use. In work with *Eucalyptus globulus*, Dean et al. (1991) showed the inheritance of pulp yield directly (see Chap. 6.4). Despite fairly good heritabilities, they stated that selection for pulp yields can be used for a rough screening, but it is not good enough to select individual trees for use in seed orchards.

There is always concern about whether superior wood of a family when grown in one environment will still be superior when grown in a different environment. As mentioned in Chapter 2.1.1.4, and Chapter 9, the genotype × environment interaction is usually, but not always, small. This was summarized by Lima et al. (1990) for wood density of *Pinus tecunumanii* in which selection for wood specific gravity in a breeding program should be possible on the basis of average family performance over sites in widely divergent areas.

As discussed in Chapter 2, there are several kinds of selection that can be done. Based upon the large tree-to-tree variation and rather high heritabilities, individual tree selection usually is more efficient than family selection for the improvement of wood properties (Allen 1985). This observation has been made by many investigators working with the genetics of wood.

12.2.2 Opportunity for Early Selection

To make quick genetic gains for improved wood properties, it is essential to use early selection. More than for most characteristics, it is possible to predict from values obtained with young trees what the wood properties of the mature tree will be. Such juvenile to mature correlations are most valuable (early selection is covered in Chap. 3). How early to select varies with the wood characteristic. For example, we found in our work with loblolly pine that a quite accurate estimate can be made of the wood density of a 30-year-old tree by assessing its wood at 4+ years. In contrast, tracheid length of the same trees cannot be reasonably estimated until after 10 years of radial growth (Zobel et al. 1972). Each wood property is different and one of the most critical aspects of a wood genetics program is to determine methods that will enable early selection of desired wood properties.

Early selection success was reported by Vargas-Hernandez and Adams (1992) with Douglas-fir (*Psuedotsuga menziesii*) where they found selection at 7 years had a relative efficiency similar to selection at 15 years. They pointed out that the wood of the trees was still juvenile and mentioned that it was uncertain how early selection would predict mature wood from juvenile wood. Based upon the results of McKimmy and Campbell (1982) and Gonzalez and Richards (1988), selection at age 15 would be effective in improving mature wood density of Douglas-fir. As a general example, Gonzalez and Kellogg (1978) recommended that genetic selection for high wood specific gravity be a part of any tree improvement program for coniferous species.

The ability to utilize wood density in early selection is difficult and complex. This was emphasized by Vargas-Hernandez and Adams (1992), who pointed out the need for information on both the genetic and phenotypic covariances between the wood density of the sample at the age of selection and the overall density at harvest age. Such data are rarely available. Villeneuve et al. (1987), found, for example in Canada, that the prediction of mature wood density was possible at ages 6 to 7 years in jack pinc (*Pinus banksiana*) and 12 to 15 years in black spruce (*Picea mariana*). As the number of traits in the selection index increases, the weights assigned are sensitive to the correlations between the traits (Burdon 1989). A cost-gain assessment must also be made. The use of an index selection on Chinese fir (*Cunninghamia lanceolata*) was described by Zhi et al. (1993) as of possible value in breeding programs.

12.3 Choice of Wood Properties – What Should Be Included?

The first and most important decision regarding inclusion of wood properties in tree improvement is to determine which characteristics to use. Even though a number of wood properties have strong enough genetic control so that good percentage gains can be made, the choice depends upon how important the changes will be and how easily they can be made. Also, one must determine if they are correlated with other important characteristics in the breeding program.

If desired wood is not available within a species, it is sometimes possible to develop it through the production of hybrids (Chap. 11.4). Usually the hybrid has wood intermediate to that of the parents. Use of vegetative propagation with hybrids is most successful.

There is no question that for essentially all wood usages, wood density is an important characteristic. It has considerable variability and is from moderately to strongly inherited and should be included in most tree improvement programs. The effects of wood density on the final product, both solid and fiber, are well known (see Chap. 4). Yet when there is a negative correlation, such as sometimes occurs between growth rate and wood density, selection must be carefully done. An example was given by Olson et al. (1985), who found some negative correlation between growth rate and specific gravity and cellulose yield in *Populus.* They concluded that a selection index would be difficult to develop and the authors recommended that only volume should be used in a selection program because it is more important than wood density. This need to determine the economic value resulting from improvement of wood properties has been stressed by many authors, such as Burdon and Thulin (1965) for *Pinus radiata.*

The value of indirect selection has previously been mentioned. It was summarized for 8-year-old *Eucalyptus globulus* in Portugal as follows: "Selection indices constructed to maximize genetic gain in dry weight gave results which clearly favored indirect selection on height. A genetic gain of around 47% in dry weight may be expected from second generation selection of one tree in 200... ." (Borralho et al. 1992).

Although the emphasis is on wood density, as underlined in many places in this book, numerous other wood properties can be used in a tree improvement program, although they are usually not included because of the overall importance of wood density; but some of the important characteristics related to pulping such as kappa number, pulp yield, tear, and burst, have shown promise. This was summarized by Wright (1991) as: "... heritabilities for these traits in *Eucalyptus grandis* have indicated a strong degree of genetic control suggesting that selection and breeding for them will be successful."

There are instances when selection for wood improvement is not successful. One example is red pine (*Pinus resinosa*) in which tree-to-tree differences are small in relation to other hard pines (Fowler and Morris 1977). Even provenance differences are small.

12.4 Summary

Rapidly changing technologies and differing wood requirements make it difficult to suggest which wood properties should be included in breeding programs. Obviously, identifying wood properties for short rotations of 6 to 10 years is easier than for long rotations.

Two wood properties almost always emphasized are uniformity and wood density because of their importance to the final product. Spiral grain is stressed for solid wood products. To be effective, the number of wood characteristics to include in a breeding program should be kept small. When there are negative correlations between wood and growth characteristics, the choices must be carefully made so as to end up with the greatest amount of desired wood.

There is some dispute about the place of wood in a tree improvement program. We feel that it should be a secondary characteristic; i.e., emphasis in the initial selection should be on adaptability, volume, tree form, and disease tolerance. Then, within this group of trees, those with the desired wood should be used. Good wood improvement grains are possible by using this approach because of the general high heritabilities of wood and the large tree-to-tree variation usually present.

Chapter 13

Improvement in Wood by Using Genetics

13.1 Current and Future Usage of Genetics to Change Wood

Initially, only a few tree improvement programs included wood as a characteristic to be genetically manipulated. Emphases in the early programs were on adaptability, pest tolerance, growth rate, and tree form. Certainly those characteristics are of importance, but since wood is the final objective of many forestry activities, it should also be included. Initially, the huge amount of variation in wood was not well-recognized and most foresters attributed it to environmental differences, since the training of foresters emphasized the controlling effect of environment. However, as Smith (1970) pointed out, the range of variation in wood is so large, even after intensive selection for volume, form, and branching, that effective selection for wood improvement is possible (Fig. 13.1).

Genetic improvement is frequently reported as percentage gains. The use of a percentage must be done carefully and as Zobel (1964) emphasized: "Improvement over what?" Much confusion results from the use of different methods to calculate percentage gains. Therefore, any statement of percent genetic improvement must clearly indicate the base from which the improvement was calculated. Without that, percentage gain values lose most of their meaning.

Since most of the important characteristics of forest trees are inherited independently, it is possible to tailor-make trees, not only for form and growth but also for desired wood properties. This was clearly expressed by Smith (1970) as: "Maximum economic gains from first-phase selection should come from improvement in volume production, site adaptability, stem form and branching characteristics. Screening ... for conformity to minimum acceptable wood property standards is (also) considered essential. In second-stage selection, more emphasis could be placed on wood quality. ..."

It was not initially known how much, if any, of the variation in wood might be genetically controlled. Uncertainties as to what is "desired wood" added to the lack of interest for inclusion of wood in genetic programs. Despite this situation, some limited work on the genetics of wood was initiated early. The lack of inclusion of wood still continues and was expressed by Harding (1990) for *Araucaria cunninghamia* (hoop pine), *Pinus caribaea* var. *hondurensis*, *P. patula*, *P. elliottii*, and several hybrid pines in Australia when he stated: "Although basic density, spiral grain and percent latewood are capable of genetic manipulation, active breeding to modify species wood properties is not currently adopted."

Fig. 13.1. The ultimate objective of tree improvement is shown here where a land race of pine has been developed genetically to be well adapted, have good tree form, and have desired and uniform wood. This 17-year-old stand of *Pinus taeda* makes for efficient forestry and mill operations. (Zobel and Talbert 1984)

In contrast, information about the genetics of wood is being quite widely used in some tree improvement programs of the eucalypt and pine genera, especially where they are being used as exotics. Although some have been studied, many species have had only a limited emphasis on wood genetics; some essentially none. Many of the hardwoods fit this latter category, especially the tropical hardwoods. It would be correct to say that inclusion of wood genetics in tree improvement has had some impact in the past and is now having a major impact on some tree breeding programs. But relative to other tree characteristics, knowledge of wood genetics is sparse and utilization of the knowledge is not common. A major opportunity for wood improvement lies in the future.

The idea has been expressed that the technology of wood processing has become so advanced (and will continue to advance) that the genetic improvement of wood will be unimportant. This is not a new idea. For example, in 1951, the senior author was told by a pulpmill manager "Don't waste time trying to improve wood. We are wood chemists. We can make paper out of any kind of wood you supply us." While he was technically correct, he overlooked the advantage of wood uniformity for the desired product. The cost advantages and marketing advantage were being ignored. Interestingly, in the intervening 40 years we have had frequent requests from the same organization for information concerning wood quality and what could be done to improve the usefulness of the wood in manufacturing and sales. The importance of wood properties, as they can be changed by genetics, was clearly expressed by Williams (1994) as: "There is a very good correlation between raw material quality and fiber characteristics and final product quality."

Obviously, the authors of this book believe that wood properties are important and will become even more important as technologies and raw materials change. As an example, as asexual propagation methods become more generally used in the future, wood properties should become of even greater importance than they are currently. The gains in uniformity achieved through asexual propagation will extend to wood, and this will directly impact manufacturing and product quality. Asexual reproduction does not create new types of wood; it only increases the number of trees having desired wood properties.

Several reviewers have referred to the role that biotechnology will play in the future to identify wood property traits. There is no doubt that such technologies will be useful tools to the tree breeder. They do not create new or desired wood properties but they will assist in identifying the presence of desired wood properties. Only breeding programs can produce wood with new properties and develop wood uniformity. Consequently, the authors of this book are convinced that breeding to improve wood properties will continue or even increase in importance in the future. There is no question that in the manufacture of wood products, as is the case for all products, the better the raw material the easier and more efficient the production of those products becomes.

The bulk of wood research has dealt with wood density, but there are numerous other wood properties, especially in those woods used for specialty products, where genetics can make a major contribution. As stated by Zobel (1973): "All important wood morphological characteristics are under enough genetic control to obtain useful gains." This, of course, is true when proper methods of genetic improvement are used. The happy situation is that for most wood properties, gains are based upon strong additive genetic variation and thus improvement can be relatively large and rapid, using either seed or vegetative propagation (Fig. 13.2).

The largest boost to the use of genetic improvement of wood is when vegetative propagation is employed as the method of regeneration. The use of vegetative propagation in forestry, especially in the pines, has only just started, so the future holds a vast promise for a much more intensive utilization of the genetics of wood. Wood quality improvement plays a major role in increasing forest profitability as indicated by the 135% gain in profitability obtained on the eucalypts at Aracruz, Brazil, as outlined by Harding and Chu (1989).

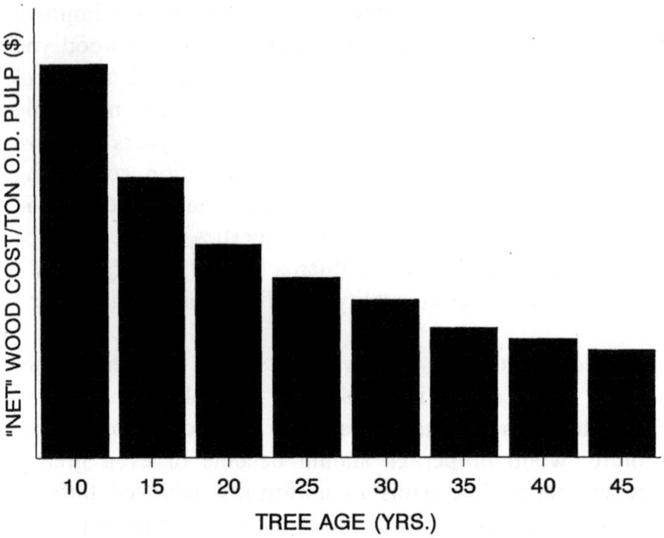

Fig. 13.2. Wood properties, when strongly inherited, can have a major effect on the economics of forestry. For the slash pine shown, the change with age is due to the amount of juvenile wood which has low density and high moisture content. Such wood characteristics can be changed by breeding. (After Howell et al. 1984)

One important wood property frequently ignored is uniformity. This could easily be cited as the most important of all wood properties and the most desired by the product manufacturer and the most closely tied to profitability. Manufacturing adaptations often can be rather easily made if the wood to be used has reasonable uniformity. However, when the usual and very large variations that naturally occur in wood are encountered by the manufacturer, there is a definite limit to the role that technological manipulation of processing can play. The real improvement and gain must be in improving the wood quality itself, the most important phase of which is through genetic manipulation. Of course, greater wood uniformity results from using plantations of single species, but genetics can produce even greater uniformity within the plantation.

In his summary of wood property research is Queensland, Australia, Harding (1990) emphasized the need to determine the potential effects of changes in genetic pedigree or growing conditions to their net economic worth. Research is needed on the genetic control of wood properties to allow the best strategies for wood improvement. Harding suggested that: "The current universal strategy employed to produce multi-purpose wood to satisfy a diversity of end uses could be modified to meet the specific needs of one or more end-users." This would necessitate somewhat more complex breeding programs but the economic gains would be considerable. Harding does emphasize that, along with the genetic manipulation, silvicultural practices and their effects on wood properties must be better known. The authors of this book endorse Harding's suggestions.

13.1.1 When to Employ Genetics

The question is often asked "When should genetics be used or when should other methods be employed to improve wood?" As usual, the answer can be simply stated as "When it is most efficient" or "When it will give the greatest gains as soon as possible and as inexpensively as possible." To this must be added "When proper forest management is used"; but a determination of efficiency is the real problem. For example, the initial cost of genetic improvement can be large but, unlike many other methods, once gain has been achieved, it usually can be carried on into future generations for little extra cost. How to make comparative costs between achieving a one-time gain by altering silviculture or manufacturing methods and a perpetual gain from genetics has not been well worked out.

In actual practice, matching the tree type to the environment will usually result in the quickest results. An example would be growing *Pinus oocarpa* (oocarpa pine) in coastal areas where the winds deform and uproot the trees. The environmental cure is to grow *P. oocarpa* where wind is minimal, and plant other species less affected by wind in the coastal areas. If the trees were crooked because of genetics, the movement into a new environment would not help, and the only method of improvement of tree form would be by selection and breeding, which would take considerable time to obtain straighter forests. Less time is required to achieve improvement in wood by genetics when asexual reproduction is used instead of sexual reproduction, but it will still take considerable time to become completely operational using the improved rooted cuttings.

Another difficult situation is the genetic correlations and/or interactions that are always present. For example, one needs to know what happens to other wood properties or tree characteristics when a change is made in a specific wood property (see Chap. 13.3). How will that change ultimately appear in the paper or boards that will be produced? In general, we in forestry are lucky because so many important tree characteristics are only weakly, or not at all, genetically correlated with each other. As discussed earlier, a good example is straightness of tree bole and wood density. Although crooked trees have more reaction wood than do straight ones, the genetics of bole straightness and wood density are essentially not correlated. One can have straight and crooked trees with either high or low wood density.

One massive study entitled *Quality control in the Forest and in the Mill* (van Buijtenen et al. 1975) was undertaken to answer the question as to the best method to alter wood in southern pines. The study investigation took several years to complete, with cooperation from the woodlands of 18 forestry organizations and 20 mill operations, in addition to help from governmental agencies. The objective of the first phase of the study was to assess the potential to change southern pine kraft linerboard by: (1) genetics of wood properties, (2) Silvicultural influence on wood properties, (3) relationship of wood properties to pulp and papermaking properties, and (4) operations research. Ten wood properties were considered, but the final assessment was based mostly on wood density.

The objective of the genetics working group was to find out how much and how easily wood properties could be changed and how expensive such a change would be. Only a few results can be mentioned here: (1) "... wood specific gravity was the most important wood property influencing the properties of unbleached southern pine kraft pulp". (2) "At short rotation ages breeding for high wood specific gravity was always desirable. ... The optimum economic strategy ... was to breed for high wood specific gravity and reduce rotation age as much as possible. ..." (3) "Modifying paper properties by refining or the use of additives is generally more expensive than achieving the same result by genetics or silviculture. ..." (4) The estimated rate of return from a standard tree improvement program was 12%, before taxes. Wood specific gravity played a most important role: "... perhaps even more so than realized. ..." (5) High wood density was the best option except in older stands where lower density was required to meet the burst specifications for linerboard. For multiwall sack paper, high wood specific gravity was even more important to obtain stronger pulp. This study was a major factor in the recognition and acceptance of the value of genetics over silvicultural or mill operations to alter wood properties.

13.2 Examples of Changes in Wood by the Use of Genetics

There are many examples of gains and improvements through use of wood in breeding programs mentioned earlier in this book, scattered throughout the literature, or now being used, even though they may be unpublished. It has been impossible to cite more than a limited number of these. For the hardwoods in general, Smith (1967) believed that moderate improvement in fiber length and diameter could be expected from crossing and superior tree selection. He goes on to say that if tracheid length and microfibril angle were included in breeding programs for conifers, improvement of these traits was feasible. For solid wood products, spiral grain is considered to be of key importance.

We have chosen a few specific studies with limited details to represent varied species and conditions. These have been summarized in Table 13.1, to give the reader a sample of what has been done and the benefits from applying genetic improvement to wood.

We made some calculations (unpublished) in 1975 that an improvement in yield of 1% from the pine wood used in the pulp mill (either from better wood properties, from genetics, or from improved technology) would result in increased income of over \$1000000/year in a kraft pulp mill with a capacity of 1200 tons/day. The improvement from genetics does not have to be great to have a significant effect on profitability. In calculations of possible gains from breeding for wood density in slash pine, Hodge et al. (1992) summarized as follows: "Assuming an increase in whole-tree density from 0.50 to 0.52 g/cm^3, this is a four percent increase in dry wood per unit volume. If pulp or biomass is the production objective, this gain from increased density would be roughly equiva-

Table 13.1. A few examples of results from inclusion of wood in conifer tree breeding programs

Species	Reference	Action and/or results
Araucaria cunninghamia	Eisemann et al. 1990	An increase of 33% in wood density was obtained with corresponding small reductions in compression wood and spiral grain
Cryptomeria japonica	Ohta and Fujisawa 1992	The broad sense heritability of the modulus of elasticity was 0.60 to 0.86, indicating that this characteristic is highly suited to clonal forestry
Cupressus lusitanica *Pinus patula* *P. radiata*	Paterson 1967b	With all factors combined for these three species, an increase of 5 to 10% in cellulose production/acre and an increase of 15% in roundwood value can be obtained by selection and breeding. Compression wood will be reduced by as much as 4 to 13%
Pines in general	van Buijtenen 1967b	As early as 1967, van Buijtenen predicted a genetic improvement of 10% above the average from the inclusion of wood. This rate can be maintained for several cycles of selection
Southern pines	van Buijtenen et al. 1974, 1975	The benefits from improving wood specific gravity were high in linerboard and essential in multi-sack paper. Five percent of the 12% gain from tree improvement could be traced to wood quality improvement
Pinus echinata	Wiselogel and Tauer 1982	Heritabilities were high enough for both wood density and tracheid length for substantial genetic gains through selection
P. elliottii	Einspahr et al. 1964	It appears possible to predict pulp yield by determining lignin percentage. Gains were greatest for cell strength and zero span tensile strength. Reduction in undesirable components (lignin, extractives) was considerable
	Hodge and Purnell 1993	When mature and juvenile wood densities are increased by selection, whole tree density can be increased by 200 kg/m^3
P. kesiya	Guldager 1972	Tests indicate that wood density considerations from provenance tests and seed orchard selections should have a high priority
P. radiata	Birt and Harris 1970	Wood density could profitably be used as a criterion for tree selection

Table 13.1. (*contd.*)

Species	Reference	Action and/or results
	Burdon and Young 1991a	There are good prospects for altered resin content, heart-wood percent, compression wood, and wood density from intensive selection. There are no strong indications that intensive selection for growth and form has had an adverse effect on wood properties
P. taeda	Goggans 1962	Latewood tracheid length, percentage of summerwood, and specific gravity are the three characteristics with which progress through selection would be the easiest
	van Buijtenen 1962	Assuming a heritability of $h^2 = 0.50$, progress by selection was 3.8% with a one-standard deviation selection differential
	van Buijtenen 1963a	The strong inheritance of wood density, along with moderate height and diameter improvement, results in an increase of dry wood production of 2 tons/acre when selecting for wood specific gravity alone or 7 tons/acre when including growth rate
	Zobel et al. 1969	An increase in specific gravity of 0.02 will mean a weight difference of 0.1 ton of dry wood per acre per year
	Zobel 1973	Pulp yields/unit volume of wood are increased; 1 to 7% greater yields can be obtained per unit weight of dry wood by increasing wood density. Pulp and paper qualities are most affected by changes in cell wall thickness
	Blair et al. 1974	Bole straighness greatly affected pulp yields and tear strength of paper by decreasing compression wood. Smaller limbs resulted in better tear because of less limbwood and associated compression wood. The study enabled placing economic values for yield and quality of pulp on genetic improvement of tree form
	Zobel et al. 1978	Producing trees from parents with higher juvenile wood density gave a gain in 10-year-old progeny of 32 kg $/m^3$ (2 lbs/ft^3). Families with a high density wood produced 5% greater yields/unit volume and an additional gain of 1% per unit dry weight of wood. The tear factor was better from trees which had high density juvenile wood

Table 13.1. (*contd.*)

Species	Reference	Action and/or results
	Jett and Talbert 1982	Gain in wood density from selection for advanced generations was 2.6%. The method used shows that parental values are enough and extensive family information is not needed
	McKinley et al. 1982	At a low level of selection intensity, the genetic gain in wood density was nearly 3% using family selection
	Talbert et al. 1982b	Working with 20-year-old trees from genetic tests, the heritabilities and variances indicate that substantial genetic gain can be made by selection for wood specific gravity in loblolly pine
Pseudotsuga menziesii	Anonymous 1989–1990	An improvement of 5.7% was obtained in wood density at 15 years of age when the best 20% of the trees were selected

lent to a four percent increase in volume with no change in density." In *Pinus taeda*, van Buijtenen (1963b) stated that an increase in dry wood production of 2 tons/acre is feasible when wood density is included in the selection criteria.

13.2.1 The Hard Pines

Other than for the eucalypts, the most intensive use of information about the genetics of wood in operational forestry has been with the group of trees loosely referred to as the hard pines. Inclusion of some wood properties in tree improvement programs has now become standard. Part of the recognition of the increased importance has resulted from better knowledge of the effect of wood quality on the final product, whether it be for pulp or for solid wood products. Although increased knowledge of genetics of wood has contributed to increased inclusion of wood in genetics programs, the economic pressure to shorten rotation ages has also played a major role.

The broad general usage of genetic manipulation of wood is exemplified by the early North Carolina State-Industry Pine Cooperative. Wood properties were initially considered important in the operational programs of the member companies of the Cooperative. Broad separations in specific gravity were made and trees were selected and established in the seed orchards by category of wood desired by each member of the cooperative. In all instances, greater uniformity, in addition to wood density, was a priority objective. Broad wood density classifications used were as follows:

1. High density wood was the objective for the seed orchards of those companies making board and box products, which required a high tearing strength.
2. Low density wood was used by those companies making newsprint, writing paper, tissue, and products that required good printability.
3. Average density was desired for those who could make no decision about the product to be produced. Here the emphasis was to avoid the extreme density trees in order to secure more uniform wood with a density similar to the average in the established forest then being harvested.

Each member of the cooperative was required to have a signed statement (by a vice president or above) as to the company's desire for wood density considered to be best for its product. This requirement stimulated a rash of studies, some on a mill scale, that added greatly to the information about wood desirability and the benefits if wood properties could be altered. These studies clearly showed the great superiority of wood density over other wood properties in affecting the final product. Unfortunately, many of the studies were never published, being considered proprietary information by the companies. The result was, however, that all members of the Cooperative included wood properties in their first generation tree improvement programs, with wood density being primary, but with other properties like cell length sometimes being included.

The effect of wood properties on the final product (i.e., what wood and tree properties to include in a breeding program) has elicited controversy. For example, some companies and researchers questioned the value of breeding for flatter-angled and smaller limbs. Although branch characteristics do not have the same importance as tree straightness, limb size improvement was shown to be especially related to tearing strength in paper made from loblolly pine (*Pinus taeda*) (Blair et al. 1974). For groundwood from loblolly pine, Schafer and Pew (1958) reported that knotty wood required more power than clear wood to produce an equivalent product. They also showed that straight trees produced better paper color and strength than did the crooked trees, both of which illustrated the improvement in wood by breeding for better tree form. Similar results were found for other products such as sawn boards (Bridgwater 1984) and plywood (Zobel et al. 1977). Pulping tests showed that high specific gravity juvenile wood had characteristics similar to mill-run chips from normal wood (Zobel 1973) and the results of selection and breeding for higher density juvenile wood were encouraging.

In their study of slash pine, Einspahr et al. (1964) pointed out a number of improvements in paper properties that resulted from genetic manipulation. One that they emphasized as having immediate practical importance was: "... the highly significant correlation between lignin and pulp yield suggests that this measurement might provide a suitable way of predicting the pulp yield of immature trees." The authors included a short table showing expected gains; this is reproduced as Table 13.2.

Table 13.2. Calculation of expected gains in slash pine through genetic improvement of wood quality (Einspahr et al. (1964), Table 7)

Wood or fiber characterisitics	Parental increase by selection[a] Average values	Actual	Percent	Tentative estimates of expected percent gains[b] Broad sense heritability	Veg. propagated individuals	Seed progeny
Specific gravity,	0.56	0.07	12.5	0.49	6.1	3.7
Fiber length	2.46	0.58	23.6	0.75	17.7	10.6
Weighted av., mm						
Fiber strength	54.10	11.50	21.2	0.84	17.9	10.7
Zero-span tensile, lb/in						
Pulp yield, %	39.40	3.10	7.9	0.53	4.3	2.5
Lignin, %	28.10	−1.60	−5.7	0.72	−4.1	−2.5
Extractives, %	6.20	−4.30	−69.3	0.25	−17.0	−10.1
Summerwood, %	49.10	11.00	22.4	0.53	11.9	7.1

[a] Expected increase if only those individuals two standard deviations above the mean are selected.
[b] Tentative gains expected using gross heritability figures for vegetatively propagated individuals and assuming narrow sense heritability equal to approximately 60% of the gross heritability.

The largest study on the wood of a hard pine (*P. taeda*) was the huge cooperative effort described earlier in reports by van Buijtenen et al. (1974, 1975). In their summary, it was stated that first-generation seed orchards are producing volume gains of 10 to 20%. Additionally, specific gravity improvement will add another 5%, along with undefined gains in wood quality from straighter stems and smaller limbs. It was found that the best way to improve yields and quality of linerboard was through the application of genetics in tree improvement programs.

Considerable work has been done on the wood of *Pinus radiata*, with positive improvements being evident. In addition to the effects of genetic selection on wood density, other important effects have been recorded. For example, Paterson (1967b) working with three species, including radiata pine, found that selection for tree form would reduce the amount of compression wood by about 4 to 13% but that it would have little effect on stem taper, which has $h^2 = 0.07$ (see Table 13.1). Some of the many wood studies on radiata pine have been put into operational programs with considerable success, but there have been too few of these applications. One reason is that the wood of *P. radiata* is generally considered to be acceptable, and there have been no strong pressures for change, except in a few instances, where greater wood density or less spiral grain have been desired.

Because of the prevalence of juvenile wood in the fast-growing young plantations now being harvested, there has been a concentration on improving juvenile wood properties. Success has been considerable such as that previously quoted for *Pinus taeda* by Zobel et al. (1978). In radiata pine, Nicholls et al. (1980) calculated an expected gain of 6.4% in juvenile wood density per generation using a practical selection intensity for high wood density of one in five trees.

There is a rapidly increasing amount of data being obtained on the gains in wood properties possible with the tropical pines. In nearly every instance, the gains were large enough to be of good economic value if incorporated into tree improvement operations.

It is clear from the few studies listed, as well as from the many other studies available, that breeding to include wood has had a considerable effect on product quality. Yet there are always exceptions. For example, Snyder (1976), reporting on the importance of wood density in a breeding program for longleaf pine (*P. palustris*) stated: "... high economic values would have to be assigned to specific gravity before it would add enough dry weight equivalent to qualify as a trait for selection. The view ... that specific gravity be dropped as a conscious trait in selection seems correct."

13.2.2 Other Conifers

Improvement for *Araucaria cunninghamia* (hoop pine) from Australia was obtained with a substantial response for basic density by Eisemann et al. (1990). They found that a one-in-ten selection resulted in an increase in wood density of

8.4%. When tree growth and form were included in the selection index, there was a 33% gain in wood density, a 2% reduction in compression wood, and a 17% reduction in spiral grain. The authors summarized the use of the selection index as follows: "A selection index combining growth, straightness and wood properties was efficient in selections for improved growth ... the assessed population of hoop pine would respond favorably to selection to improve growth and stem straightness without adverse effects on wood properties."

Considerable information about wood is known for Douglas-fir (*Pseudotsuga menziesii*). In a recent publication, Vargas-Hernandez and Adams (1991) recognized the need for inclusion of wood in a genetics program. The following recommendation was made: "... wood density should not be ignored when selections are made. ... Early testing for wood density should be a strong consideration as a component of advanced-generation breeding programs in Douglas-fir." Also for Douglas-fir, McKimmy (1987) made it clear that the influence of heredity on wood density is sufficient to warrant serious efforts to incorporate wood into a tree breeding program.

Tracheid length is sometimes considered to be critically in need of improvement. For Norway spruce (*Picea abies*), Ujvári and Szönyi (1973) reported that 60-year-tracheid length can be grown in 40 years if the trees with the longest tracheids are selected as parents. This can be reduced to 20 years if an intensive program is used, especially if vegetative propagation is employed.

13.2.3 Temperate Hardwoods

Research in the genetic manipulation of wood properties for the temperate hardwoods has shown that definite improvements are possible. The information has been used only sparingly in operations. Most has been accomplished with the diffuse-porous species, especially in the genus *Populus*. In a discussion of selection traits to be used with hybrid poplars, Mirza and Strobl (1991) emphasized the importance of specific gravity and reported very high heritabilities within and between families. Less has been done with the ring-porous hardwoods, but one example is green ash (*Fraxinus pennsylvanica*), as reported by Lowe and Greene (1990). They concluded that: "... a tree improvement program could be expected to produce moderate gains in specific gravity. Among-family variation in fiber length was too small to warrant attempts at genetic improvement".

Despite relatively low heritabilities for wood density of diploid *Populus tremuloides*, van Buijtenen et al. (1962) found enough genetic control to allow gains from selection for fiber length. The tearing strength and breaking length showed differences between clones sufficiently large to enhance fiber strength and burst. They stated: "It is likely that considerable improvement of the pulp and papermaking properties might be achieved by breeding. ..." In contrast, although there were large progeny differences in the wood of *Betula pendula* (European white birch), Nepveu and Velling (1983) reported no differences in pulp yields.

With the highly heritable fiber length, and even though fiber length variation was small, Boyce and Kaeiser (1961) found that the maximum average length of fibers that can be expected after several generations of selection in poplar will be less than 2.0 mm for trees 20 years of age (the average of indigenous stands at that age varied from 0.85 to 1.2 mm), and results in product quality from such small gains would not be very marked. However, recent work with the eucalypts indicate that small changes in fiber length will be valuable (Campinhos and da Silva 1990).

A 9% gain for wood density from improved *Populus trichocarpa* was reported by Reck (1974). He recommended a relatively low selection intensity for wood compared to growth but believed good gains could be obtained from a step-by-step selection program with volume, wood density, and fiber length.

Working with sycamore (*Platanus occidentalis*), Huber and Bongarten (1981) reported that substantial gains in specific gravity could be obtained by selection for ray cells and fiber content and reduction in vessel content in the wood, although the changes in tissue proportions were probably too small to affect pulping seriously.

13.2.4 Tropical Hardwoods

With the exception of the eucalypts, little is known about the inheritance of wood properties in the tropical hardwoods and what is known is rarely used in applied tree breeding programs. This deficiency must be overcome if the tropical hardwoods are to assume the importance they should have. It is true that there is a small start in using improved wood from *Gmelina*, *Tectona*, *Bombacopsis*, *Acacia*, *Swietenia*, and some others, but few have had their operational success reported (Fig. 13.3).

13.2.4.1 Eucalypts

Eucalypts are discussed separately because of the relatively heavy genetic manipulation with their wood. A whole series of studies have been made on the eucalyptus and a number of these have been incorporated into operational programs. These were summarized by Dean et al. (1990) as: "Research into the inheritance of wood and pulping properties has demonstrated that these are generally highly heritable, and that variations have a high impact on the overall cost of production ... breeding for improved properties is very worthwhile." A similar summary was made by Campinhos and Claudio-da-Silva (1990), who stated that selection provides substantial improvements in pulp yield. Additional gains can be obtained through further screening of the selected trees.

The Aracruz story, described in earlier chapters, was chosen to illustrate the work on the wood of *Eucalpytus* because of its thorough studies, strik-

Fig. 13.3. In a few instances, intensive breeding has been done with some tropical hardwoods. One example is *Bombacopsis quinata* in Colombia by the Pizano Company. This species has excellent wood quality. Breeding is underway to adjust wood density and, hopefully, towards achieving the desired color of the wood

ing results, huge size, degree of inclusion of the genetics of wood, and ultimately the effect of improved wood on the profitability of the company. Initially, Aracruz did not intensively include wood in their genetic improvement program. However, in a short time, especially after rooted cuttings began to be used operationally, genetic consideration and studies of wood became a major part of silviculture for the company. Significantly, wood properties were included among the important characteristics in the breeding program. Aracruz's operation became very effective, not only because of the efforts of the tree improvers but because of full interest and participation by the mill personnel. Many of the studies became joint ventures between forest and mill research.

Because the administration was convinced of the value of including wood and, since vegetative propagation was being used, genetic findings were rapidly incorporated into the operational program. Wood requirements were closely tailored to mill tests to produce a uniform, high quality pulp; one of the most desired in the paper industry. The application of genetics to wood is used on a large scale with currently 30 to 40 million rooted cuttings being planted annually, each with a rigorous control of the wood that will be produced. The main wood properties emphasized were uniformity, desired wood density, and increased cellulose yields. Some of the earlier results were summarized by Brandáo in 1984, as follows in Table 13.3. The following are a few comments.

1. The improvement and greater uniformity in wood density would be expected with intensive selection and choice of a narrow range best for the desired product.
2. The improved pulp content partially results from higher density but also from use of clones with high cellulose yield. This latter phase of the gain is possible using vegetative propagation and is a novel use of genetics in wood improvement.

Table 13.3. Results from the Aracruz Forestry Program on forest productivity and pulpwood quality

	Initial forest 7 years of age	Today's forest 7 years of age	Change Quality	percent
Average yield (m^3/ha/yr)	33	70	+37	+112
Range in basic density	(300–900)	(500–600)[b]		
Av. basic density (kg/m^3)	460	575	+115	+25
Pulp yield (%)	48	51[b]	+3	+6
Pulp content (kg pulp/m^3)	238	293[b]	+55	+23
Specific consumption (m^3/ton of wood)	4.20	3.41[b]	−0.79	−19
Forest productivity (tons pulp/ha/yr)	7.85	18.45[b]	+10.60	+135

[a] After Brandao (1984). These show the results from intensive forest management using rooted cuttings and from use of genetically improved wood.
[b] Wood without bark.

3. Much of the improvement in pulp yields/hectare of the forest was from greater growth but a significant amount can be credited to higher density and better cellulose yields.

The total improvement of forest productivity of over 135% is an impressive gain indeed, a portion of which can be credited to the use of knowledge about the genetics of wood.

For the eucalypts, Harding and Chu (1989) developed indexes to rank plus trees according to the predicted wood properties of their progenies at harvest age. This enabled screening to remove quickly the trees with obviously undesired wood from the breeding programs.

Other recent studies on the eucalypts have shown the strong influence of genetics. For example, Borralho et al. (1992) showed that pulp yields for 8-year-old *E. globulus* in Portugal had a moderate genetic component with $h^2 = 0.21$. Clarke (1990), reporting on *E. grandis* in South Africa, found high heritabilities for fiber yield, burst strength, and tear strength but low heritabilities for pulp yield, kappa number, and paper brightness. He concluded: "Substantial improvement in fibre yield per tree can be expected through conventional breeding with gains of the order of 27% over the trial mean (136 kg) predicted per generation." Put another way, an increase in just 1% in pulp yield would produce an extra 3 tons of product, (an increase of 11% in profit per hectare) at no extra cost (Fig. 13.4).

13.3 Improving Wood When There Is a Negative Correlation with Growth Rate

For some species, or for some provenances or in some environments, there is a negative relationship of growth rate to wood properties and especially to wood density (see Chap. 10.2). In species with a negative genetic correlation between growth and wood properties, intensive selection for growth results in lower wood density, shorter cells, or other undesired characteristics. Thus, if gains in both wood and volume are to be obtained at the same time, special breeding programs must be used. Only one of the many designs available will be mentioned here because each varies with the species and the unique conditions extant.

Independent culling levels are frequently employed to achieve gains in both volume and wood density when the two are negatively correlated. Results have shown that a reduction in density was avoided using this breeding method (Jefferson and Yanchuk 1985). The authors hypothesized that it will require more than two breeding cycles to counteract the expected deterioration in density in white spruce (*Picea glauca*).

An example of the problem of a negative correlation and how it was handled will illustrate the difficulty and gains possible. In a 15-year-old test of Douglas-fir (*Pseudotsuga menziessii*), Vargas-Hernandez and Adams (1991) reported

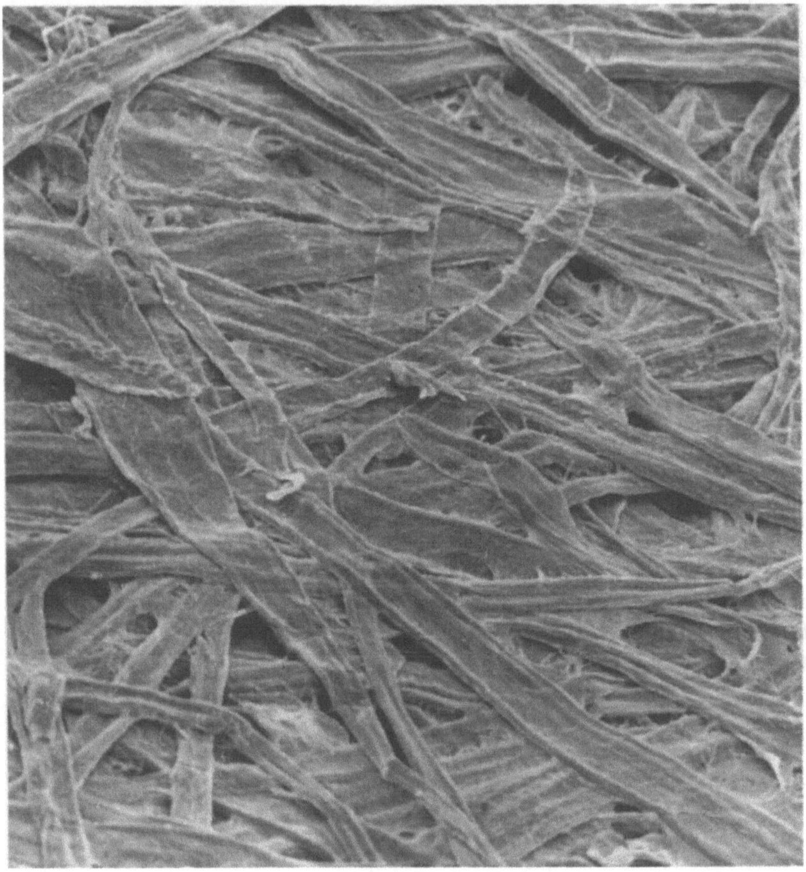

Fig. 13.4. The final product can be altered by breeding and by improving mill technologies. An example is shown of the fiber of *Gmelina arborea*, magnified 352× after 10 min beating. Wood from genetically developed high or low wood density trees would considerably alter the pattern and quality of the fiber mat shown. (Courtesy Companhia Florestal Monte Dourado, Jari, Brazil)

the following for juvenile wood of the species, (juvenile wood is formed to the 10th to 20th annual ring from tree center in this species). "Selection for volume alone without regard for wood density ... will produce the maximum estimated increase in volume growth (13.37%), ... but it will reduce overall core density (−2.41%). ... Although ... a reduction of 2.41% in average wood density might appear to be unimportant, especially when compared with gains obtained in bole volume, the impact of this reduction is substantial in terms of dry weight of fiber per hectare." If selection for core density alone were made, wood density would be increased by 5.66% and a reduction in volume growth of −8.47% so wood weight would be reduced by 3.29% relative to the original population. Maximum gains on both traits cannot be obtained at the same time (Vargas-Hernandez and Adams 1991). The same problem for Douglas-fir was covered by King et al. (1988), where index selection was used to assess the tradeoffs. They reported: "Conservative options would restrict the loss in wood density or seek to improve both traits at the expense of maximizing gain in one trait. Less conservative options would allow that a loss in wood density was acceptable to gains in volume and overall dry weight."

Correlations between a negative growth rate and wood density are frequently found for *Pinus radiata*. In their assessment of 20-year-old selected families, Burdon and Young (1991) stated that selection for overall growth rate must result in some lowering of density, but the wood production is greater in fast-growing trees. They commented that it will be difficult to simultaneously achieve appreciable gains in both traits even when using a multiple-trait selection index and recommended emphasizing growth rate because: "Restricting gains in basic density to zero in a combined selection index will result in appreciable gains in basal area, which is generally the more important trait to improve."

Although there were some slightly negative correlations between growth and wood density, Yanchuk and Kiss (1992) found that selection of white spruce families for both positive height growth and positive wood density is quite feasible. Similarly, faster growth has frequently been reported as detrimental to tree form resulting in more compression wood. An example was given by Burdon and Young (1991b) where, in 20-year-old radiata pine, more compression wood was found in the fast-growing trees. Although McKinley et al. (1982) showed that selection for growth alone would result in a negative response for wood specific gravity in loblolly pine in Texas, they concluded that gains were still feasible for total dry wood production if proper selections for both growth and wood density are made.

The situation was well summarized by van Buijtenen (1963b) for loblolly and slash pines. He found some negative correlations between growth and wood density and warned that indiscriminate selection for either growth or specific gravity might not result in the desired improvement in the total amount of dry matter produced. He found some families that had both high specific gravity and high growth rate, making a twofold improvement possible.

13.4 Summary

Initially, it was generally accepted that wood variation existed because of environmental factors affecting tree growth. This is certainly a major reason for wood variability, but extensive studies, especially with the conifers, the poplars, and the eucalypts, have shown frequent and strong genetic control of wood properties. This is especially true for the most important wood property, wood density. This knowledge is now sometimes being used in some operational programs and large, economically valuable improvements in yield and quality have resulted.

However, the value of including wood in a genetics program is not always accepted. Some researchers suggest that the gains from including wood are so small that they do not pay. In this vein Zobel (1965a) summarized for fiber length in the hardwoods: "Even if heritabilities are strong, the total amount of gain in absolute fiber length increase will not be large when compared to conifers ... even though the gains may be small, they may result in meaningful sheet property improvement if paper is made from them." Others point out that sometimes there is a negative relationship between wood or growth characteristics and desired wood properties and that, if gains are to be made, a complex index breeding system would need to be developed. Despite the objections, however, improvement in both yield and quality of final product can be related to the genetic improvement of wood. In some instances it has proven to be of secondary importance

Fig. 13.5. Breeding for wood improvement has traditionally been for fiber or commercial solid wood products, but the most serious needs and uses of wood are for energy. Small genetic programs are now being established for the production of better wood for firewood. (Photo from Guatemala, after Zobel and Talbert 1984)

to volume. Breeding for improvement of wood properties should definitely be a major part of tree improvement programs.

Genetic gains in wood properties are obtained, as for other characteristics, through selection and breeding made possible by the relatively high heritabilities and large selection differentials. The use of vegetative propagation will make breeding for and use of improved wood much more efficient. This has already been shown by Kellison (1981) and Ikemori et al. (1986). Vegetative propagation makes possible the rapid production of large quantities of highly desirable wood on a large industrial scale.

Improvement of wood through breeding should be achieved as part of general tree improvement programs. Usually, improvement of wood is secondary to increasing volume growth and having well-adapted land races with good tree form. Most wood improvement by genetics has been for pulp or commercial solid wood products, but consideration is now being given to developing wood for specialty uses, such as for energy or for decorative woods (Fig. 13.5).

References

Abdel-Gadir AY, Krahmer RL 1992 Genetic variation in age of demarcation between juvenile and mature wood in Douglas-fir. Wood Fiber Sci 25: 384–394

Abdel-Gadir AY, Krahmer RL, McKimmy MD 1993a Intra-ring variations in mature Douglas-fir trees from provenance plantations. Wood Fiber Sci 25: 170–181

Abdel-Gadir AY, Krahmer RL, McKimmy MD 1993b Relationships between intra-ring variables in mature Douglas-fir trees from provenance plantations. Wood Fiber Sci 25: 182–191

Adams WT, Howe GT 1985 Stem sinuosity measurement in young Douglas-fir progeny tests. Proc IUFRO Breed Strat for Douglas-fir Vienna, Austria, 147–149

Akachuku AE 1984 The possibility of tree selection and breeding for genetic improvement of wood properties in Gmelina arborea. For Sci 30: 275–283

Allen PJ 1977 Genotypic and phenotypic correlations of wood and tree characteristics. Variation of growth, stem quality and wood properties in relation to genetic improvement in tropical forest trees. Topic 3 IUFRO Workshop Brisbane, Australia, 13 pp

Allen PJ 1985 Estimation of genetic parameters for wood properties in slash pine in southeast Queensland. Res Note 41 Dept For Queensland, Australia, 15 pp

Almeida CM, Brune A, Silva JC, Oliveira LM 1981 Estimativas de herdabilidades e correlacoes em progenies jovenus de Eucalyptus citriodora. Rev Arvore 5: 250–268

Amaral AC, Ferreira M, do Couto HT 1977 Methods to determine the basic density in wood of tropical pines. Institute de Pesq E Estudos Glorestais Curitiba, Brazil No. 15, 47–67

Anonymous 1962 The influence of environment and genetics on pulpwood quality – an annotated bibliography. TAPPI Monograph Series 24, Appleton, WI, 316 pp

Anonymous 1963 Existing methods of value for small sample measurements of wood and fiber properties. Tappi 16: 150–156

Anonymous 1968a New methods of measuring wood and fiber properties in small samples. Tappi 51: 75A–80A (For Biol Subcomm No. 2 of TAPPI)

Anonymous 1968b Population studies of Norway spruce in Poland. Polish For Res Inst, Poland, 180 pp

Anonymous 1988 Annual Report, SAPPI Forests Research Division Ann Rep Johannesburg, South Africa, 21, 22

Anonymous 1989–1990 Genetics of Wood Quality. Ann Rep Pacific Northwest Tree Improv Res Coop, Oregon State Univ Corvallis, OR, 2–14

Anonymous 1990 Genetics of wood quality. Ann Rep Pacific Northwest Tree Impr Res Coop, Oreg State Univ Corvallis, OR, 3–17

Anonymous 1991 Genetic parameters and gains expected from selection for dry weight in Eucalyptus globulus. CELBI Co (Mimeo Rep) 29 pp

Arbez M, Baradat Ph, Birot Y, Azoeuf P, Hoslin R 1978 Variabilité et hérédité de l'angle du fil du bois mesuré à l'aide d'un traceur radioactif chez le pin maritime et le pin lariccio de calabre. Can J For Res 8: 280–289

Arbez M, Millier C 1972 Variability, heritability and correlations between characters in young Calabrian pines. IUFRO-SABRAO Joint Symposium, Tokyo, p 31

Arita M 1967 Studies on the crown broadness in Cryptomeria japonica in view of genetical and ecological characteristics. Res Bull Fac Agr Gifu Univ, Japan, 25: 36

Armstrong JE, Funk DT 1979 Genetic variation in the wood of Fraxinus americana. Wood Fiber 12: 112–120
Avanzo E 1974 Relative importance of basic density and basal area in the selection of clones of Aigeiros. Cellul Carta 25: 29–39
Baas P 1982 New perspectives in wood anatomy. W Junk, The Hague, 252 pp
Bailey IW 1920 The cambium and its derivative tissues II: Size of cambial initials in gymnosperms and angiosperms. Am J Bot 7: 355–367
Bailey JK, Feret PO, Bramlett DL 1974 Crown character differences between well-pruned and poorly-pruned Virginia pine trees and their progeny. Silvae Genet 23: 181–184
Bailey LF 1948 Figured wood: a study of methods of production. J For 46: 119–125
Bakulin VT 1980 Wood anatomy of induced polyploids of poplars. Biol Nauk 1: 25–30
Balatinecz JJ, Kennedy RW 1968 Mechanism of earlywood-latewood differentation in Larix decidua. Tappi 51: 414–422
Bamber RK, Burley J 1983 The wood properties of radiata pine. Common Agr Bur, Oxford, England, 84 pp
Banks CH 1967 Spiral grain and its effect on the quality of South African timber. 14th IUFRO Meet Munich, 1–26
Bannan MW 1967 Anticlinal divisions and cell length in conifer cambium. For Prod J 17(6): 63–69
Bannister MH 1970 P. radiata progeny trial, Pigeon Valley, Nelson. Rep For Res Inst for 1970 NZ For Ser, Rotorua, NZ, 22, 23
Bannister MH, Burdon R 1969 Genetical survey of P. radiata. Rep For Res Inst New Zealand, Rotorua, NZ, 20, 21
Bannister MH, Vine MH 1981 An early progeny trial in Pinus radiata 4. Wood density. N Z J For Sci 11: 221–243
Bannister MH, Williams AL, McDonald R, Forde MB 1962 Variation of turpentine composition in five population samples of Pinus radiata. N Z J Sci 5: 489–495
Baradat Ph 1976 Use of juvenile mature relationships and information from relatives in combined multitrait selection. Proc IUFRO Joint Meet Genetic Work Parties Adv Breed Bordeaux, France, 121–138
Baradat MA, Birot Y, Azoeuf R, Hoslin R 1978 The variability and heredity of the grain angle of wood measured by radioisotopes of maritime and laricio pine of Calabria. Can J For Res 8: 280–289
Barbour RJ, Bailey RE, Cook JA 1992 Evaluation of relative density, diameter growth, and stem form in a red spruce (Picea rubens) stand 15 years after precommercial thinning. Can J For Res 22: 229–238
Barefoot AC 1963 Selected wood characteristics in young yellow poplar (Liriodendron tulipifera). For Prod J 13: 233–239, 443–448
Barefoot AC, Hitchings RG, Ellwood EL 1964 Wood characteristics of kraft paper properties of four selected loblolly pines. Tappi 47: 343–356
Barefoot AC, Hitchings RG, Ellwood EL, Wilson E 1970 The relationship between loblolly pine fiber morphology and kraft paper properties. Bull NC Agr Exp Stn Tech Bull 202 NC State Univ Raleigh, NC, 88 pp
Barker JA 1973 Location effects on heritability estimates and gain predictions for ten-year-old loblolly pine. PhD Thesis, School For Res NC State Univ Raleigh, NC, 105 pp
Barnes RD, Gibson GL, Bardey MA 1983 Variation and genotype – environment interaction in international provenance trials of Pinus caribaea var. hondurensis and implications for population improvement strategy. Silvicultura 8 Sáo Pedro, Brazil, 35–43
Barnes RD, Woodend MA, Schweppenhauser MA, Mullin LJ 1977 Variation in diameter growth and wood density in six-year-old provenance trials of Pinus caribaea on five sites in Rhodesia. Silvae Genet 26: 163–167
Barr NA 1983 Sawing qualities of Eucalyptus saligna as a guide to selection of trees for breeding and seed collection. Fast Growing Trees Silvicult 32: 800–801

Barrett JP, Bengston GW 1964 Oleoresin yields for slash pines from seven seed sources. For Sci 10: 159–164

Barrichelo LE, Brito JO 1979 Radial variability in the wood of Pinus caribaea var. hondurensis. IPEF 18: 81–102

Bauch J 1980 Variation of wood structure due to secondary changes. IUFRO Conf Div 5 Oxford, England, p 16 (Abstr)

Bauer E 1975 Method for removing wood samples from trees. For Sci 21: 51–52

Beaudoin M, Hernàndez RE, Koubaa A, Poliquin J 1992 Interclonal, intraclonal and within-tree variation in wood density of poplar hybrid clones. Wood Fiber Sci 24: 147–153

Beaulieu J, Corriveau A 1985 Variabilité de la densité du bois et de la production des provenances d'épinette blanche 20 ans après plantation. Can J For Res 15: 833–838

Bendtsen BA 1978 Properties of wood from improved and intensively managed trees. For Prod J 28(10): 61–72

Bernard-Dagon C, Fillon C, Pauly G, Baradat P, Illy G 1971 Les terpines du pin maritime: espects biologigues et génétiques. I Variabilité de la composition monoterpenique dans un individu, entre individus et entre provenances. Ann Sci For 28: 223–258

Bertolucci FL, Demuner BJ, La Torraca SM 1992 Estimativas de parâmetros genéticos e fenotípicos para características de madeira e polpa de eucalypto. Proc D Encontro Nacional da Teenicelepa, Figueira da Fáz 13: 9–25

Bethel JS 1977 Wood as a renewable material. SAF Ann Meet – Forests for America's future: Beyond the bicentennial New Orleans, LA, 77–80

Bhat KM, Bhat KV 1983 Wood properties of one-year-old Eucalyptus tereticornis. IUFRO Div 5 Conf Madison, WI, 4 pp

Bhat KM, Bhat KV, Dhamodaran TK 1990 Wood density and fiber length of Eucalyptus grandis grown in Kerala, India. Wood Fiber Sci 22: 54–61

Birot YM, Arbez M, Azoef P, Hoslin R 1979 Phenotypic variation of spiral grain along the stem in Corsican pine and Douglas-fir. Ann Sci For 36: 165–173

Birot Y, Nepveu G 1979 Variabilité clonale et liaisons ortets-ramets dans un population d'Epicéa. Silvae Genet 28: 37–47

Birot Y, Dufour J, Ferrandes P, Teissier du Cross E, Azoeuf P, Hoslin R 1980 Variabilité de l'angle du fil du bois chez quelques feuillus: hêtre, chêne et Eucalyptus dalrympleana. Ann Sci For 37: 19–36

Birt DV, Harris JM 1970 Preliminary assessment of wood density in provenance trials of Pinus taeda, P. elliottii and P. caribaea. For Prod Rep 295 NZ For Ser Rotorua, New Zealand, 7 pp

Bissett IJ, Dadswell HE 1949 The variation of fibre length within one tree of Eucalyptus regnans. Aust For 8: 86–96

Bissett IJ, Dadswell HE, Wardrop AB 1951 Factors influencing tracheid length in conifer stems. Aust For 15: 17–30

Blair R, Olson J 1984 Influence of forest management practices on wood quality. Symp: Utilization Changing Wood Resources of the Southern US Raleigh, NC (mimeo), 16 pp

Blair RL, Zobel BJ, Barker J 1975 Predictions of gain in pulp yield and tear strength in young loblolly pine through genetic increases in wood density. Tappi 58(1): 89–91

Blair RL, Zobel BJ, Franklin EC, Djerf AC, Mendel JM 1974 The effect of tree form and rust infection on wood and pulp properties of loblolly pine. Tappi 57(7): 46–50

Blanchette RA, Biggs AR 1992 Defence mechanisms of woody plants against fungi. Springer, Berlin Heidelberg New York, 458 pp

Borralho NM, Cotterill PP, Kanowski PJ 1992 Genetic parameters and gains expected from selection for dry weight in Eucalyptus globulus ssp. globulus in Portugal. For Sci 38: 80–94

Bosshard HH 1951 Variability of the elements of ash wood as a function of cambial activity. Schweiz Z Forstw 102: 648–665

Boulding KE 1977 Wood in the energy economy. SAF Ann Meet – Forests for America's future: Beyond the Bicentennial New Orleans, LA, 1–7
Boyce SG, Kaeiser M 1961 Environmental and genetic variability in the length of fibers in eastern cottonwood. Tappi 44: 363–366
Boyd JD 1974 Relating lignification to microfibril angle differences between tangential and radial faces of cell wall layers in wood cells. Drev Vysk 19: 41–54
Boyd JD, Foster RC 1975 Microfilbrils in primary and secondary wall growth develop trellis configurations. Can J Bot 53: 2687–2701
Boyd L 1968 Genetic variation in packing density within a selected population of loblolly pine (Pinus taeda). For Prod J 18(11): 65
Boyle TJ, Balatinecz JJ, McCaw PM 1987 Genetic control of some wood properties of black spruce. 21st Can Tree Impr Assoc Truro, Nova Scotia, p 198
Brandáo LG 1984 The new eucalypt forest. Marcus Wallenberg Symp Proc 1 Falun, Sweden, 3–15
Brazier JD 1965 An assessment of the incidence and significance of spiral grain in young conifer trees. For Prod J 15: 308–312
Brazier JD 1967 Timber improvement I. A study of the variation in wood characteristics in young Sitka spruce. Forestry 4: 117–128
Brazier JD 1969 Some considerations in appraising within-ring density. 2nd World Cons For Tree Breed IUFRO Washington, DC, 9 pp
Brewbaker JL, Hutton EM 1979 Leucaena-versatile tropical legume. New Age Crops – Am Assoc Adv Sci Press, Boulder, CO, 207–259
Bridgwater FE 1984 The impact of genetic improvement of stem straightness on yield and value of lumber. Symp Utilization Changing Wood Resource in South US Raleigh, NC, 80–87
Bridgwater FE, Talbert JT, Jahromi S 1983 Index selection for increased dry weight in a young loblolly pine plantation. Silvae Genet 32: 157–161
Britt KW 1967 Wood and fiber properties: measurement and interpretation. Proc 4th For Biol Conf TAPPI, Appleton, WI, 84–98
Brown CL, Klein J 1961 Observations on inheritance of wood specific gravity in seedling progeny of loblolly pine. J For 59: 898–899
Brown GA 1971 A statistical analysis of density variation in Pinus caribaea grown in Jamaica. Selective Breeding to Improve Some Tropical Conifers 15th IUFRO Congress Sect 22 Gainesville, FL, 17 pp
Brown HP, Panshin AJ 1940 Commercial timbers of the United States. McGraw-Hill, New York, 554 pp
Bruckman G, Walters CS 1964 A comparison of methods for determining volume of increment cores. J For 62: 172–177
Bunn EH, 1981 The nature of the resource. N Z J For 26: 162–169
Burdon RD 1970 Clonal replication trial in P. radiata. Ann. Rept. For Res Inst N Z For Ser Rotorua, NZ, p 23
Burdon RD 1975 Compression wood in Pinus radiata clones on four different sites. N Z J For Sci 3: 286–303
Burdon RD 1989 Early selection in tree breeding: principles for applying index selection and inferring input parameters. Can J For Res 19: 499–504
Burdon RD 1992 Testing and selection: strategies and tactics for the future. Breeding Tropical Trees, IUFRO Meet Cartegena and Cali, Colombia, 1 p
Burdon RD, Gaskin RE, Zabkiewicz JA, Low CB 1992a Genetic survey of Pinus radiata. 7: Variation and inheritance of pinene composition in wood oleoresin. N Z J For Sci 22: 246–256
Burdon RD, Harris JM 1973 Wood density in radiata pine clones on four different sites. N Z J For Sci 3: 286–303
Burdon RD, Harris JM, Bannister MH, Birt DV, Low CB 1972 Variation between and

within populations of Pinus radiata in wood density, ring width and bark thickness. New Zealand For Ser FRI Gen Tree Impr Rep 60, Rotorua, 13 pp
Burdon RD, Low CB 1992 Genetic survey of Pinus radiata. 6: Wood properties: variation, heritabilities and interrelationships with other traits. N Z J For Sci 22: 228–245
Burdon RD, Thulin IJ 1965 The improvement of Pinus radiata. N Z For Ser FRI Symp No. 6 Rotorua, New Zealand, 92 pp
Burdon RD, Young GD 1991a Preliminary genetic parameter estimates for wood properties from top-ranked Pinus radiata progenies and comparisons with controls. Proc 11th Meet Represen Res Work Grp No. 1 (Forest Genetics) Australian For Council Coonawarra, South Australia, 137–140
Burdon RD, Young GD 1991b Some wood properties in four Pinus radiata provenances at Kaingaroa Forest, rings 1–20 from pith-pilot results. Proc 11th Meeting Representative Research Working Group No. 1 (Forest Genetics) Australian For Council Coonawarra, South Australia, 141–143
Burdon RD, Young GD, Cown DJ 1992b Evaluation of wood properties in highly select Pinus radiata progenies and preliminary estimates of genetic parameters. Proc Div 5 IUFRO Conf Forest Production vol I Nancy, 168–169
Burley J, Nikles DG 1973 Selection and breeding to improve some tropical conifers. Dept For Queensland, 456 pp
Burley J, Nuevo CC, Franklin PG 1983 Preliminary estimates of heritability of wood density parameters in Pinus caribaea var. hondurensis from Fiji and Queensland, Australia. Conf IUFRO Division 5 Madison, Wl, 5 pp
Burley J, Press A, Morgan JF 1967 Pine wood quality studies in Central Africa. 2. Development of laboratory techniques. A. Spiral grain. IUFRO Congress Sect 41 Munich, pp 1–17 (Symposium on Spiral Grain)
Byram TD, Lowe WJ 1988 Specific gravity variation in a loblolly pine seed source study in the Western Gulf Region. For Sci 34: 798–803
Byram TD, Lowe WJ, McKinley CR, Robinson JF, Stander AF, van Buijtenen JP 1984 Thirty-second progress report of the cooperative tree improvement program. Texas A&M University College Station, TX, 27 pp
Byrd VL 1964 An investigation of the effect of wood chemical constituents on kraft paper properties of four selected loblolly pines. MS Thesis NC State Univ Raleigh, NC, 113 pp
Byrd VL, Ellwood EL, Hitchings RG, Barefoot AC 1965 Wood characteristics and kraft paper properties of four selected loblolly pines. II: Wood chemical constituents and their relationship to fiber morphology. For Prod J 15: 313–320
Callaham RZ, Liddicoet AR 1961 Altitudinal variation at 20 years in ponderosa and Jeffrey pines. J For 59: 814–820
Cameron JF, Berry PF, Phillips EW 1959 The determination of wood density using beta rays. Holzforschung 13: 78–84
Campbell RK 1965 Phenotypic variation and repeatability of stem sinuosity in Douglas-fir. Northwest Sci 39(2): 47–59
Campinhos E 1980 More wood of better quality through intensive silviculture with rapid-growth improved Brazilian Eucalyptus. Tappi 63: 145–147
Campinhos E, Claudio-da-Silva 1990 Development of the eucalyptus tree of the future. ESPRA Conf Seville, Spain, 22 pp
Carson SD, Carson MJ 1991 Realizing gains in resistance to Dothistroma. Proc 11th Meet Rep Res Work Grp No. 1 Australian For Council Coonawara, South Australia, 71–75
Carson MJ, Inglis CS 1988 Genotype and location effects on internode length of Pinus radiata in New Zealand. N Z J For Sci 18: 267–271
Cave ID, Walker JC 1994 Stiffness in wood in fast-grown plantation softwoods: the influence of microfibril angle. For Prod J 44(5): 43–48
Cech MY, Kennedy RW, Smith JH 1960 Variation in some wood quality attributes of one-year-old black cottonwood (Populus trichocarpa). Tappi 43: 857–859

CELBI 1979 Heredity studies of Eucalyptus globulus. Tech Consul Fast Growing Plantation Broadleaved Trees for Mediterranean and Temperate Zones. FAO, Lisbon, Portugal, pp 1–15
Chafe SC 1985 The distribution and interrelationship of collapse, volumetric shrinkage, moisture content and density in trees of Eucalyptus regnans. Wood Sci Technol 19: 329–345
Champion HG 1927 On the twisting of tree trunks. Indian For 53: 236–237
Champion HG 1929 More about spiral grain in conifers. Indian For 55: 57–58
Champion HG 1930 Second interim report on the progress of investigations into the origin of twisted fibre in Pinus longifolia. Indian For 56: 511–520
Champion HG 1945 Genetics of forestry. QJ For 39: 74–81
Chaperon H, Raox H, Siohan A, Alazard P 1988 Genetic variability of the technological characteristics of maritime pine wood. AFOCEL Ann Rech Sylvicol, 327–345
Chen YT, Zhi JW, Guang N 1994 Combining ability for main characters of wood qualities in Chinese fir. For Res 7(2): Mimeo (in press)
Chiu CM, Lee CH 1994 Wood-bark grain spirality correlations in *Calocedrus formosana.* IAWA Bull: 17 pp, Mimeo (in press)
Chowdhury KA 1931 Anatomical studies on the wood of a hybrid larch. J For 29: 797–805
Chudnoff M 1980 Tropical timbers of the world. US For Serv FPL, Madison WI, 826 pp
Chudnoff M, Geary TF 1973 On the heritability of wood density in Swietenia macrophylla. Turialba, Costa Rica, 23: 359–361
Clark A, Saucier JR 1991 Influence of planting density, intensive culture, geographic location, and species on juvenile wood formation in southern pine. Georgia For Res Pap 85, Georgia For Comm, 13 pp
Clark JD 1961 Convenient projector for fiber measurements. Tappi 44(1): 64
Clarke CH 1990 The estimation of genetic parameters of pulp and paper properties in Eucalyptus grandis and their implications for tree improvement. MS Thesis, Univ Natal Dept Gen Pietermaritzburg, South Africa, 107 pp
Clauson ML, Wilson JB 1991 Comparison of video and x-ray for scanning wood density. For Prod J 41(3): 58–62
Coggeshall MV, Land SD, Ammon VD, Cooper DT, McCracken FI 1981 Genetic variation in resistance to canker disease of young American sycamore. Plant Dis 65: 140–142
Colley J 1973 Influence of vessel elements on the picking tendency of eucalypt pulps. Pap Technol Ind 14: 293–296
Comerford NB, Leaf AL 1981 The mean accuracy and precision of using stem disc moisture contents to estimate stem dry weight of Pinus resinosa. For Ecol Managem 3: 329–334
Conkle MT 1980 Amount and distribution of isozyme variation in various species. 17th Can Tree Impr Assoc Newfoundland, Canada, 109–117
Conrad J 1977 Ein Indiz für die Vererbbarkeit gewisser Holzstukturen. Forst Holzwirt 32: 217–218
Corea E 1989 Evaluation of a provenance trial of Pinus oocarpa and Pinus patula ssp. tecunumanii in four sites in Costa Rica. MS Thesis Turrialba, Costa Rica, 179 pp
Corriveau A, Beaulieu J, Mothe F 1987 Wood density of natural white spruce populations in Quebec. Can J For Res 17: 675–682
Corriveau A, Beaulieu J, Mothe F, Polequin J, Doucet J 1990 Densité et largeur des cerves des populations d' Epinettes blanches de la région forestière des Grands Lacs et du Saint-Laurent. Can J For Res 20: 121–129
Cown DJ 1971 Wood properties of Pinus nigra. In 1970 Rep N Z For Ser For Res Inst Rotorua, New Zealand, 7 pp
Cown DJ 1992 Corewood (juvenile wood) in Pinus radiata – should we be concerned? N Z J For Sci 22: 87–95
Cown DJ, McConchie DL, Young GD 1983 Wood properties of Pinus caribaea var. hondurensis grown in Fiji. FRI Bull 17 N Z For Serv Rotorua, New Zealand, 54 pp

Cown DJ, McConchie DL, Young GD 1991a Radiata pine – wood properties survey. FRI Bull No. 50 Rotorua, New Zealand, 50 pp

Cown DJ, Parker ML 1979 Densitometric analysis of wood from five Douglas-fir provenances. Silvae Genet 28: 48–53

Cown DJ, Young GD, Burdon RD 1992 Variation in wood characteristics of 20-year-old half-sib families of Pinus radiata. N Z J For Sci 22: 1–10

Cown DJ, Young GD, Kimberly MO 1991b Spiral grain patterns in plantation-grown Pinus radiata. N Z J For Sci 21: 206–216

Crawford DL, Pettey TM, Antai SP 1982 Genetic manipulation of lignin degrading Streptomyces by protoplast fusion and transformation. Res Dev Div TAPPI Conf Atlanta, GA, 249–257

Cregg BM, Dougherty PM, Hennessey TC 1988 Growth and wood quality of young loblolly pine trees in relation to stand density and climatic factors. Can J For Res 18: 851–858

Crist JB, Dawson DH 1975 Anatomy and dry weight yields of two Populus clones grown under intensive culture. North Cent For Exp Stat US For Ser, St. Paul, MN, 6 pp

Crow AB 1962 Site, silviculture and wood quality. 11th Ann For Symp Louisiana State Univ Baton Rouge, LA, 16–34

Cubbage FW 1990 Economic impacts of southern pine wood quality changes. Proc South Plant Wood Quality Workshop Athens, Georgia, US For Ser Tech Rep SE-63, New Orleans, LA, 95 pp

Curró P 1972 Variations of basic density in wood of Populus × euramericana (P. × canadensis). Cellul Carta 23: 41–49

Dadswell HE 1951 The influence of wood structure on the quality of eucalypt groundwood. Aust Pulp Pap Ind Tech Assoc Proc 5: 1–18

Dadswell HE 1957 Tree growth characteristics and their influence on wood structure and properties. 7th British Common For Conf CSIRO, Melbourne, Australia, 19 pp

Dadswell HE 1958 Wood structure variations occurring during tree growth and their influence on properties. J Inst Wood Sci 1: 1–24

Dadswell HE 1960 Tree growth-wood property interrelationships. Proc Spec Field Inst For Biol Sch For NC State Univ Raleigh, NC, 86 pp

Dadswell HE 1972 The anatomy of eucalypt wood. CSIRO Melbourne, Australia, 36 pp

Dadswell HE, Fielding JM, Nicholls JW, Brown AG 1961 Tree to tree variations and the gross heritability of wood characteristics of Pinus radiata. Tappi 44: 174–179

Dadswell HE, Nicholls JW 1959 Assessment of wood qualities for tree breeding I. Pinus elliottii var elliottii in Queensland. Aust CSIRO Div For Prod Technol Pap 4: 1–16

Dadswell HE, Wardrop AB 1959 Growing trees with wood properties desirable for paper manufacture. J Aust Pulp Pap Ind Tech Assoc 12: 129 136

Dadswell HE, Wardrop AB 1960 Some aspects of wood anatomy in relation to pulping quality and to tree breeding. J Aust Pulp Pap Ind Tech Assoc 13(5): 161–173

Dadswell HE, Watson AJ 1961 Influence of the morphology of woodpulp fibres on paper properties. Tech Sect British Pap Board Makers Assoc London, England, 537–572 (Symposium Formation and Structure of Paper)

Dadswell JW, Dadswell HE, Perry DH 1963 Assessment of wood qualities for tree breeding. Silvae Genet 12: 105–110

Darrow WK 1992 Heart shake in South African-grown Pinus elliottii. Wood Fiber Sci 24: 241–251

Davidson J 1972 Variation, association and inheritance of morphological and wood characters in an improvement program for Eucalyptus deglupta. PhD Thesis, Australian National Univ Canberra, Australia, 263 pp

Davidson J 1983 Program in breeding Eucalyptus deglupta. Fast Growing Trees, Silvicult 31: 520–533

Davis EM 1961 The effect of density upon wood properties. Wood Work Dig 63(9): 52–53

Dean CA, Cotterill PP, Burdon RD 1991 Early selection for growth of radiata pine. I. Trends over time in additive and dominance genetic variances and co-variances. Silvae Genet 40(5): 145–153

Dean GH 1990 An investigation into Kajaani fibre length statistics for use in tree breeding (E. nitens, Hampshire S.S.O., 8 Y.O.) APPM Internal Report (unpublished), Mimeo 12 pp

Dean GH, French J, Maddern KN 1991 Breeding eucalypts for manufacture of pulp and paper. Proc 11th Meet Res Work Grp No. 1 Australian For Council Coonawarra, Australia, 125–128

Dean GH, French J, Tibbits WN 1990 Variation in pulp and paper making characteristics in a field trial of Eucalyptus globulus. Appita

de Guth EB 1980 Relationship between wood density and tree diameter in Pinus elliottii of Missiones Argentina. IUFRO Conf Div 5 Oxford, England, 1 p (Summary)

Delaporte P 1983 The variability in inter- and intraprovenances of Picea sitchensis. Improvement strategy by vegetative propagation. AFOCEL 283–337

Derr HJ, Enghardt H 1960 Is geographic seed source of slash pine important? South Lumberman 201(2513): 95–96

De Souza VR, Carpim MA, Barrichelo LE 1986 Wood basic density of different provenances of Eucalyptus grandis and E. saligna from different diameter classes and different heights along the trees. IPEF 33: 65–72

Dietrichson J 1963 Some results from an anatomic investigation on Norway spruce provenances in Sweden and Norway. 1st World Consultation Forest Genetic Tree Improvement Stockholm, Sweden, 9 pp

Dietrichson J 1964 The selection problem and growth-rhythm. Silvae Genet 13: 178–184

Dillner B, Ljunger A, Herod OA, Thune-Larssen E 1970 The breeding of Eucalyptus globulus on the basis of wood density, chemical composition and growth rate. Production Industrial Utilization. Eucalyptus, Lisbon, vol 23, 120–155

Dinwoodie JM 1961 Tracheid and fibre length in timber: a review of literature. Forestry 34(2): 125–144

Dodd RS, Power AB 1989 Specific gravity and fibre length variation among Eucalyptus trees. Tech Rep 35·01·446 For Prod Lab Univ California, Berkeley, 12 pp

Doede D 1993 Genetics of stem volume, stem form, and branch characteristics of sapling Noble fir. MS Thesis Oregon State Univ Corvallis, OR, 40 pp

Donaldson LA 1992 Within- and between-tree variation in microfibril angle in Pinus radiata. N Z J For Sci 22: 77–86

Donaldson LA 1993 Variation in microfibril angle among three genetic groups of Pinus radiata trees. N Z J For Sci 23: 90–100

Dorn D 1969 Relationship of specific gravity and tracheid length to growth rate and provenance in Scotch pine. 16th Northeastern For Tree Improv Conf Quebec, Canada, 8 pp

Durand PY 1983 Variation in the technological quality of the wood of framire in plantations in the Ivory Coast. Fast Growing Trees Silvicult 32: 811–815

Dvorak WS, Kellison RC 1991 Annotated bibliography on the wood properties of Pinus tecunumanii. CAMCORE Bull 9 NC State Univ Raleigh, NC, 16 pp

Dvorak WS, Wright JA 1993 Parent-offspring correlations for wood specific gravity of Pinus tecunumanii between native and exotic environments. Res Rep Smurfit Carton de Colombia Cali, Colombia, 8 pp

Echols RM 1958 Variation in tracheid length and wood density in geographic races of Scotch pine. Yale Univ Sch For Bull 64, New Haven, CT, 52 pp

Echols RM 1959 Estimation of pulp yield and quality of living trees from paired-core samples. Tappi 42: 875–877

Echols RM 1961 Liquid dispersion method for mounting wood fibers on slides. For Sci 7: 369–370

Echols RM 1969 Powered drive for large increment cores. J For 67: 123–125
Echols RM 1973 Effects of elevation and seed source on tracheid length in young ponderosa pine. For Sci 19: 47–49
Echols RM, Conkle MT 1971 The influence of plantation and seed source elevation on wood specific gravity of 29-year-old ponderosa pines. For Sci 17: 388–393
Echols RM, Dorman KW 1962 Heritability of specific gravity and summerwood percent and relation to other factors in slash pine. Tappi 45: 599–601
Eguiluz-Piedra T, Zobel BJ 1986 Geographic variation in wood properties of Pinus tecunumanii. Wood Fiber Sci 18: 68–75
Ehrenberg C 1970 Breeding for stem quality. Unasylva 24(3): 23–31
Einspahr DW, Benson MK, Peckham JR 1967 Variation and heritability of wood and growth characteristics of five-year-old quaking aspen. Inst Pap Chem Gen Physiol Note 1, Madison, WI, 1–6
Einspahr DW, Goddard RE, Gardner HS 1964 Slash pine wood and fiber property heritability study. Silvae Genet 13: 103–109
Einspahr DW, van Buijtenen JP, Peckham JR 1963 Natural variation and heritability in triploid aspen. Silvae Genet 12: 51–58
Einspahr DW, van Buijtenen JP, Peckham JR 1969 Pulping characteristics of ten-year-old loblolly pine selected for extreme wood specific gravity. Silvae Genet 18: 57–61
Eisemann RL, Harding KJ, Eccles DB 1990 Genetic parameters and predicted selection response for growth and wood properties in a population of Araucaria cunninghamii. Silvae Genet 39(5–6): 206–216
Elliott GK 1958 Spiral grain in second-growth Douglas-fir and western hemlock. For Prod J 8: 205–211
Ellis JC 1971 Use of X-rays in measuring ring widths from increment borings. N Z For Ser Reprint 534, 223–230
Engler A 1905 Einfluss der Provenenz des Samens auf die Eigenschaften der forstlichen Holzgewächse II. Mitt Schweiz Centralanst Forstl Versuchswes 8: 81–236
Ericson B 1959 A mercury immersion method for determining the wood density of increment core sections. Rapp avd Skogsprod Skogsforskn Inst No. 1 Stockholm, Sweden, 31 pp
Ericson B 1960a Studies on the genetical wood density variation in Scots pine and Norway spruce. Stat Skogsforskn Inst Stockholm, Rapp 4, 52 pp
Ericson B 1960b Latewood percentage, density and volumetric shrinkage in wood of Picea abies f. virgata. A comparison with Picea abies. For Res Inst Sweden 2, Stockholm, 15 pp
Ericson B 1961 Forest tree breeding with a view to raising the yield of pulp. Some preliminary results. Medd Statens Skogsforskningsinst 32: 194–263
Ericson B 1966 Determination of basic density in small wood samples. An examination of some sources of error. Rapp Uppsats Inst Skogsprod Skoshogsk Stockholm, Sweden, No. 9, 1–34
Eriksson, K-EL, Dinus RJ 1991 Progress towards making trees easier to pulp and bleach. Pulp Pap Mag, 40–44
Ezell AW, Lowe WJ, Murphy WK, Wright JA 1983 Plantation culture for energy production – the importance of within species selection. Wood Fiber Sci 15: 69–73
Falconer DS 1960 Introduction to quantitative genetics. Roland, New York, 365 pp
Falconhagen ER 1979 Provenance variation in growth, timber and pulp properties of Pinus caribaea in South Africa. Bull 39 South Afr For Res Inst Dept For Pretoria, South Africa
Farmer RE 1970 Variation and inheritance of eastern cottonwood growth and wood properties under two soil moisture regimes. Silvae Genet 19: 5–8
Farmer RE, Wilcox JR 1966 Variation in juvenile growth and wood properties in half-sib cottonwood families. US For Ser Res Pap NC-6, Eulfport, MI, 4 pp
Farmer RE, Wilcox JR 1968 Preliminary testing of eastern cottonwood clones. Theor Appl Gen 38: 197–201

Ferguson RB, Land SB, Cooper DT 1977 Inheritance of growth and crown characters in American sycamore. Silvae Genet 25: 180–182
Ferrari G 1992 Wood quality in stump shoots of some species of Eucalyptus. Proc Div 5 IUFRO Conference Forest Production vol I Nancy, 241, 242
Ferrari G, Scaramuzzi G 1980 Investigations on the effect of site on wood characteristics in Populus × euroamericana. IUFRO Conf Div 5 Oxford, England, 40 (Summary)
Fielding JM 1960 Branching and flowering characteristics of Monterey pine. For Timber Bureau Bull 37 Canberra, Australia, 59 pp
Fielding JM 1967 Spiral grain in Pinus radiata plantations in the Australian Capital Territory. Common Aust Dept Nat Dev For Res Inst Leaflet 103: 5–28
Foelkel CE, Ferreira M, Nehring JH, Rolim MB 1975 Variability of slash pine wood in the radial direction. IPEF 10: 1–11
Foulger AN 1968 Effect of aphid infestation on properties of grand fir. For Prod J 18: 43–47
Foulger AN 1969 Through-bark measurements of grain direction: preliminary results. For Sci 15: 92–94
Fowler DP, Morris RW 1977 Genetic diversity in red pine: Evidence for low genetic heterozygosity. Can J For Res 7: 343–347
Frampton LJ, Jett JB 1989 Juvenile wood specific gravity of loblolly pine tissue culture plantlets and seedlings. Can J For Res 19: 1347–1350
Franklin EC 1974 Phenotypic and genetic variation of sulfate naval stores yields in loblolly pine. TAPPI For Biol Conf Seattle, WA 99–102
Franklin EC 1979 Model relating levels of genetic variance to stand development of four North American conifers. Silvae Genet 28: 207–212
Franklin EC, Callaham RZ 1970 Multinodality, branching and forking in lodgepole pine (Pinus contorta var. murrayana) Silvae Genet 19: 180–184
Franklin EC, Squillace AE 1973 Wood parameters: necessity for comparing them on a unit volume basis. Tappi 57(11): 118–120
Freese F 1967 Elementary statistical methods for foresters. Ag Handbook 317 USDA Forest Service, New Orleans, LA, 86 pp
Fries A 1986 Volume growth and wood density of plus tree progenies of Pinus contorta in two Swedish field trials. Scand J For Res 1: 403–419
Gabriel WJ 1956 Preliminary report on clonal differences in the wood and phloem of Populus deltoides and P. trichocarpa. 3rd Northeast For Tree Improv Conf Ithaca, NY, 33–35
Gansel CR 1965 Inheritance of stem and branch characters in slash pine and relation to gum yield. 8th South For Tree Improv Conf Savannah, GA, 57–63
Garcia de Leon JP 1982 Genetic variation of wood properties among populations of Pinus caribaea. PhD Thesis Univ Oxford, 248 pp
Garges LD, Labosky P, Blankenhorn PR, Rishel LE 1984 Lumber recovery from gypsy moth-killed red and white oak trees. For Prod J 34(7–8): 45–50
Garrett PW, Randell WK, Shigo AL, Shortle WC 1979 Inheritance of compartmentalization of wounds in sweetgum (Liquidambar styraciflua) and eastern cottonwood (Populus deltoides). USDA For Ser Res Pap NE 443, 4 pp
Garrett PW, Shigo AL, Carter J 1976 Variation in diameter of central columns of discoloration in six hybrid poplar clones. Can J For Res 6: 475–477
Geary TF 1967 Variation in density and diameter between open-pollinated families of Pinus khasya. Rhod Zambia Malawi J Agric Res 5: 297–300
Geyer WA, Gilmore AR 1965 Effect of spacing on wood specific gravity in loblolly pine in southern Illinois. Agric Exp Stn Note 113 Univ Illinois, Urbana, IL, 5 pp
Ghisi G 1969 A comparison of some methods for determining the basic density of poplar wood for industrial use. Publ Cent Speriment Agric E For 10: 238–248
Gibson GL 1982 Genotype-environment interaction in Pinus caribaea. Comm For Inst Oxford Interim Rep, Oxford, England, 112 pp

Gilmore AR 1968 Geographic variation in specific gravity of white pine and red pine in Illinois. For Prod J 18: 49–51

Gilmore AR, Boyce SF, Ryker RA 1966 The relationship of specific gravity of loblolly pine to environmental factors in southern Illinois. For Sci 12: 399–405

Gladstone WT, Barefoot AC, Zobel BJ 1970 Kraft pulping of earlywood and latewood from loblolly pine. For Prod J 20(2): 17–24

Gladstone WT, Gray RL 1972 Effects of forest fertilization on wood quality. South For Res Cen Weyerhaeuser Co Hot Springs, Arkansas, 18 pp

Goddard RE, Cole DE 1966 Variation in wood production of six-year-old progenies of select slash pines. Tappi 49: 359–362

Goddard RE, Strickland RK 1964 Crooked stem form in loblolly pine. Silvae Genet 13: 155–157

Goggans JF 1961 The interplay of environment and heredity as factors controlling wood properties in conifers with special emphasis on their effects on specific gravity. Tech Rep 11 Sch For NC State Coll, Raleigh, 56 pp

Goggans JF 1962 The correlation, variation and inheritance of wood properties in loblolly pine. Tech Rep 14 School For Res NC State Univ Raleigh, NC, 155 pp

Goggans JF 1964 Correlation and heritability of certain wood properties in loblolly pine (Pinus taeda). Tappi 47: 318–322

Göhre K 1960 The distribution of density in poplar stems. Wiss Abh Dtsch Akad Landwiss Berl 44: 51–79

Göhre K, Götze H 1956 Untersuchungen über die Rohwichte des Rotbuchenholzes. Arch Forstwes 5: 716–748

Goldstein IS 1980 New technology for new uses of wood. Tappi 63: 105–108

Gonzalez JS 1987 Circumferential variability in the relative density of discs taken at breast height of lodgepole pine plus trees. Can Tree Improv Assoc Proc Truro, Nova Scotia, p 199

Gonzalez JS 1989 Circumferential variation in the relative density of lodgepole pine. Can J For Res 19: 276–279

Gonzalez JS, Kellogg RM 1978 Evaluating wood specific gravity in a tree-improvement program. Can Dept Environ West For Prod Lab Vancouver, British Columbia, 22 pp

Gonzalez JS, Richards JE 1988 Early selection for wood density in young coastal Douglas-fir trees. Can J For Res 18: 1182–1185

Govette RL, Sinclair SA, Bowyer JL 1983 Lumber grade yields from spruce budworm damaged balsam fir. For Prod J 33(5): 46–50

Grattapaglia D 1994 Eucalypts catch up with the genome research wave. Dendrome 1(1): 5–7

Green HV 1965 Wood characteristics IV: the study of wood characteristics by means of a photometric technique. Tech Rep Series 419 Woodlands Res Index No 167 Montreal, Canada, 18 pp

Greene JT 1966 Selection for tracheid length in loblolly pine. Georgia For Res Council Rep No 18, Macon GA, 30 pp

Guldager P 1972 Breeding for density in Pinus kesiya. Zambia Minist Lands Nat Resour Div For Res Res Note 8, Kitwe, Zambia, 15 pp

Haigh RW 1961 The effect of provenance and growth rate on specific gravity and summerwood percentage in young Douglas-fir. Univ Brit Col For Club Res Commit Vancouver Canada 19, 57 pp

Hale JD 1955 Thickness and density of bark. Pulp Pap Mag Can: 3–7

Hale JD 1959 Physical and anatomical characteristics of hardwoods. Tappi 42: 670–677

Hall PJ 1988 Using the pilodyn tester in selecting plus trees. North J App For 5(3): 170–171

Hamilton JR, Cech FC, Ammons FL 1984 A tool for obtaining large cores from standing trees. For Prod J 34: 33–34

Hamilton JR, Wendel GW 1967 Specific gravity and fiber length of some hybrid poplars growing in West Virginia. W Va Agric Exp Stn Bull 556T: 6

Hamrick JL 1962 Environmental and genetic control of bark thickness in loblolly pine (Pinus taeda). NC State Univ (Mimeo rep), Raleigh, NC, 7 pp

Hamrick JL, Metton LB, Linhart YB 1979 Levels of genetic variation in trees: Influence of life history characteristics. Symp Isozymes N Am For Trees Berkeley, CA 35–41

Hamza KF, Lewark S 1992 Sampling in studies of wood properties in four Eucalyptus species grown in Ruvu, Tanzania. Proc Div 5 IUFRO Conf Forest Production vol I Nancy, p 156

Hancock WV 1962 The assessment of wood quality of trees selected for breeding purposes. 8th Meet Comm For Tree Breed in Canada, Fredericton, N.B. Part II, K1–K3

Hans AS 1976 Variation in wood density of Eucalyptus grandis and E. tereticornis. J Sci Technol 1(4): 109–112

Harding KJ 1990 Queensland wood property research during the 1980's. Appita 43: 155–157

Harding KJ, Chu LQ 1989 Wood quality improvement in the tree breeding program at Dongmen. 4th Tech Exchange Seminar China-Australia Afforestation Proj Dongmen State Forest Farm Guangxi Zhuang Autonomous Region, PR China, 11 pp

Harding KJ, Chu LQ, Crane RA 1989 Basic density and other wood property investigations at Dongmen. 4th Tech Exchange Seminar China-Australia Afforestation Project, Dongmen State Forest Farm Guangxi Zhuang Autonomous Region, PR China, 20 pp

Harding KJ, Kanowski PJ, Woolaston RR 1991 Preliminary genetic parameter estimates for wood quality traits of Pinus caribaea var. hondurensis in Queensland, Australia. Silvae Genet 40: 152–156

Harding KJ, Woolaston RR 1991 Genetic parameters for wood and growth properties of Araucaria cunninghamii. Silvae Genet 40: 232–237

Harris G 1993 Comparison of northern softwood and southern pine fiber characteristics for groundwood publication papers. Tappi 76(6): 55–61

Harris JM 1954 Heartwood formation in Pinus radiata. NZ For Serv Tech Pap No 1 N Z For Ser, Rotorua, N.Z., 31 pp

Harris JM 1961 A survey of the wood properties of radiata pine grown in the Kaingaroa Forest. For Prod Rep 76 For Res Inst N Z For Ser, Rotorua, N.Z., 1–16

Harris JM 1965a The heritability of wood density. IUFRO Sec 41 Melbourne, Australia, 20 pp

Harris JM 1965b A survey of wood density, tracheid length and latewood characteristics in radiata pine grown in New Zealand. N Z For Serv Tech Rep 47, Rotorua, N.Z., 31 pp

Harris JM 1966 Specific gravity and summerwood percent. N Z For Serv For Res Inst FRI Rotorua, N.Z., Symp 6: 34–36

Harris JM 1969 The use of beta rays in determining wood properties – parts 1–5. N Z J Sci 12: 395–451

Harris JM 1970 Breeding to improve wood quality. Unasylva 24(23): 32–36

Harris JM 1977 Note on wood density of Pinus caribaea grown under temperate, subtropical and tropical conditions. Variation of growth, stem quality and wood properties in relation to genetic improvement of tropical forest trees. IUFRO Workshop Brisbane, Australia, 9 pp

Harris JM 1983 Variation in wood quality: research requirements and research strategy. IUFRO Conf Div 5 Madison, WI, 117–134

Harris JM 1984 Non-destructive assessment of spiral grain in standing trees. N Z J For Sci 14: 395–399

Harris JM 1989 Spiral grain and wave phenomena in wood formation. Springer, Berlin, Heidelberg, 214 pp

Harris JM, Birt DV 1972 Use of beta rays for early assessment of wood density development in provenance trials. Silvae Genet 21: 21–25

Harris JM, James RN, Collins MJ 1976 Case for improving wood density in radiata pine. N Z J For Sci 5: 347–354

Harris JM, Meylan BA 1965 The influence of microfibril angle on longitudinal and tangential shrinkage of Pinus radiata. Holzforschung 19: 144–153
Harris JM, Polge H 1967 A comparison of X-ray and beta ray techniques for measuring wood density. J Inst Wood Sci 19: 34–42
Hart CA, Hafley WL 1967 Estimation of wood fiber length from increment cores. Tappi 50: 615–618
Hart GA, Harding KJ 1990 Microcomputer assisted measurement of wood properties. Wood Struc Lab Queensland Dept For Indooroopilly, Queensland, 10 pp (Mimeo Rep) Proc IAWA/IUFRO Joint 2nd Pacific Regional Wood Anatomy Conf, Los Banos, Philippines, 12 pp
Harzmann L 1965 Information on the phenomenon of crossed spiral grain in the wood of Liquidambar formosana. Arch Forstwes 14: 801–806
Hawthorn JC 1961 An assessment of the use of growth rate as an indicator of specific gravity in hemlock and balsam. Thesis, School For Univ Brit Columbia Vancouver, Canada
Hedden RL 1991 Relation of Nantucket pine tip moth attack and fusiform rust infection in loblolly pine families. SJAF 15(4): 204–208
Heger L, Parker ML, Kennedy RW 1974 X-ray densitrometry: a technique and an example of application. Wood Sci 7: 140–148
Heikinheimo O 1951 Experiences in the growing of curly birch. Comm Inst For Fenn 39.5, Helsinki, Finland, 53 pp
Hejnowicz Z, Zagorska-Marek B 1974 Mechanism of changes in grain inclination in wood produced by storeyed cambium. Acta Soc Bot Pol 43: 381–398
Henderson J, Petty JA 1972 A comparison of wood properties of coastal and interior provenances of lodgepole pine, Pinus contorta. Forestry 45: 50–57
Herpka I 1979 Genetic variability of basic density of juvenile wood in Populus deltoides and its significance for selection. IUFRO Meet Conc Poplars, France & Belgium, Wageningen, Netherlands, 12 pp
Hildebrandt G 1960 The effect of growth conditions on the structure and properties of wood. 5th World For Congr Seattle, WA, 12 pp
Hiley WE 1955 Quality in softwoods. Q J For 49: 159–164
Hiller CH 1954 Variations in fibril angles in slash pine. USFS Rep 2003 FPL Madison, WI, 6 pp
Hiller CH 1964 Estimating the size of the fibril angle in latewood tracheids of slash pine. J For 62: 249–252
Hiller CH 1968 Trends of fibril angle variation in white ash. USFS Res Pap FPL 99 Madison, WI, 8 pp
Hirai 1958 Studies on weight growth of forest trees (VI) Chamaecyparis obtusa of the Tokyo University Forest in Chiba. Bull Tokyo Univ For 54: 199 217
Hodge GR, Purnell RC 1993 Genetic parameter estimates for wood density, transition age and radial growth in slash pine. Can J For Res 23: 1881–1891
Hodge GR, White T, Powell G 1992 Genetics of wood density characteristics in slash pine. Coop For Gen Res Prog 34th Prog Rep Univ Florida Gainesville, FL, 12–20
Holst MJ 1960 Forest tree breeding and genetics at the Petawawa Forest Experiment Station. 7th Meet Commit For Tree Breed Vancouver, Canada, K1-K27
Honda S 1912 Black heartwood of Cryptomeria. Dai-nippon Saurin Kaihoo (Bull Jpn For Assoc) 35: 50–51
Horn RA, Coens CL 1970 Rapid determination of the number of fibers per gram of pulp. Tappi 53: 2120–2122
Howard ET 1973 Heat of combustion of various southern pine materials. Wood Sci 5: 194–197
Howe GE 1969 Early results of a sugar maple improvement study. 11th Meeting Forest Tree Breeding For Branch Dept Fish For Canada Ottawa, Canada, p 286

Howell RA, Leach GN, Nutter DE, Steltenkamp MS 1984 Some age-associated trends of planted slash pine as related to kraft pulping. Symposium Utilization Changing Wood Resources Southern US, NC, 243–260

Hsu JK, Walters LS 1975 Effect of irrigation and fertilization on selected physical and mechanical properties of loblolly pine (Pinus taeda). Wood Fiber 7: 192–206

Huber DA, Bongarten B 1981 Specific gravity as a selection criterion in sycamore. 16th Southern Forest Tree Improvement Conference Blacksburg, VA, 320–327

Ifju G, McLain TE 1982 Quantitative wood anatomy based on stereological methods and its use for predicting paper properties. TAPPI Research Development Division Conference, Atlanta, GA 15–28

Ikemori YK, Martins FC, Zobel BJ 1986 The impact of accelerated breeding on wood properties. 18th IUFRO World Congr Ljubljana, Yugoslavia, 359–368

Isebrands JG, Hunt CM 1975 Growth and wood properties of rapid-grown Japanese larch. Wood Fiber 7: 119–128

Iwakawa M, Watanabe M, Sato T, Mikami S, Inuma M, Kida S 1967 Inheritance of some characters in open pollinated progenies of Pinus densiflora. Bull For Exp Stn Meguro, Japan 207: 31–67

Jackson LW 1959 Loblolly pine tracheid length in relation to position in tree. J For 57: 366-367

Jackson LW 1964 Fibril angle variation in open- and control-pollinated progeny of slash and loblolly pine. Bull Georgia Acad Sci 22(1): 4 pp

Jackson LW, Greene JT 1957 Methods for fiber measurements. J For 55: p 300

Jackson LW, Greene JT 1958 Tracheid length variation and inheritance in slash and loblolly pines. For Sci 4: 316–318

Jackson LW, Strickland RE 1962 Geographic variation in tracheid length and wood density of loblolly pine. GA Res Counc 8: 1–4

Jackson LW, Warren BJ 1962 Variation and inheritance in specific gravity of slash and loblolly pine progeny. Res Pap GA For Res Counc 14: 4 pp

Jayme G, Krause T 1963 Über die Packungsdichte der Zellwände in Laubhölzern. Holz Roh-Werkst 21: 14–19

Jeffers JW 1959 Regression models of variation in specific gravity in four provenances of Sitka spruce. J Wood Sci 4: 44–59

Jefferson PA, Yanchuk A 1985 Heritability and genetic relationships for wood quality characteristics – the key. Forintek Canada Corp Workshop Quebec, Canada, 1–9

Jesus RM, Vital BR 1986 Comparacao entre densidades de matrizes e de clones de Eucalyptus grandis. Rev Arvore 10(1): 82–90

Jett JB, McKeand SE, Weir RJ 1991 Stability of juvenile wood specific gravity of loblolly pine in diverse geographic areas. Can J For Res 21: 1080–1085

Jett JB, Talbert JT 1982 Place of wood specific gravity in the development of advanced generation seed orchards and breeding programs. South J Appl For 6: 177–180

Jett JB, Weir RJ, Barker JA 1977 The inheritance of cellulose in loblolly pine. TAPPI For Biol Comm Meet Madison, WI, 4 pp

Johansson K 1993 Influence of initial spacing and tree class on the basic density of Picea abies. Scand J For Res 8: 18–27

Johnson H 1960 Delineation of improvement objectives and their possible attainment in growth and stem form. 5th World For Cong Seattle, WA, 716–721

Johnson IG, Colley R 1991 Diameter-wood density relationships in radiate pine at three sites. Proceedings Meeting Representative Research Workers Group no. 1 (Forest Genetics) Australian For Council Coonawarra, South Australia, 141–143

Johnson JW, McElwee RL 1967 Geographic variation in specific gravity and three fiber characteristics of sweetgum. 9th Southern Forest Tree Improvement Conference Knoxville, TN, 8 pp

Johnsson H 1950 Offspring of curley birch. Årsberätt för Växtförädl Skogsträd, 18–29

Jourdain CJ, Olson JR 1982 Selection for wood quality in five-year-old American sycamore. North American Forest Biology Workshop Univ Kentucky Lexington, KY, 125–130

Jozsa LA, Brix H 1989 The effects of fertilization and thinning on wood quality of a 24-year-old Douglas-fir stand. Can J For Res 19: 1137–1145

Julien LM, Edgar JC, Conder TM 1971 A density-gradient technique for obtaining wood and bark chip density. US For Ser USDA North Central For Exp Stat Note NC-112 St Paul, MN, 4 pp

Kadambi K, Dabral SN 1955 On twist in Chir (Pinus longifolia). Indian For 81: 58–64

Kallmes OJ, Bernier GA 1963 Estimating the thickness of pulped wood fibres. Nature 197(4874): 1330

Kandell ES 1971 Polynomial models to study and present within-tree variation of wood properties. Wood Fiber 3: 106–111

Kane M, Urueña H, Dvorak W, Atehortua C 1993 The potential of Bombacopsis quinata as a commercial plantation species. For Ecol Manage 56: 99–112

Kanô T 1957 Forest-biological studies on wood quality on the dimension of bulk-density of wood and its distribution in the stem of Todo-fir. Bull Gov For Exp Stn Tokyo 101: 73–99

Kanô T 1961 The position in stems indicating the average values of bulk density. Bull Gov For Exp Stn Meguro, Japan 134: 141–148

Kanô T, Saito H 1970 On the selection of the basic density in the Japanese red pine (Pinus densiflora) wood. Mok. Gakk. 16: 305–309

Kanowski PJ 1986 Characterization and interpretation of variation in forest trees: a reappraisal based on a progeny test of Pinus caribaea. PhD Thesis, Oxford Univ, 221 pp

Kanowski PJ, Mather RA, Savill PS 1991 Genetic control of oak shake: some preliminary results. Silvae Genet 110: 166–168

Karki L 1978 The saw-log quality of pine cultivations can be best improved by breeding. Metsa Ja Puri: 20–25

Karki L 1980 Genetically narrow crowned and fine-branched trees are valuable in forestry. Found For Tree Breed Finland, Helsinki, Finland, Info Sheet 3, 3pp

Keating WG, Bolza E 1982 Characteristics, properties and uses of timbers. Texas A&M Univ Press College Station, TX

Keays JL, Hatton JV 1971 Complete-tree utilization studies II. Yield and quality of kraft pulp from the components of Picea glauca. Tappi 54: 1721–1724

Keiding H, Olsen HC 1965 Assessment of stem form in clones and in progenies of larch. Silvae Genet 14: 115–122

Keith CT, Kellog RM 1985 Wood quality considerations in tree improvement programs. Workshop Forintek Canada Corp Quebec, Canada, 77 pp

Keller R 1973 Caractéristiques du bois de pin maritime. Variabilité et transmission héréditaire. Ann Sci For 30: 31–62

Keller R, Azoef P, Hoslin R 1974 Tracer method for the measurement of spiral gain in standing trees. Ann Sci For 31: 161–169

Keller R, Thiercelin F 1975 The effect of the big wood rays upon some properties of beech wood. Ann Sci For 32: 113–129

Kellison RC 1967 A geographic variation study of yellow-poplar (Liriodendron tulipifera) within North Carolina. Tech Rep 33 NC State Univ School For Res Raleigh, NC, 41 pp

Kellison RC 1981 Characteristics affecting quality of timber from plantations, their determination and scope for modification. IUFRO Div 5, 17 World Congress Kyoto, Japan, 77–88

Kellogg RM 1990 An integrated forest management and resource quality research effort for coastal Douglas-fir in Canada. Proceedings Southern Plantation Wood Quality Workshop, Athens, GA, US For Ser Tech Rep SE-63, 75–81

Kellogg RM, Thykeson E, Warren WG 1975 The influence of wood and fiber properties on kraft converting-paper quality. Tappi 58: 113–116

Kellogg RM, van Buijtenen JP, Hatton JV, Wahlgren HE, Brink DL 1982 New methods of measuring wood and fiber properties in small samples. TAPPI Res and Dev Div Conf Asheville, NC, 1–2

Kennedy RW 1966 Intra-increment variation and heritability of specific gravity, parallel-to-grain tensile strength, stiffness and tracheid length in clonal Norway spruce. Tappi 49: 292–295

Kennedy RW 1969 Wood formation in jack pine seedlings of various provenances. 11th Comm For Tree Breed Canada Part 2 Ottawa, Canada, 101–109

Kennedy RW 1970 Specific gravity of early- and late-flushing Douglas-fir trees. Tappi 53: 1479–1481

Kennedy RW 1971 Influence of provenance and test location on wood formation in Pinus banksiana seedlings. Silvae Genet 20: 60–63

Kennedy RW, Smith JH 1959 The effects of some genetic and environmental factors on wood quality in poplar. Res Note No 19, Faculty For Univ Brit Columbia, 2 pp

Kennedy RW, Warren WG 1969 Within-tree variation in physical and chemical properties of Douglas-fir. 2nd World Consultation Forest Tree Breeding IUFRO, Washington DC, 19 pp

Kennedy RW, Wilson LW 1954 Studies on smooth and corkbark alpine fir II. Specific gravity comparison. Pulp Pap Mag Can, 3

Khalil MA 1985 Genetics of wood characters of black spruce (Picea mariana) in Newfoundland, Canada. Silvae Genet 34: 221–230

Kibblewhite RP, Bawden AD 1992 Kraft fibre qualities of Pinus radiata toplogs, thinnings and slabwood, and a "genetic misfit". N Z J For Sci 22: 96–110

Kibblewhite RP, Lloyd JA 1983 Interrelations between the wood, fibre, and kraft pulp properties of new-crop radiata pine corewood and slabwood. FRI Bull No 45 N Z For Ser Rotorua, New Zealand, 26 pp

King JN 1986 Selection of traits for growth, form and wood quality in a population of coastal Douglas-fir (Pseudotsuga menziesii var. menziesii) from British Columbia. PhD Thesis Univ Alberta, Edmonton

King JN, Yeh FC, Heaman J, Dancik BP 1988 Selection of wood density and diameter in controlled crosses of coastal Douglas-fir. Silvae Genet 37: 152–157

King JN, Yeh FC, Heaman JC, Dancik BP 1992 Selection of crown form traits in controlled crosses of coastal Douglas-fir. Silvae Genet 41: 362–370

King JP 1967 Seed source variation in tracheid length and specific gravity of five-year-old jack pine seedlings. 8th Lake States Forest Tree Improvement Conference US For Ser, St. Paul, Minnesota Res Pap NC–23, 5–9

King JP 1980 Variation in specific gravity in 3-year-old coppice clones of Eucalyptus saligna growing in Hawaii. Aust For Res 10: 295–299

Kininmonth JA, Williams DH 1977 Measuring the moisture content of wood. N Z For Ser Info Ser 60, 12 pp

Kinloch BB 1963 Observations on the effects of fusiform rust on specific gravity and fiber length in the wood of loblolly pine. Mimeo Rep School For NC State Univ Raleigh, NC, 4 pp

Kirk DB, Breeman LG, Zobel BJ 1972 A pulping evaluation of loblolly pine wood. Tappi 55: 1600–1604

Klein GS 1965 Variations in specific gravity of foreign softwood species and Norway spruce from south and west Norway. Medd Norsk Skogforsøksves 20(74): 137–169

Klein GS 1966 Increment cores as a basis for determining a number of properties of Picea abies. Norsk Skogbr 12: 448–449

Kleinschmit J, Knigge W 1967 Durch Umwald und Erbanlagen bedingte Variation der Trockensubstanzerzeugung, Struktur und Rohdichte junger Fichten (Picea abies). Allg Forst-Jagdztg 138: 189–198

Klem G 1942 Effect of planting space (Plantevstandens unflytelse pa granvedens og sulfitcellulosens kvalitet). Medd Norsk Skogforsøksves 28: 293 pp
Klem GA 1957 The quality of Norway spruce (Picea abies) of Norweign and German origin. Medd Norsk Skogforsøksves 48: 291–314
Klem GS 1965 Variation in specific gravity of foreign softwood species and Norway spruce from south and west Norway. Medd Norsk Skogforsøksves 20(74): 137–169
Klumpers J, Janin G, Becker M, Levy G 1993 The influence of age, extractive content and soil water on wood color in oak; the possible genetic determination of wood color. Am Sci For 50(Suppl 1): 4035–4095
Knigge W 1993 Giant sequoia (Sequoiadendron giganteum) in Europe. Holz Roh-Werkst 51: 145–155
Knigge W, Koltzenburg C 1965 The influence of timber qualities and ecological conditions on the cell sizes and on the proportions of types of cells in hardwoods in the temperate zones. IUFRO Sect 41 Committee Fibre Characteristics Melbourne, Australia, 2, 51 pp
Knudsen MV 1958 A comparative study of some technological properties of Norway spruce in a provenance test. IUFRO 12 Congr Proc 4: 219–228
Koch P 1972 Utilization of the southern pines. Vol I The Raw Material. Vol II Processing. US For Ser Agr Handb 420, Supt of Documents, Washington, DC, 1663 pp
Koeler A 1939 Heredity versus environment in improving wood. J For 37: 683–687
Koltzenburg C 1966 Die Abhängigkeit der Holzeigenschaften der Rotbuche (Fagus sylvatica) von Lichtgenuss, sociologischer Stellung und anderen Wuchsbedingungen. Thesis, Forst Fak Univ Göttingen, Hann Münden, 121 pp
Koo KM, Hong SM 1967 Tracheid length and compression strength of X Pinus rigitaeda and its back-cross hybrid. Inst For Genet Suwon, Korea, Res Rep 5: 85–90
Kozlowski TT 1963 Growth characteristics of forest trees. J For 61: 655–661
Krahl-Urban 1960 Some results of the investigation into spiral grain in beech. Silvae Genet 9: 139
Krahmer RL, Snodgrass JD 1967 Selected physical properties of wood in a forest. For Prod J 17(3): 21–29
Krempl H 1965 Spiral growth in the log and branchwood of spruce. IUFRO Sect 41 Melbourne, Australia, 1–9
Krempl H 1970 Analyses of spiral gain in spruce. Mitt Forst. Bundes-Versuchsanst., Wien, 116 pp
Kretschmann DE, Moody RC, Pellerin RF, Bendtsen BA, Cahill JM, McAllister RH, Sharp DW 1983 Effect of various proportions of juvenile wood on laminated veneer lumber. U S For Ser Res Pap FPL-RP-521, Madison, WI 100 pp
Kristinic A 1968 Variability and heritability of height and number of branches in Salix alba populations at Bakovci and Lipovljani. Part II Sum List 92(1/2): 27–49
Kromhout CP, Toon RE 1978 Variation of wood properties of some tropical species in southern Africa. Workshop Genetic Improving Tropical Trees Brisbane, Australia, 8–45
Ku Y, Su Y, Kuo Y 1991 Wood characteristics and pulping potential of *Gmelina arborea*. Proc Tropical Pulp Conf Jakarta, Indonesia 1: 17–31
Kucêra B 1994 A hypothesis relating current annual height increment of juvenile wood formation in Norway spruce. Wood Fiber Sci 26: 152–167
Kurinobu S 1990 Ann. Rep.: Kanto Forest Tree Breeding Institute, Japan 22: 41–48
Kurinobu S, Yomogida H, Yamaguchi K 1990 Ann. Rep. For Tree Breed Inst 24 Mito, Japan, 7–24
LaBorde AP 1966 The effect of geographic seed source on the specific gravity of planted loblolly pine in southeastern Louisiana. MF Thesis, Louisiana State Univ Baton Rouge, LA, 21 pp
Lacaze J, Polge H 1970 Relations phénotypiques au stade juvénile entre la densité du bois et divers caractères phénologiques et de vigeur chez Picea abies. Ann Sci For 27: 231–242

Ladell JL 1959 A method of measuring the amount and distribution of cell wall material in transverse microscope sections of wood (Pinus strobus). J Inst Wood Sci 3: 43–45
Ladrach W 1984 Growth of Leucaena in the Cauca Valley and the Atlantic Coast after three years. Investigation Forestal Carton de Colombia Cali, Colombia, 6 pp
Ladrach WE 1986 Wood quality of Pinus kesiya. Carton de Colombia Cali, Colombia, Res Rep 111, 11 pp
Ladrach WE, Lambeth C 1991 Growth and heritability estimates for a seven-year-old open-pollinated Pinus patula progeny test in Colombia. Silvae Genet 40:169–173
Ladrach W, Zobel B 1986 Avances de la silvicultura en plantaciones forestales y su impacto en el desarrollo economica de los paises. Centro de Convenciones y Exposiciones Gonzalo jiménez de Quesada. Bogota, Colombia, 12 pp (mimeo)
Lahiri FA 1959 Possibilities of improving wood quality through breeding. Indian For 85: 327–334
Lamb GN 1940 Figure in wood. Northern Nut Growers Assoc Ann Rep, Lafayette, Indiana, 28–33
Lambeth CC 1980 Juvenile-mature correlations in Pinaceae and implications for early selection. For Sci 26: 571–580
Land SB, Dicke SG, Tuscan GA, Patterson PE 1983 Genetic, site and within-tree variation in specific gravity and moisture content in young sycamore trees. Tappi 66(?): 149–155
Land SB, Lee JC 1981 Variation in sycamore wood specific gravity. Wood Sci 13: 166–170
Lange PW 1959 The morphology of hardwood fibers. Tappi 42:786–792
Langlet O 1959 A cline or not a cline – a question of Scots pine. Silvae Genet 8: 13–22
Lannan PT 1960 Pulpwood properties: response of processing and of paper quality to their variation. Tappi 43: 40–47
Lantz CW, Hofmann JG 1969 Geographic variation in growth and wood quality of loblolly pine in North Carolina. 10th Southern Conference Forest Tree Improvement Houston, TX, 175–188
Larson PR 1960 A physiological consideration of the springwood-summerwood transition in red pine. For Sci 6: 110–122
Larson PR 1963 Stem form development of forest trees. For Sci Mono 5, 42 pp
Larson PR 1973 The physiological basis for wood specific gravity in conifers. IUFRO Div 5 Meet Brisbane, Australia 2: 672–680
Lee CH 1974 Geographic variation of growth and wood properties in eastern white pine – 15 year results. 21st Northeast Forest Tree Improvement Conference Univ New Brunswick Fredericton, New Brunswick, 36–41
Lee CH 1975 Geographic variation of growth and wood properties in Japanese larch in southwestern lower Michigan. 12th Lake States Forest Tree Improvement Conference, Michigan, 35–46
Lee CH 1979 Absence of growth-wood property correlation in twenty-seven black pine seed sources. Wood Fiber 11: 22–28
Lee CH, Wahlgren HE 1979 Specific gravity sampling with minimal tree damage – a study of red pine. Wood Sci 11: 241–245
Levi MP 1978 Blue-flecked panneling: a new market for southern pine beetle-killed trees. South Lumberman 237(2944): 70–71
Levi MP 1981 A guide for using beetle-killed southern pine based on tree appearance. South pine beetle handbook US For Serv Agric Handbook New Orleans, LA, 572, 18 pp
Lewark S 1980 Vergleichende Untersuchung der Jahrringbreite und Rohdichte von Stecklingen und Sämlingen der Lärche. Forstarchiv 51: 47–54
Lewark S 1981 Variabilität der Ästigkeit und Rundholzqualität der Fichte. Holz-Zbl 107: 275–277
Lewark S 1982a Studies of wood characteristics of young Norway spruce from clonal experiments. Forstarchiv 53: 14–21

Lewark S 1982b Variability, heritability and response to selection of wood characteristics in a Norway spruce clonal experiment. IUFRO Jt Mtg Work Parties Genetic Breeding Strategy Including Multiclonal Varieties. Sensenstein, West Germany, 25 pp

Lewark S 1982c Untersuchungen von Holzmerkmalen junger Fichten aus Stecklingsklonen. Forstarchiv 53: 14–21

Lewark S, Hamza KF 1992 Juvenile wood of four Eucalyptus species grown at Ruvu Tanzania. Proc Div 5 IUFRO Conf Forest Production, Vol I, Nancy, France, p 196

Liese W, Ammer U 1958 Investigations on the length of wood fibers in poplars. Holzforschung 11:169–174

Liese W, Ammer U 1962 Anatomical studies on extreme twisted pine-wood. Holz Roh-Werkst. 20: 339–346

Lima RT 1987 Stability of genotypes of Pinus oocarpa for specific gravity in three different environments in South America. MS Thesis Coll For Res NC State Univ Raleigh, NC, 48 pp

Lima RT, Jett JB, Dvorak WS 1990 Family stability of wood specific gravity in Pinus tecunumanii established on three sites in South America. New For 3: 311–316

Littleford TW 1961 Variation of strength properties within trees and between trees in a stand of rapid-growth Douglas-fir. For Prod Lab Can Vancouver, Canada

Loo-Dinkins JA, Fowler DP, Schneider MH 1982 Geographic variation in specific gravity among Japanese larch from different provenances. Wood Fiber 14: 281–286

Loo-Dinkins JA, Gonzalez JS 1991 Genetic control of wood density profile in young Douglas-fir. Can J For Res 2: 935–939

Loo-Dinkins JA, Tauer CC, McNew RW 1985 Genetic variation in the time of transition from juvenile to mature wood in loblolly pine (Pinus taeda). Silvae Genet 34: 14–19

Loo-Dinkins JA, Tauer CG, van Buijtenen JP 1984 Juvenile-mature relationships and heritability estimates of several traits in loblolly pine (Pinus taeda). Can J For Res 14: 822–825

Loo-Dinkins JA, Ying CC, Hamm EA 1991 Stem volume and wood relative density of a non-local Douglas-fir provenance in British Columbia. Silvae Genet 40: 29–35

Lowe WJ, Greene TA 1990 Geographic variation in specific gravity and fiber length of green ash (Fraxinus pennsylvanica) in East Texas. Silvae Genet 39: 194–198

Lowerts GA, Kellison RC 1981 Genetically controlled resistance to discoloration and decay in wounded trees of yellow poplar. Silvae Genet 30: 98–101

Lowery DP 1966 A spiral grain classification system and its application. For Prod J 16(1): 47–50

MacDonald RG, Franklin JN 1969 Pulp and paper manufacture. 2nd edn, vol 1: The pulping of wood. McGraw-Hill, New York, 769 pp

MacLennan L 1990 Annual research report. Inst For Comm For Res Univ Natal Pietermaritzburg, Rep South Africa, p XVIII

Malan FS 1988 Genetic variation in some growth and wood properties among 18 full-sib families of South African grown Eucalyptus grandis. A preliminary investigation. S Afr For J 146: 38–43

Malan FS 1989 The wood properties of SA grown E. grandis. Wood S Afr. 14: 61–71

Malan FS 1993 The quality and wood properties of four provenances of South African grown Pinus tecunumanii. Mimeo. Paper, CSIR, Pretoria, S. Africa, 22 pp. Submitted anales des Sciences Forestieres

Marian JE, Stumbo DA 1960 A new method of growth ring analysis and the determination of density by surface texture measurements. For Sci 6(3): 276–291

Martin JW 1984 Forest management practices that will influence product characteristics of the changing wood resource. Symposium Utilization Changing Wood Resources Southern US NC State University, Raleigh, NC, 9 pp

Marton J 1967 Determination of lignin in small pulp and paper samples using the acetyl bromide method. Tappi 50: 335–337

Marton R, Stairs GR, Schreiner EJ 1967 Influence of growth rate and clonal effects on wood anatomy and pulping properties of hybrid poplars. TAPPI 4th For Biol Conf, Appleton, WI 183–201
Marts RO 1955 Wood and fiber structure by incident fluorescence microscopy. J Biol Photogr Assoc 23(4): 151–155
Matheson AC, Turner CH, Dean GH 1986 Genetic variation in the pulp qualities of Eucalyptus obliqua. Appita 39: 205–212
Matziris DI, Zobel BJ 1973 Inheritance and correlation of juvenile characteristics in loblolly pine (P. taeda). Silvae Genet 22: 38–45
May MN 1958 Some considerations regarding the problem of defining wood quality objectives for tree improvement programs. 34th Forest Biology Committee Meeting TAPPI Madison, WI, 5 pp
McAlister RH, Clark A 1991 Effect of geographic location and seed source on the bending properties of juvenile and mature loblolly pine. For Prod J 41(9): 39–42
McAlister RH, Clark A 1992 Shrinkage of juvenile and mature wood of loblolly pine from three locations. For Prod J 42(7/8): 25–28
McBride CF 1967 Some economic losses in producing lumber from spiral grained logs. IUFRO Sect 41 München, Germany, 1–12
McCutchan BG 1982 Estimated gain in form, quality and volumetric traits in American sycamore in response to the selection for tree dry weight. MS Thesis School For Res NC State University, Raleigh, NC, 82 pp
McElwee RL 1963 Genetics in wood quality improvement. Southern Conference Tree Improvement. Gulfport, Mississippi, 21–24
McGee CE 1968 Northern red oak seedlings growth varies by light intensity and seed source. US For Ser Res Note SE-90, Asheville, NC, 4 pp
McKimmy MD 1959 Factors related to variation in specific gravity in young-growth Douglas-fir. Oregon For Prod Res Cent Bull 8, 52 pp
McKimmy MD 1966 A variation and heritability study of wood specific gravity in 46-year-old Douglas-fir from known seed sources. Tappi 49: 542–549
McKimmy MD 1978 Heritability and prediction of genetic gain for average wood density in Douglas-fir. 32nd Annual Meeting for Forest Product Research Society Corvalles, OR, OP, p 34
McKimmy MD, Campbell RK 1982 Genetic variation in the wood density and ring width trends in coastal Douglas-fir. Silvae Genet 31: 43–55
McKimmy MD, King JP 1978 Strength properties in young ponderosa pine of known parentage. 32nd Annual Meeting Forest Products Research Society Atlanta GA Session 8, p 8
McKimmy MD, King JP 1980 Strength relationships in young ponderosa pine of known parentage. Wood Sci 12: 165–167
McKimmy MD, Nicholas DD 1971 Genetic differences in wood traits among half-century-old families of Douglas-fir. Wood Fiber 2: 347–355
McKimmy RJ 1987 Wood properties of Douglas-fir grown in Victoria. Aust For 50: 81–85
McKinley CR, Lowe WJ, van Buijtenen JP 1982 Genetic improvement of wood specific gravity in loblolly pine (P. taeda) and its relation to other traits. TAPPI Research and Development Conference Asheville, NC, 153–158
McMillan CW 1973 Fibril angle of loblolly pine wood as related to specific gravity, growth rate and distance from the pith. Wood Sci Technol 7: 251–255
McNab WH 1983 Total tree and product weight of beetle-killed loblolly pines in northeast Georgia. GA For Res Pap 42: 11 pp
Megraw RA 1985 Wood quality factors in loblolly pine. TAPPI Press, Atlanta, GA, 89 pp
Megraw RA 1986 Effect of silvicultural practices on wood quality. TAPPI Research and Development Conference Raleigh, NC, 27–34
Meier RJ, Goggans JF 1977 Heritabilities of height, diameter and specific gravity of young Virginia pine. For Sci 23: 450–456

Mergen F 1954 Inheritance of oleoresin yield in slash pine. Naval Stores Review Lake City, FL (Dec Issue): 2 pp

Mergen F, Burley J, Yeatman CW 1964 Variation in growth characteristics and wood properties of Norway spruce. Tappi 48: 499–504

Mergen F, Furnival GM 1960 Discriminant analysis of Pinus thunbergii × Pinus densiflora hybrids. Proc Soc Am For, Washington, DC 36–40

Mergen F, Hoekstra PE, Echols RM 1955 Genetic control of oleoresin yield and viscosity in slash pine. For Sci 1: 19–30

Meyer-Uhlenreid KH 1958 Investigations on the inheritance of an anatomical characteristic in crosses of poplars from different sections. Züchter 28: 209–216

Meyer-Uhlenreid KH 1959 Inheritance of wood-fibre lengths in various species of Populus. Züchter 29: 117–123

Meylan BA, Probine MC 1969 Microfibril angle as a parameter in timber quality assessment. For Prod J 19(4) 30–34

Mikami S 1973 Preliminary studies on the genetic improvement in spiral grain in Japanese larch. IUFRO Meet Div 5 Brisbane, Australia, pp 746–751

Miler Z, Miler A, Pasternak B 1979 Wood fibre length in pine provenance samples. Pr Komisji Nauk Rolniezychi Komisji Nauk Lesnych 48: 95–101

Miller SR 1961 Variation in bark thickness of slash pine by site, tree size and distance up stem. Woodland Res Note No 10 Union Bag Camp Paper Corp Savannah, GA, 3 pp

Mirza A, Strobl S 1991 Specific gravity – another selection trait for a hybrid poplar clonal screening program? Can Tree Impr Assoc Maintaining Diversity Ottawa, Canada, 13, 14

Mitchell HL 1955 Breeding for high-quality wood. USFS – For Prod Lab No 2050 Madison, WI, 13 pp

Mitchell HL 1960 Development of an adequate concept of wood quality for the guidance of geneticists and forest managers. 5th World For Congr Seattle, WA, 1341–1348

Mitchell HL 1963 Specific gravity variation in North American conifers. IUFRO Sect 41 Committee on Fiber Characteristics Melbourne, Australia, vol 2, 15 pp

Mitchell KJ 1961 Apical growth in relation to the transition from springwood to summerwood in conifers. Univ Brit Col For Club 19, Vancouver, B.C., 71 pp

Mohrdiek O 1979 Holzdichte und Faserlänge bei Populus-Leuce-Arten. Holzzucht 33(3/4): 23–26

Mork E 1928 Die Qualität des Fichtenholzes unter besonderer Rücksichtnahme auf Schleif- und Papierholz. Papier-Fabr 26: 741–747

Mothe F 1983 Study of the genetic variability between and within populations of wood quality and growth in common spruce. Centre National de Recherches Forestières Univ Nancy, France, 111 pp

Mottet A 1965 Considerations on the relationship between certain anatomical characteristics and density in tropical woods. IUFRO Sect 41 Melbourne, Australia, 15 pp

Müller-Stoll WR 1947 Photometric investigation of wood structure. Part I. The determination of the annual ring structure and the proportion of latewood by photometric methods. USDA translation No 20, 22 pp

Nanson A, de Jamblinne A, Istas JR, Schalk J 1975 Inheritance of wood density in Pinus nigra. Bull Soc R For Belg 82: 68–79

Navarro de Andrade E 1961 O Eucalipto. 2nd edn FAO, World Conference on Eucalyptus Saõ Paulo, Brazil, 664 pp

Nebgen RJ, Lowe WJ 1982 Inheritance of growth, branch angle and specific gravity in three American sycamore populations. Silvae Genet 31: 86–89

Nelson ND, Burk T, Isebrands JG, Ek AR 1979 Clone and spacing effects on branch characteristics of short rotation intensively cultured Populus. 1st North Central Tree Improvement Conference Madison, WI, 74–85

Nepveu G 1976 Wood quality juvenile-mature correlations. Tree Breed Sta Bordeaux, France, 16 pp

Nepveu G 1984a Contrôle héréditaire de la densité et de la rétractibilité du bois de trois espèces de chêne (Quercus petraea, Quercus robur et Quercus rubra). Silvae Genet 33: 110–115
Nepveu G 1984b Determinisme genotypique de la structure anatomique du bois chez Quercus robur. Silvae Genet 33: 91–95
Nepveu G, Birot Y 1979 Les corrélations phénotypiques juvénile-adulte pour la densité du bois et al vigueur chez l'épicéa Ann Sci For 36: 125–149
Nepveu G, Keller R, Tessier du Cros E 1978 Sélection juvénile pour la qualité du bois chez certains peupliers noirs. Ann Sci For 35: 69–72
Nepveu G, Teissier du Cros E 1976 Juvenile selection for wood quality in the euramerican poplars. IUFRO Meeting Genetic Improvement Bordeaux, France, 20 pp
Nepveu G, Velling P 1983 Variabilité génétique individuelle dans la qualité du bois chez Betula pendula. Silvae Genet 32: 37–49
Nicholls JW 1963 A note on the heritability of heartwood formation in Pinus radiata. Lab Rep No 1 Study on microscopic and sub-microscopic structure of wood in relation to properties. Wood Prod Lab CSIRO Melbourne, Australia, 6 pp
Nicholls JW 1965 Preliminary observations on the change with age of the heritability of certain wood characteristics in *Pinus radiata* clones. Silvae Genet 16: 18–20
Nicholls JW 1966 Preliminary observations on some aspects of the heritability of wood characteristics. Mimeographed Report, CSIRO, Melbourne, Australia, 16 pp
Nicholls JW 1967a Assessment of wood qualities for tree breeding IV. Pinus pinaster grown in western Australia. Silvae Genet 16: 21–28
Nicholls JW 1967b Selection criteria in relation to wood properties. Australian For Res Work Group No 1 Forestry House Canberra, Australia
Nicholls JW 1967c Preliminary observations on the change with age of the heritability of certain wood characters in Pinus radiata clones. Silvae Genet 16: 18–20
Nicholls JW 1967d Spiral grain in wood. British Columbia Lumberman, Vancouver BC pp 1–3
Nicholls JW 1967e Preferred methods for the investigation of spiral grain. Mimeo Paper, CSIRO, Melbourne, Australia, 5 pp
Nicholls JW 1967f Some aspects of the genetic and environmental control of spiral grain in Pinus species. Mimeo Rep For Prod Lab CSIRO Melbourne, Australia, 1–10
Nicholls JW 1977 Notes on a field method for measuring wood density. Breeding Pinus radiata IUFRO Work Party, Melbourne, Australia Newsletter No 1, 4–5
Nicholls JW, Brown AG 1971 The ortet-ramet relationship in wood characteristics of Pinus radiata. Appita 25: 200–209
Nicholls JW, Brown AG 1974 The inheritance of heartwood formation in Pinus radiata. Silvae Genet 23: 138–141
Nicholls JW, Dadswell HE 1961 Structure in relation to silvicultural treatment of Pinus radiata. CSIRO Div Tech Pap For Prod, CSIRO Melbourne, Australia, 11–12
Nicholls JW, Dadswell HE, Fielding JM 1964 The heritability of wood characteristics of Pinus radiata. Silvae Genet 13: 68–71
Nicholls JW, Dadswell HE, Perry DH 1963 Assessment of wood qualities for tree breeding. Silvae Genet 12: 105–110
Nicholls JW, Morris JD, Pederick LA 1980 Heritability estimates of density characteristics in juvenile radiata wood. Silvae Genet 29: 54–61
Nicholls JW, Pawsey CK, Brown AG 1976 Further studies on the ortet-ramet relationship in wood characteristics of Pinus radiata. Silvae Genet 25: 73–79
Nicholson J 1967 Repeatability of tracheid characteristics in Douglas-fir seedlings grown under two photoperiods. 14th Northeastern Forest Tree Improvements Conference Univ Toronto, Canada, 43–49
Nilsson B 1963 Wood density in progenies after hybridization within Picea abies. World Consultation Forest Genetic Tree Improvement Stockholm, Sweden, Sect 7/8: 1–4
Nobori Y, Fukazawa K 1988 The analysis of density variation for one stem from high

age planted stand of larch by X-ray densitometric system. Tech Note No 259 Oji Inst For Tree Improv Kuriyama Hokkaido, Japan, 15–18

Noh ER, Lee SK, Koo YB 1986 Compartmentalizing ability of poplar clones for discoloration and decay associated with artificial wounds. Res Rep Inst For Genet Suwon, Korea 22: 21–25

Noh ER, Lee SK, Koo YB, Kim CS 1987 Compartmentalizing ability of a hybrid poplar, Populus nigra × P. maximowiczii F_1 clones for wood discoloration and decay associated with artificial wounds. Res Rep Inst For Genet Suwon, Korea 23: 16–22

Noh ER, Hyun SK, Jo JM, Cho RM, Lee SK, Ahn JK, Kim JJ 1984 Activities relative to poplar breeding, cultivation, exploitation and utilization. (1980–1984). Res Rep Inst For Genet Suwon, Korea 20: 16–45

Northcott PL 1957 Is spiral grain the normal growth pattern? For Chron 33: 335–352

Northcott PL 1958 Spiral grain in wood. British Columbia Lumberman, 3 pp

Noskowiak AF 1963 Spiral grain in trees: a review. For Prod J 13: 266–275

Noskowiak AF 1968 Spiral grain patterns from increment cores. For Prod J 18(4): 57–60

Nylinder P 1953 Variation in density of planted spruce. Medd Statens Skogsforskningsinst 43(3): 1–43

Nylinder P 1965 Non-destructive field sampling systems for determining the wood density of standing timber over large areas, variation within and between species, and the influence of environmental and other measurable factors on wood density. IUFRO Meet Sect 41 Melbourne, Australia, 1–25

Ohkura S 1960 An approach to the causes of spiral grain in trees. Bull Shinshu Univ For 2 (Engl Summary Univ Calif School For 25: 11–12)

Ohta S, Fujisawa Y 1992 Wood characteristics and genetic variation in sugi (Cryptomeria japonica). Proc Div 5 IUFRO Conf For Prod vol I Nancy, France, p 254

Okkonen EA, Wahlgren HE, Maeglin RR 1972 Relationships of specific gravity to tree height in commercially important species. For Prod J 22(7): 37–41

Okwuagwu CO, Guries RP 1980 Estimates of general and specific combining ability and heritability for juvenile wood specific gravity and tracheid length in jack pine. 27th Northwestern Forest Tree Improvement Conference, Moscow, Idaho, 128–137

Olesen PO 1971 The water displacement method. A fast and accurate method of determining the green volume of wood samples. For Inst The Royal Veterin-Agric Univ of Copenhagen, 57 pp

Olesen PO 1977 The variation of the basic density level and tracheid width within the juvenile and mature wood of Norway spruce. Forest Tree Improvement 12, Royal Veterin-Agric Univ Copenhagen, 21 pp

Olson JR, Jordain CJ, Rousseau RJ 1985 Selection for cellulose content, specific gravity and volume in young Populus deltoides clones. Can J For Res 15: 393–396

Orme RK 1992 Improvement of Eucalyptus globulus and Eucalyptus nitens for pulpwood production. Prod Tech for Genetically Improved Fast Growing For Tree Species AFOCEL-IUFRO Bordeaux, France, 36–42

Osorio LF, Dvorak WS 1993 Volume and wood density results for Pinus tecunumanii at 8 years of age in Colombia. Southern Forest Tree Improvement Committee Atlanta, GA, 348–356

Otegbeye GO, Kellison RC 1980 Genetics of wood and bark characteristics of E. viminalis. Silvae Genet 29: 27–31

Oteng-Amoako A, de Zeeuw CH, Meyer RW 1983 Variation of selected anatomical properties and specific gravity in the stem of naturally grown and plantation grown Triplochiton scleroxylon from Ghana. IUFRO Conf Div 5 Madison, WI, 45–46

Palmer ER 1977 The potential of fast growing tropical pines for pulp and paper. Trop Sci 19: 127–140

Palmer ER, Johnson JS, Gibbs JA, Ganguli S, Dutta AP 1984 Pulping characteristics of Gmelina arborea grown in plantations in the Solomon Islands. Trop Dev Res Inst L65, London 21 pp

Panshin AM, de Zeeuw CH 1980 Textbook of wood technology, 4th edn, McGraw-Hill, New York, 722 pp
Parker ML, Jozsa LA 1973 X-ray scanning machine for tree ring width and density analyses. Wood Fiber 5: 192–197
Parker ML, Kennedy RW 1973 The status of radiation densitometry for measurement of wood specific gravity. IUFRO Meet vol 2 Brisbane, Australia, 882–893
Parrot L 1960 Genetic variation in the density of Picea abies wood. Ann Ec Nat Eaux For 17: 267–334
Paterson DN 1967a Spiral grain in East African exotic softwood forest plantations. IUFRO Conf München, Germany, 14 pp
Paterson DN 1967b Control of wood quality and quantity in the East African exotic softwood tree breeding programme. IUFRO Conf München, Germany, 25 pp
Paul BH 1953 Forest genetics in relation to wood quality. Proc Lake States Forest Genetic Conference Misc Rep 22: 55–59, Saint Paul, MN
Paul BH 1956 Specific gravity of Populus species and hybrids. US For Serv, FPL Rep N 2060 Madison, WI, 17 pp
Pawsey CK 1965 Timber sampling from standing trees. Aust For Res 1(3): 33–36
Pearson FG, Fielding HA 1961 Some properties of individual growth rings in European larch and Japanese larch and their influence upon specific gravity. Holzforschung 15: 82–89
Pearson FG, Weir RJ, Smith WD 1980 Utilization of pine thinnings. Southern Forest Economics Worker Conference Univ South Mississippi Long Beach, MS, 27 pp
Pearson RG, Gilmore RC 1980 Effect of fast growth rate on the mechanical properties of loblolly pine. For Prod J 30(5): 47–54
Pederick LA 1970 Variation and inheritance of stem form and bark thickness in young loblolly pine. Tech Rep 41 NC State Univ School For Raleigh, NC, 44 pp
Pederick LA 1971 Inheritance of spiral grain in young radiata pine. IUFRO Conf Sect 41 Gainesville, FL, 9 pp
Pelle DN 1967 A special style for spiral grain estimation of Pinus radiata. Aust For Res 3(1): 54
Pereira H 1992 The raw-material properties of very young and fast grown eucalypts. Proc Div 5 IUFRO Conf For Prod vol I Nancy, France, p 250
Perry TO 1960 The inheritance of crooked stem form in loblolly pine (Pinus taeda). J For 58: 943–947
Persson A 1972 Studies on the basic density in mother trees and progenies of pine. Studia Forestalia Suecica No. 96 Skogshøgsh, Stockholm, 37 pp
Persson A, Downie B 1992 Variation in bark thickness of young Pinus contorta var. latifolia in Sweden. Scand J For Res 7: 99–106
Peterson TA 1968 Variation in radial growth patterns and specific gravity of red pine (Pinus resinosa). For Prod J 18(11): 63
Petrini S 1959 The two oldest Swedish provenance tests in Scots pine. Medd Statens Skogsforskningsinst 48(11): 1–49
Pew JC, Knechtges RG 1989 Cross-sectional dimensions of fibers in relation to paper-making properties of loblolly pine. Pap Trade J 109: 1–3
Phelps JE, Isenbrands JG, Teclaw RM 1983 Raw material evaluations of short cultured coppice. IAWA Bull 4: 10 (Summary)
Phelps JE, Rink G, Workman EC 1992 Color studies of black walnut: past and recent efforts. Proc Div 5 IUFRO Conf For Prod vol I Nancy, France, p 203
Phelps JE, Workman EC 1992 Vessel area studies in black walnut (Juglans nigra). Wood Fiber Sci 24: 60–67
Phillips EW 1963 Timber improvement by tree selection and breeding. 1st World Consul For Gen Tree Improv Stockholm, Sweden, 5 pp
Phillips EW 1965a A consideration of the use of the beta-ray technique in determining cell dimensions. IUFRO Meet Sect 41 Melbourne, Australia, 7 pp

Phillips EW 1965b Methods and equipment for determining the specific gravity of wood. IUFRO Meet Sect 41 Melbourne, Australia, 14 pp
Pillow MY 1954 Specific gravity relative to characteristics of annual rings in loblolly pine. US For Ser FPL No 1989 Madison, WI, 21 pp
Pillow MY, Bray MW 1935 Properties and sulphate pulping characteristics of compression wood. Pap Trade J 1–3
Pillow MY, Chidester GH, Bray MW 1941 Effect of wood structure on properties of sulphate and sulphite pulps from loblolly pine. South Pulp Pap J 4(7): 6–12
Plank ME 1984 Lumber recovery from insect-killed lodgepole pine in the northern Rocky Mountains. US For Ser Res Paper PNW-320, Portland, Oregon, 12 pp
Plumptre RA 1977 Variation of growth, stem quality and wood properties in relation to genetic improvement of tropical forest trees. 3rd World Consul Forest Tree Breeding Canberra, Australia, 28 pp
Polge H 1963a Density analysis with radiograms. Ann Ec Nat Eaux For 20: 533–581
Polge H 1963b Contribution to the study of the quality of wood of the principal exotic coniferous species used in French reforestation. Ann Ec Nat Eaux For, 20(3): 403–460
Polge H 1967 Determination of fiber length using 5 mm increment cores. Tappi 50: 460–462
Polge H 1968 Tests précoces de la qualité du bois sur 25 provenances d'Abies grandis. Ann Sci For 25: 3–23
Polge H 1980 Wood quality appraisal through non-destructive testing. IUFRO Meet Sect 41 Oxford, England, p 52
Polge H, Garros S, Michel P 1971 Inheritance of specific gravity in four-year-old seedlings of silver fir. Ann Sci For 28: 185–194
Polge H, Illy G 1967 Observations sur l'anisotropie du pin maritime des Landes. Ann Sci For 24: 205–231
Polge H, Illy G 1968 Study on inheritance of specific gravity and correlations with growth by non-destructive tests on four-year-old seedlings. Silvae Genet 17: 173–181
Polge H, Keller E 1970 First appraisal of wood quality (in situ) by using torque measurements. Ann Sci For 198–223
Polge H, Keller R 1973 Qualité du bois et largeur d'accroissements en forêt de Trançais. Ann Sci For 30: 91–125
Polk RB 1972 Heritabilities of some first-order branching traits in Pinus banksiana. 8th Central States Forest Tree Improvement Conference Columbia, MO, 33–39
Posey CE 1969 Phenotypic and genotypic variation in eastern cottonwood in the southern great plains. 10th Southern Conference Forest Tree Improvement Houston, TX, 130–135
Posey CE, Bridgwater FE, Walker N 1970 Effect of seed origin on tracheid length, specific gravity and volume of shortleaf pine in Oklahoma. For Sci 16: 66–71
Powers HR 1992 Loblolly pine seed sources differ in susceptibility to the southern pine beetle in South Carolina. South J Appl For 16: 169–174
Powers HR, Mathews FR 1987 Five fusiform rust-resistant seed sources in coastal South Carolina: a field comparison. South J Appl For 11: 198–201
Prestemon DR 1965 Improving the power increment borer for hardwoods. J For 63: 763–765
Prestemon DR 1967 Variation in heart stain and density in tanoak. For Prod J 17(7): 33–41
Pryor LD 1950 A hybrid Eucalyptus. Proc Linn Soc N S W 75(1–2): 96–101
Pryor LD 1951 A genetic analysis of some Eucalyptus species. Proc Linn Soc N S W 76(3–4): 140–147
Pryor LD, Chattaway MM, Kloot HH 1956 The inheritance of wood and bark characteristics in Eucalyptus. Aust J Bot 4: 216–239
Purkayastha SK, Tandon RD, Rao KR 1973 A note on the variation in wood density in some 36-year-old teak trees from different seed origins. Indian For 99: 215–217

Purnell RC 1988 Variation in wood properties of Eucalyptus nitens in a provenance trial on the Eastern Transvaal Highveld in South Africa. S Afr For J 144: 8–22
Pyszynki W 1977 Complex wavy grain in the stem of Aesculus. Acta Soc Bot Pol 46: 231–249
Quirk JT 1975 Dot-grid integrating eyepiece; 10 sampling techniques for estimating cell wall areas. Wood Sci 8: 88–91
Quirk JT, Smith D 1975 Comparison of dual linear and dot-grid eyepiece methods for estimating wood properties of Douglas-fir. Wood Sci 8: 92–96
Reck SG 1974 Growth and wood characteristics in clones of Populus trichocarpa grown in West Germany and their significance for selection. Proc IUFRO Meet Session II Stockholm, Sweden, 129–146
Rees LW, Brown RM 1954 Wood density and seed source in young plantations of red pine. J For 52: 662–665
Reid JS 1963 Wood quality in rapidly growing conifers planted in New Zealand. IUFRO Sect 41 Committee on Fibre Characteristics, Melbourne, Australia, 1–38
Remröd J 1976 Choosing Scots pine (Pinus sylvestris) provenances in Northern Sweden – analysis of survival, growth and quality in provenance experiments planted in 1951. Skogshögsk Stockholm Inst Skogsgen Rapp Upps 19, 132 pp
Rendle BJ, Phillips EW 1958 The effect of growth (ring width) on the density of softwoods. Forestry 31: 113–120
Richardson SD 1960 The role of physiology in forest tree improvement. 5th World Forestry Congress Seattle, WA, 733–741
Rink G, Thor E 1973 Inheritance and gains of three wood properties in Virginia pine. 12th Southern Forest Tree Improvement Conference Baton Rouge, LA, 24–30
Roberts LW, Gahan PB, Aloni R 1988 Vascular differentiation and plant growth regulators. Springer, Berlin Heidelberg New York, 154 pp
Rone VM 1970 Variation in wood density and tracheid length in Picea abies progeny. Lesovedenie 5: 78–82
Rowan SJ 1970 Fusiform rust gall formation and cellulose, lignin and other wood constituents of loblolly pine. Phytopathology 60: 1216–1220
Ruden T 1954 Brown curly birch and some other abnormal wood formations in birch. Medd Nor Skogforsöksves 12: 451–505
Rudman P, DaCosta EW, Gay FJ 1967 Wood quality in plus trees of teak (Tectona grandis). Silvae Genet 16: 102–105
Rudman P, Higgs M, Davidson J, Malajczuk N 1969 Breeding eucalypts for wood properties. 2nd World Conference Forest Tree Breeding, Washington, DC, 9 pp
Rydholm SA 1965 Pulping processes. Interscience, New York, 1269 pp
Sachsse H 1965 The effect of the rate of growth on the occurance of spiral grain. Proc IUFRO Meet Melbourne, Australia, 15 pp
Santamour FS 1979 Inheritance of wound compartmentalization in soft maples. J Arboric 5: 220–225
SAPPI 1993 Annual report forest research. Pulping Studies, Howick, South Africa, 32, 33
Saucier JR, Clark A 1992 New evidence on the role of environmental factors associated with the length of juvenile wood formation in loblolly pine. Proc Div 5 IUFRO Conf For Prod vol I Nancy, France, 158–159
Saucier JR, Taras MA 1966a Specific gravity and fiber length variation within annual height increments in red maple. For Prod J 16(2): 33–36
Saucier JR, Taras MA 1966b Wood density variation among six longleaf pine seed sources grown in Virginia. J For 64: 463–465
Saucier JR, Taras MA 1967 Wood density and percent summerwood variation among nine loblolly pine seed sources grown in Alabama. 9th Southern Conference Forest Tree Improvement Knoxville, TN, 115–119
Savill PS, Kanowski PJ, Gourlay ID, Jarvis AR 1993 Genetic and intra-tree variation in the number of sapwood rings in Quercus robur and Quercus petrea. Silvae Genet 42: 371–375

Scallan AM, Green HV 1974 A technique for determining the transverse dimensions of the fibres in wood. Wood Fiber 5: 323–333

Scaramuzzi G 1958 Variation of the volume proportions of wood elements within the stem in Populus × euramericana. Guinier Pap 4: 242–251

Scaramuzzi G 1960 Technological investigations on the wood of some Euroamerican poplar hybrids. Publ Cent Sper Agr For Roma 3: 193–216

Scaramuzzi G 1966 A comparison of some methods for determining the basic density of increment cores of Euramerican poplar hybrids. 6th Congr For Mundial, Madrid, 1–11

Schafer ER, Pew JC 1958 Grinding of loblolly pine. Relation of wood properties and grinding conditions to pulp and paper quality. US For Serv For Prod Lab Rep 1163: 1–18

Schmidtling RC, Amburgey TL 1982 Genetic variation in decay susceptibility and its relationship to growth and specific gravity in loblolly pine. Holzforschung 36: 159–161

Schniewind AP 1962 Horizontal specific gravity variation in tree stems in relation to their support function. For Sci 8: 111–118

Schoenike RE 1962 Natural variation in jack pine (Pinus banksiana) PhD Thesis University of Minnesota, St Paul, MN, 233 pp

Schönbach H 1960a Ergebnisse weiterer untersuchungen über Cellulosegehalt und Rohwichte Graupaer Pappelklone. Wiss Abh Deutsch Akad Landwiss Berlin 44(4): 81–88

Schönbach H 1960b Results of further studies in cellulose content and specific gravity of gray poplar clones. Cont to Poplar Res No IV Inst For Sci, German Acadamy Agriculture, Berlin 44: 83–98

Schreiner EJ 1935 Possibilities of improving pulping characteristics of pulpwoods by controlled hybridization of forest trees. Pap Trad J Tech Sec C: 105–109

Schreiner EJ 1958 Possibilities for genetic improvement in the utilization potentials of forest trees. Silvae Genet 7: 122–128

Schultz-Dewitz GW 1965 Variation in cell diameter and cell wall thickness within and between trees of northern coniferous species. IUFRO Conf Melbourne, Australia, 12 pp

Schütt P 1958 Schwankungen in Zellulose- und Ligningehalt bei einigen in Westdeutschland angebauten Pinus contorta-Herkünften. Silvae Genet 7: 65–69

Schütt P 1962 Individual variations and variations stemming from differences in place of origin in wood density and fiber length on Pinus contorta. Papier 16: 671–675

Schwartz SL, Bray MW 1941 Suitability of high-density, high summerwood slash pine for kraft and high-grade papers. Pap Trade J 113: 1–7

Scott CW, MacGregor WD 1952 Fast grown wood, its features, and value with special reference to conifer planting in the United Kingdom since 1919. 6th British Common For Conf Canada, Ontario 20 pp

Senft JF, Bendtsen BA, Galligan WL 1985 Weak wood. J For 83: 476–485

Senft JF, Quanci MJ, Bendtsen BA 1986 Property profile of 60-year-old Douglas-fir. Juvenile wood – what does it mean to forest management and forest products? For Prod Res Soc Coop Tech Workshop Proc 47309, 17–28

Sesbou A 1981 Study of the genetic variability of quality of wood and collapse in Eucalyptus camaldulensis. PhD Thesis, Université de Nancy, France, 145 pp

Sesbou PA, Nepveu G 1991 Infraspecific variability in Eucalyptus camaldulensis coppice and effect of the first cutting on wood density, pulp yield and fiber length. Silvae Genet 40: 173–179

Shelbourne CJ 1963 Growth rates, stem form, and branching characteristics of five provenances of P. khasya grown under plantation conditions in N. Rhodesia. Commonw For Rev 42: 334–336

Shelbourne CJ 1969 Predicted genetic improvement from different breeding methods. 2nd World Consul Forest Tree Breeding Washington DC 6 pp

Shelbourne CJ, Thulin IJ, Scott RM 1972 Variation, inheritance, and correlation amongst growth, morphological and wood characters in radiata pine. N Z For Ser Prod For Div Gen Tree Improve Rep Rotorua N Z 61: 26
Shelbourne CJ, Zobel BJ, Stonecypher RW 1969 The inheritance of compression wood and its genetic and phenotypic correlations with six other traits in 5-year-old loblolly pine. Silvae Genet 18: 43–47
Shigo AL, Marx HG 1977 Compartmentalization of decay in trees. US For Ser Agr Inf Bull Asheville, NC 405, 73 pp
Shigo AL, Shortle WC, Garrett PW 1977 Genetic control suggested in compartmentalization of discolored wood associated with tree wounds. For Sci 23: 179
Shukla GL 1972 Some statistical aspects of partitioning genotype-environmental components of variability. Heredity 29: 237–245
Siddiqui KM, Khan M, Akhtar S 1979 Results of 10-year-old Eucalyptus camaldulensis provenance study at Peshawar. Silvae Genet 28: 24–26
Silen RR 1964 Regeneration aspects of the 50-year-old Douglas-fir heredity study. Proc 1964 Annual Meeting Western Reforestation Coordinating Committee Corvallis, Oregon
Simpson WT 1993 Specific gravity, moisture content, and density relationships for wood. US For Ser, Madison WI FPL-GTR-76, 13 pp
Sinclair SA, Ifju G 1979 Lumber quality of beetle-killed southern pine in Virginia. For Prod J 29(4): 18–22
Singh T 1986 Wood density variation of six major tree species of the Northwest Territories. Can J For Res 16: 127–129
Smith DM 1955 A comparison of two methods for determining the specific gravity of small samples of second-growth Douglas-fir. US For Ser Rept No 2033 Madison, WI, 21 pp
Smith DM 1961 Method of determining specific gravity of small wood chips. US For Ser Rep 2209 FPL Madison, WI, 10 pp
Smith DE 1962 An investigation into the specific gravity relationship between limb sections and the bole or increment core of young loblolly pine (Pinus taeda) trees. MS Thesis, School of For NC State Univ Raleigh, NC, 58 pp
Smith DM 1962 Patterns of variation of wood structure and properties within trees of southern yellow pine and the influence of environmental factors on these problems. Symposium wood quality in relation to the southern pulp and paper industries. Raleigh, NC, 14–16
Smith DM 1967 Microscopic methods for determining cross-sectional cell dimensions. US For Ser Res Paper FPL-79, Madison, WI, 20 pp
Smith JH, Kozak A 1967 Thickness and percentage of bark of the commercial trees of British Columbia. Univ British Columbia, Fac For, Vancouver, (mimeo rep) 33 pp
Smith JH, Rumma U 1971 Specific gravity and fiber length of hybrid poplar. J For 69: 34
Smith WB 1992 Determining moisture content in eastern redcedar. For Prod J 42(7/8): 67–69
Smith WJ 1967 The heritability of fibre characteristics and its application to wood quality improvement in forest trees. Silvae Genet 16: 41–50
Smith WJ 1970 Improvement in yield and wood quality through selection criteria and indices. 2nd Work Group No 1 Research Committee Australian For Council Beerwah, Queensland, 1–40
Smith WJ 1977 Variation of growth, stem quality and wood properties in relation to genetic improvement of tropical forest trees. (Variation in wood quality and productivity of some Queensland plantation grown softwoods). IUFRO Workshop Brisbane, Australia, 26 pp
Snyder B 1976 Selecting for specific gravity and growth in longleaf pine. US For Ser South Inst For Gen Biloxi, MS, Final Report (Mimeo)

Sohn SI, Goddard RE 1975 Genetic study of wood specific gravity in slash pine. 22nd Northeastern Forest Tree Improvement Conference Syracuse, New York, 61–72
Son DS, Chung SB 1970 Fibre length and specific gravity of Populus alba × glandulosa hybrid. Inst For Genet Suwon, Korea Res Rep 8: 1–5
Sprague JR, Talbert JT, Jett JB, Bryant RL 1983 Utility of the pilodyn in selection for mature wood specific gravity in loblolly pine. For Sci 29: 696–701
Spurr SH, Hyvärinen MJ 1954 Wood fiber length as related to position in tree and growth. Bot Rev 20: 561–575
Squillace AE, Bengston GW 1961 Inheritance of gum yield and other characteristics of slash pine. Proc 6th Southern Forest Tree Improvement Conference Gainesville, FL, 85–96
Squillace AE, Echols RM, Dorman KW 1962 Heritability of specific gravity and summerwood per cent and relation to other factors in slash pine. Tappi 45: 599–601
Squillace AE, Gansel CR 1974 Juvenile-mature correlations in slash pine. For Sci 20: 225–229
Squillace AE, Silen RR 1962 Racial variation in ponderosa pine. For Sci Monograph No 2, 27 pp
Stahl EG, Persson A, Persson B 1986 Timber quality in northern latitudes – implications for forest tree improvement. Sveriges Lantbruksuniversitet, Rep 6 (Frans Kempe Symp in Umeå), 149–177
Stairs GR 1969 Seed source and growth rate effects on wood quality in Norway spruce. 11th Meeting Committee Forest Tree Breeding in Canada Ottawa, Canada, 231–236
Stairs GR, Marton R, Brown AF, Rizzo M, Petrik A 1966 Anatomical and pulping properties of fast- and slow-grown Norway spruce. Tappi 49: 296–300
Stevens M, Lejeune J, Schalck J 1983 Variation in specific gravity and tension wood in fast growing poplar hybrids. IUFRO Conf Div 5 Madison, WI
Stonecypher RW, Cech FC 1960 An efficient method of taking large increment cores. J For 58: 644–645
Stonecypher R, Cech FC, Zobel BJ 1964 Inheritance of specific gravity in two- and three-year-old seedlings of loblolly pine. Tappi 47: 405–407
Stonecypher R, Zobel BJ 1966 Inheritance of specific gravity of five-year-old seedlings of loblolly pine. Tappi 49: 303–305
Stonecypher RW, Zobel BJ, Blair R 1973 Inheritance patterns of loblolly pines from a non-selected natural population. NC Agric Exp Stn Tech Bull 200: 1–60
Strauss SH, Critchfield WB 1982 Inheritance of β-pinene in xylem oleoresin of knobcone × Monterey pine hybrids. For Sci 28: 687–696
Strauss SH, Lande R, Namkoong G 1992 Limitations of molecular-marker-aided selection in forest tree breeding. Can J For Res 22: 1050–1061
Strickland RK 1960 Geographic variations in specific gravity and tracheid length of loblolly pine. MS Thesis Univ Georgia Athens, GA, 68 pp
Strickland RK, Goddard RE 1965 Inheritance of branching and crown characteristics in slash pine. 8th Southern Conference Forest Tree Improvement Savannah, GA, 57–63
Strickland RK, Goddard RE 1966 Correlation studies of slash pine tracheid length. For Sci 12: 54–62
Stringer JW, Olson JR, Carpenter SD 1982 Variation in wood properties of nine black locust clones. 7th N American Forest Biology Workshop Lexington, KY, p 73
Struve DK, Talbert JT, McKeand SE 1984 Growth of rooted cuttings and seedlings in a 40-year-old plantation of Eastern white pine. Can J For Res 14: 462–464
Sudo S 1983 Assessment for wood quality improvement of akamatsu (Pinus densiflora). IUFRO Conf Div 5 Madison, WI, p 53 (abstract)
Syzmanski MB 1991 Loblolly pine provenance variation in age of transition from juvenile to mature wood specific gravity. For Sci 37: 160–174
Talbert JT, Jett JB 1981 Regional specific gravity values for plantation grown loblolly pine in the Southeastern United States. For Sci 27: 801–807

Talbert JT, Jett JB, Bridgwater FE, Jahromi ST 1982a Genetic parameters of wood specific gravity in a control pollinated loblolly pine genetic test. TAPPI Research Development Division Conference Asheville, NC, 179–182
Talbert JT, Jett JB, Bryant RL 1982b Inheritance of wood specific gravity in an unimproved loblolly pine population. 20 years of results. Silvae Genet 32: 33–37
Taras AM, Wahlgren HE 1963 A comparison of increment core sampling methods for estimating tree specific gravity. US For Ser USDA Res Pap SE 7, Athens, GA, 16 pp
Tauer GG, Loo-Dinkins JA 1990 Seed source variation in specific gravity of loblolly pine grown in a common environment in Arkansas. For Sci 36: 1133–1145
Taylor F 1965 A study of the natural variation of certain properties of the wood of yellow poplar (Liriodendron tulipifera) within trees, between trees and between geographic areas. PhD Thesis NC State Univ Raleigh, NC, 190 pp
Taylor FW 1969 The effect of ray tissue on the specific gravity of wood. Wood Fiber 1: 142–145
Taylor FW 1977a A note on the relationship between branch and stemwood properties of selected hardwoods growing in the midsouth. Wood Fiber 8: 257–261
Taylor FW 1977b Variation in specific gravity and fiber length of selected hardwoods. For Sci 23: 190–194
Taylor FW 1981 Rapid determination of southern pine specific gravity with a pilodyn tester. For Sci 27: 59–61
Teissier du Cross E, Kleinschmit J, Azoeuf P, Haslin R 1980 Spiral grain in beech, variability and heredity. Silvae Genet 29: 5–13
Tellerup E 1953 Individual differences in the shape of wood rays in Fagus sylvatica. A wood anatomical investigation. Royal Vet Agr Col Yearbook, Copenhagen, Denmark, 147–157
Thiercelin F 1970 Tardivité du débourrement et densité dans une population adulte de Picea abies. Ann Sci For 27: 243–254
Thor E, Brown SJ 1962 Variation among six loblolly pine provenances tested in Tennessee. J For 60: 476–480
Thorbjornsen E 1961 Variation in density and fiber length in wood of yellow poplar. Tappi 44: 192–195
Timell TE 1969 The chemical composition of tension wood. Svensk Paperstidn 72: 173–181
Timell TE 1986 Compression wood in gymnosperms. Springer, Berlin Heidelberg New York, 2150 pp (3 vols)
Toda R 1970 Abstracts of Japanese literature in forest genetics and related fields vols IA, IB. Gov For Exp Stn Meguro, Japan, 917 pp
Todorovski S 1969 Distribution of weight in the sessile oak (Quercus sessifolia) trunk. School Agr For Univ Skopje Terezije Belgrade, 28 pp (Trans for USDA)
Tsoumis GT 1991 Science and technology of wood-structure, properties, utilization. van Nostrand Reinhold, New York, 494 pp
Turnbull RF 1965 Review of current attitudes on growth stresses. IUFRO Sect 46 Melbourne, Australia, 8 pp
Turner CH, Balodis V, Dean GH 1983 Variability in pulping quality of E. globulus from Tasmanian provenances. Appita 36: 371–376
Ujvári É, Szönyi L 1973 Expectable gain breeding long fibre Norway spruce. Erdész Kut 69(2): 93–99
Valentine FA 1962 Natural variation in specific gravity in Populus tremuloides in northern New York. 9th Northeastern Forest Tree Improvement Conference, Syracuse, NY, 17–24
Valley RB, Morse TH 1965 Measurement of fiber length using a modified coulter particle counter. Tappi 48: 372–376
van Buijtenen JP 1955 Summerwood formation in loblolly pine. PhD Thesis, Texas A & M Univ College Station, TX, 83 pp

van Buijtenen JP 1956 Experimental control of environmental factors and their effect upon some aspects of wood anatomy in loblolly pine. Inst Paper Chem Appleton, WI, 21 pp (mimeo rep)
van Buijtenen JP 1960 Sampling fiber lengths in quaking aspen. 5th World Forestry Congress, Seattle, WA, 3: 1367–1370
van Buijtenen JP 1962 Heritability estimates of wood density in loblolly pine. Tappi 45: 602–605
van Buijtenen JP 1963a Heritability of wood properties and their relation to growth rate in Pinus taeda. 1st World Consultation Forest Genetic Tree Improvement Stockholm, Sweden, 13 pp
van Buijtenen JP 1963b Tree to tree variation in physical and anatomical properties of wood of loblolly and slash pines. Symposium Wood Quality Relationship to Southern Pulp Paper Industry, Raleigh, NC, 12 pp (mimeo)
van Buijtenen JP 1964 Anatomical factors influencing wood specific gravity of slash pines and the implications for the development of a high-quality pulpwood. Tappi 47: 401–404
van Buijtenen JP 1965 Inheritance of fiber properties in North American conifers. IUFRO Section 41, vol 2. Melbourne, Australia, 12 pp
van Buijtenen JP 1967a Quality control begins in the woods. Tappi 4th Forest Tree Biology Conference, Appleton, WI, 102–110
van Buijtenen JP 1967b Pulpwood properties and tree breeding: a synthesis. 14th IUFRO Congress Sect 41 München, Germany, 243–262
van Buijtenen JP 1969a Controlling wood properties by forest management. Tappi 52: 257–259
van Buijtenen JP 1969b Relationships between wood properties and pulp and paper properties. 2nd World Consultation Forest Tree Breeding, Washington, DC, 1–13
van Buijtenen JP 1978 Response of "lost pines" seed sources to site quality. 5th North American Forest Biology Workshop, Gainesville, FL, 7 pp
van Buijtenen JP 1982 Fiber for the future. Tappi 65(8): 10–12
van Buijtenen JP, Alexander SD, Einspahr DW, Ferrie AE, Hart T, Kellogg RM, Porterfield RL, Zobel BJ 1974 Quality control in the forest and in the mill. Tappi CA Rep 54 Atlanta, GA, 96 pp
van Buijtenen JP, Alexander SD, Einspahr DW, Ferrie AE, Hart T, Kellogg RM, Porterfield RL, Zobel BJ 1975 How will tree improvement and intensive forestry affect pulp manufacture? (Digest of CA Rep 54). Tappi 58(9): 129–134
van Buijtenen JP, Einspahr DW, Joranson PN 1959 Natural variation in Populus tremuloides. Tappi 42: 819–823
van Buijtenen JP, Einspahr DW, Peckham JR 1962 Natural variation in Populus tremuloides II. Variation in pulp and papermaking properties. Tappi 45: 58–60
van Buijtenen JP, Einspahr DW, Peckham JR 1968 Micropulping loblolly pine grafts selected for extreme wood specific gravity. Silvae Genet 17: 15–19
van Buijtenen JP, Joranson PN, MacLaurin DJ 1961 Pulping southern pine increment cores by means of a small scale kraft procedure. Tappi 44: 166–169
van Buijtenen JP, Zobel BJ 1994 Genetics and breeding of wood. Chapter 13 Tree Breeding (Mandal AK) Inter Book Distrib and Publ, Dehra Dun, India (in press)
van der Sijde HA, Shaw MJ, van Wyk G 1985 Reaction wood in Pinus taeda – a preliminary report. S Afr For J: 27–32
van Wyk G 1977 Progress with the Eucalyptus grandis breeding program in the Republic of South Africa. 3rd World Consultation Forest Tree Breeding, Canberra, Australia, 3 pp
van Wyk G, Hodgson LM 1983 Clonal seed orchards of Eucalyptus grandis in South Africa. Fast Growing Trees Silvicult 31: 566–568
Vargas-Hernandez J 1990 Genetic variation of wood density components in coastal Douglas-fir and their relationship to growth rhythm. PhD Thesis, Oregon State Univ Corvallis, OR, 113 pp

Vargas-Hernandez J, Adams WT 1991 Genetic variation of wood density components in young coastal Douglas-fir: implications for tree breeding. Can J For Res 21: 1801–1807

Vargas-Hernandez J, Adams WT 1992 Age-age correlations and early selection for wood density in young coastal Douglas-fir. For Sci 38: 467–478

Vargas-Hernandez J, Adams WT 1994 Genetic relationships between wood density components and cambial growth rhythm in young coastal Douglas-fir. Can J For Res (in press) 20 pp (Pap # 2978 For Res Lab Oregon State Univ)

Vargas-Hernandez J, Adams WT, Krahmer RL 1994 Family variation in age trends of wood density in young coastal Douglas-fir. Wood Fiber Sci (in press) (Pap # 2934 For Res Lab Oregon State Univ) 17 pp

Veal MA, Blair RL, Jett JB, McKean WT 1974 Impact of fusiform rust on pulping properties of young loblolly and slash pine. TAPPI Biological Conference Seattle, WA, 47–58

Velling P 1974 Phenotypic and genetic variation in wood basic density of Scots pine (Pinus silvestris). Folia For 188: 29 pp

Velling P 1983 Individual genetic variability of wood quality in Betula pendula. Folia For 21 pp

Vidakovic M, Ahsan J 1970 The inheritance of crooked bole in Shisham (Dalbergia sissoo). Silvae Genet 19: 94–98

Villeneuve ME, Morganstern EK, Sebastian LP 1987 Variation patterns and age relationships of wood density in families of jack pine and black spruce. Can J For Res 17: 1219–1222

Vital BR, Lucia RM 1980 Procedéncia de semintes e qualidade da madeira de Eucalyptus grandis. Rev Arvore 4(2): 170–178

von Pechmann H 1958 Die Auswirkung der Wuchsgeschwindigkeit auf die Holzstruktur und die Holzeigenschaften einiger Baumarten. Schweiz Z Forstwes 109: 615–647

von Wedel K 1962 Hydrostatic methods for the determination of the density of wood samples. Holz Roh- Werkst 20: 360–364

von Wedel KW, Zobel BJ, Shelbourne CJ 1967 Knotwood in loblolly pine. Tech Rep #36 School For NC State Univ Raleigh, NC, 43 pp

von Wedel, Zobel BJ, Shelbourne CJ 1968 Prevalence and effects of knots in young loblolly pine. For Prod J 18: 97–103

Wagner FG, Taylor FW 1993 Low lumber recovery at southern pine sawmills may be due to misshapen sawlogs. For Prod J 43(3): 53–55

Wakeley PC 1969 Effects of evolution on southern pine wood. For Prod J 19(2): 16–20

Wall RE 1971 Variation in decay in aspen stands as affected by their clonal growth pattern. Can J For Res 1: 141–146

Walters CS 1951 Figured walnut propagated by grafting. J For 49: 917

Wang H, Cai X, Zang D, Zheng Y 1991 Provenance variation in the growth rate and wood properties of Eucalyptus grandis in China. Res Inst Chinese Acad For Beijing, China, 6 pp

Wang S, Littell RC, Rockwood DL 1984 Variation in density and moisture content of wood and bark among twenty Eucalyptus grandis progenies. Wood Sci Technol 19: 97–100

Wangaard FF 1958 Research on quality of wood in relation to its utilization as fiber for papermaking. Norsk Skogind 8: 1–10

Wangaard RR, Kellogg RM, Djerf AD 1967 Prediction of whole tree pulp properties for slash pine. Tappi 50: 109–113

Waterman AM 1954 Septoria canker of poplars in the United States. USDA, US For Ser Circ Philadelphia, PA 947

Watson AJ 1965 Fibre characteristics and wood properties; tropical and semi-tropical hardwoods. IUFRO Sect 41 Melbourne, Australia, 7 pp

Watson AJ, Dadswell HE 1962 Influence of fibre morphology on paper properties. Part II. Earlywood and latewood. Appita 15: 116–129

Watson AJ, Hodder IG 1954 Relationship between fibre structure and handsheet properties in Pinus taeda. Aust Pulp Pap Tech Assoc Proc 8: 290–310
Watson AJ, Wardrop AB, Dadswell HE, Cohen WE 1952 Influence of fibre structure on pulp and paper properties. Aust Pulp Pap Ind Tech Assoc Proc 6: 243–267
Waugh G 1972 Growth stresses – genetic and environmental influences. CSIRO For Prod Newsletter No 389 South Melbourne Victoria, Australia, 3 & 4
Webb CD 1964 Natural variation in specific gravity, fiber length and interlocked grain of sweetgum (Liquidambar styraciflua) in the South Atlantic States. PhD Thesis NC State Univ Raleigh, NC, 138 pp
Webb CD 1967 Interlocked grain in Liquidambar styraciflua. 14th IUFRO Conf vol 9 München, Germany, 398–412
Webb CD 1969 Variation of interlocked grain in sweetgum. For Prod J 19(8): 45–48
Webb CD, Belanger RP, McAlpine RG 1973 Family differences in early growth and wood specific gravity in sycamore (Platanus occidentalis). 12th Southern Forest Tree Improvement Conference, Baton Rouge, LA, 213–227
Wellwood RW, Smith JG 1962 Variation in some important qualities of wood from young Douglas-fir and hemlock trees. Res Pap 50 Univ Brit Columbia, Vancouver, Canada, 15 pp
Wheeler RT 1964 Equipment suitable for measuring angle of grain and for obtaining small timber samples from living trees. Comm For Rev 43: 314–318
Wheeler EY, Zobel BJ, Weeks DL 1965 Tracheid length and diameter variation in the bole of loblolly pine. Tappi 49: 484–490
White JF, Saucier JR 1965 A comparison of the specific gravity of two slash pine varieties grown in South Florida. Tappi 49: 230–232
Wilcox MD 1974 Wood brightness variation in clones of loblolly pine. TAPPI Forest Biology Conference, Seattle, WA, 71–80
Wilcox MD 1975 Wood brightness variation in clones of loblolly pine. Silvae Genet 24: 54–59
Wilcox MD 1977 Variation in growth, stem quality, and wood properties in relation to genetic improvement of tropical forest trees. IUFRO Workshop Brisbane, Australia, 9 pp
Williams CG, Neale DB 1992 Conifer wood quality and marker-aided selection: a case study. Can J For Res 22: 1009–1017
Williams MF 1994 Matching wood fiber characteristics to pulp and paper processes and products. Tappi 77(3): 227–233
Wilson JW, Ifju G 1965 Wood characteristics VII: Intra-increment physical and chemical properties. Pulp Pap Res Inst Can Tech Rep Ser 424, Montreal, Canada, 24 pp
Wiselogel AE, Tauer CG 1982 Genetic variation in wood specific gravity and tracheid length in shortleaf pine. TAPPI Research and Development Division Conference Asheville, NC, 169–178
Wodzicki T 1961 Investigations on the kind of Larix polonica wood formed under various photoperiodic conditions. II Effect of different light conditions on wood formed by seedlings in a greenhouse. Acta Soc Bot Pol 30: 111–131
Wodzicki T, Witkowska L 1961 On the photoperiodic control of extension growth and wood formation in Norway spruce. Acta Soc Bot Pol 30: 755–764
Woessner RA 1983 Variation in wood density, height and diameter of the International Provenance Trial at Jari. Silvicultura No 30 Fast Growing Trees Sáo Pedro, Brazil, 183–185
Woodcock DW 1989 Climate sensitivity of wood anatomical features in a ring-porous oak (Quercus macrocarpa). Can J For Res 19: 639–644
Wooten TE, Barefoot AC 1965 Mechanical maceration technique for isolation of individual fibers. For Prod J 25(2): 75
Worrall J 1970 Interrelationships among some phenological and wood property variables in Norway spruce. Tappi 53: 58–63

Worrall J 1975 Provenance and clonal variation in phenology and wood properties of Norway spruce. Silvae Genet 24: 2–5

Wright JA 1987 Results of micropulping wood samples of Pinus caribaea, P. elliottii, and P. patula ssp. tecunumanii in the Eastern Transvaal and Zululand. Simposio Sobre Silvicultura Y Mejoramiento de Especies Forestales. 4 CIEF Buenos Aires, Argentina, 247–256

Wright JA 1990a Provenance variation in wood properties of Pinus caribaea var. hondurensis. N Z J For Sci 20(2): 220–230

Wright JA 1990b Variation in wood properties of Pinus oocarpa and Pinus patula ssp. tecunumanii at six sites. Silvae Genet 39: 1–5

Wright JA 1991 Impact of wood quality assessments on future fibre resources in the pulp and paper-making industry. S Afr For J 157: 96–99

Wright JA, Endo M 1993 Wood density from 460 clones in sixteen trials of Eucalyptus grandis at four years of age in Colombia. Wood Fiber Sci 25: 331–338

Wright JA, Gibson GL, Barnes RD 1986a Provenance variation in stem volume and wood density of Pinus caribaea, P. oocarpa and P. patula ssp. tecunumanii in Zambia. Comm For Rev 65: 33–40

Wright JA, Gibson GL, Barnes RD 1986b Variation of stem volume and wood density of provenances of Pinus oocarpa and P. patula ssp. tecunumanii of Aqudos, São Paulo, Brazil. IPEF 32: 21–23

Wright JA, Gibson GL, Barnes RD 1986c Variation of stem volume and wood density of Pinus caribaea provenances growing at Bukit Tapah, Maylasia. Malays For 49: 176–180

Wright JA, Gibson GL, Barnes RD 1989a Variation in volume and density of the wood of eight provenances of Pinus oocarpa and P. patula ssp. tecunumanii in Conocoto Ecuador. IPEF 41/42: 5–7

Wright JA, Gibson GL, Barnes RD 1990 Variation of stem volume and wood density of Pinus caribaea provenances growing at two locations in Fiji. Appita 43: 286–288

Wright JA, Osorio LF 1993 Density variation in provenances of Pinus maximinoi at age 14.5 years in Colombia. Res Rep 147 Smurfit Carton de Colombia Cali, Colombia, 5 pp

Wright JA, Shaw MJ, Hadebe W 1989b The correlation of the pulp and paper traits between micro-pulping and full-scale pulping. Tappi 72(4): 191–194

Wright JA, Sluis-Cremer HJ 1992 Tracheid morphology and pulp and paper strength traits of Pinus taeda and P. patula at age 17 years in South Africa. Tappi: 75(5): 183–187

Yanchuk AD, Dancik BP, Micko MM 1983a Intraclonal variation in wood density of trembling aspen in Alberta. Wood Fiber Sci 15(4): 387–394

Yanchuk AD, Dancik BP, Micko MM 1983b Variation and heritability of wood density and fibre length of trembling aspen in Alberta, Canada. Silvae Genet 32: 65–74

Yanchuk AD, Kiss GK 1992 Genetic variation of growth and wood specific gravity and its utility in the improvement of interior spruce in British Columbia. Silvae Genet 42: 141–148

Yazawa K, Kurata H 1963 Study on the measurement of cell elements of wood with the dimensions calculator by facsimile process. J Jpn Wood Res Soc 7: 193–198

Yiannos PN 1964 The apparent cell-wall density of wood and pulp fibers. Tappi 47: 468–471

Yitai C, Jiwei Z, Ning G 1992 A study of the combining ability for main characters of wood qualities in Chinese fir. Res Inst Subtropical Forestry, China, 6 pp

Zabel R 1956 Decay resistant variations within northeastern forest tree species. 3rd Northeast Forest Tree Improvement Conference, Ithica NY, 13–17

Zavarin E, Mirov NT, Colling EN, Snasberk K, Costello K 1968 Chemical composition of turpentine from Pinus khasya. For Sci 14: 55–61

Zhang SY, Owoundi RE, Nepveu G, Mothe F, Dhôte JF 1993 Modelling wood density in European oak (Quercus petrea and Quercus robur) and simulating the silvicultural influence. Can J For Res 23: 2587–2593

Zhang SY, Zhong Y 1991 Effect of growth rate on specific gravity of East-Liaoning oak (Quercus liaotungensis) wood. Can J For Res 21: 255–260
Zhang SY, Zhong Y 1992 Structure-property relationships of wood in East-Liaoning oak. Wood Sci Technol 26: 139–149
Zhi JW, Chen YT, Li GX 1993 Genetic analysis and index selection on the main economic characters of clones of Chinese fir. Research Institute of Tropical Forestry, CAF China For Res 6. Suppl
Zobel BJ 1956 Genetic, growth and environmental factors affecting specific gravity in loblolly pine (Comments by Schreiner EJ). For Prod J 6: 442–337
Zobel BJ 1961 Inheritance of wood properties in conifers. Silvae Genet 10: 65–70
Zobel BJ 1963 Breeding for wood properties in forest trees. World Consultation Forest Genetics Tree Improvement Stockholm, Sweden, 31 pp
Zobel BJ 1964 Breeding for wood properties in forest trees. Unasylva 18: 89–103
Zobel BJ 1965a Inheritance of fiber characters and specific gravity in hardwoods – a review. IUFRO Conf Melbourne, Australia, 16 pp
Zobel BJ 1965b Inheritance of spiral grain. IUFRO Sec 41 Melbourne, Australia, 9 pp
Zobel BJ 1970 Developing trees with wood properties more desirable for paper. Tappi 53: 2320–2325
Zobel BJ 1971 Genetic manipulation of wood of the southern pines including chemical characteristics. Wood Sci Technol 54: 255–271
Zobel BJ 1972 Genetic effects on wood qualities. TAPPI Forest Biology Committee Meeting, Appleton, WI, 9 pp (mimeo)
Zobel BJ 1973 Improvement of trees through genetics and the resultant physical and chemical properties of fibres. TAPPI-API Management Meeting, Mobile, Alabama, 16 pp (mimeo)
Zobel BJ 1974 Improving yields from forest land. Proc Fiber Conservation Utilization, Pulp Pap. seminar: Chicago, 16 pp
Zobel BJ 1978 Changing wood by genetic manipulation. For Prod Res Soc Atlanta, GA, 34 (Abstr)
Zobel BJ 1980 Genetic improvement of forest trees for biomass production. In: Saarkanen and Tillman (eds) Progress in biomass conversion C. Academic Press, New York, 37–58
Zobel BJ 1981 Wood quality from fast grown plantations. Tappi 64(1): 71–74
Zobel BJ 1984 The changing quality of the world wood supply. Wood Sci Technol 18: 1–17
Zobel BJ, Campinhos E, Ikemori Y 1983 Selecting and breeding for desirable wood. Tappi 66(1): 70–74
Zobel BJ, Cole D, Stonecypher R 1962a Wood properties of clones of slash pine. Forest Genetics Workshop Macon, GA, 32–39
Zobel BJ, Jett JB, Hutto R 1978 Improving wood density of short rotation southern pine. Tappi 61(3): 41–44
Zobel BJ, Jett JB, McVickers GW 1977 Effects of tree form on yield and quality of plywood. Mimeo Rep NC State Univ, 11 pp
Zobel BJ, Kellison RC, Mathias M 1969 Genetic improvement in forest trees – growth rate and wood properties in young loblolly pine. 10th Southern Conference Forest Tree Improvement, Houston, TX, 57–75
Zobel BJ Kellison RC, Mathias MF, Hatcher AV 1972 Wood density of the southern pines. Tech Bull 208 North Carolina Agric Exp Stn Raleigh, NC, 56 pp
Zobel BJ, Mathias M, Kellison RC, Roberds JH 1968a Moisture content of southern pine trees. Tech Rep 37 School For Res North Carolina State University, Raleigh, NC, 44 pp
Zobel BJ, McElwee RL, Browne C 1962b Interrelationship of wood properties of loblolly pine. 6th Southern Forest Tree Improvement Committee, Gainesville, FL, 142–162
Zobel BJ, Stonecypher R, Browne C, Kellison RC 1966 Variation and inheritance of cellulose in southern pines. Tappi 49: 383–387

Zobel BJ, Stonecypher RW, Browne C 1968b Inheritance of spiral grain in young loblolly pine. For Sci 14: 376–379
Zobel BJ, Talbert J 1984 Applied forest tree improvement. Wiley, New York, 511 pp
Zobel BJ, Thorbjornsen E, Henson E 1960 Geographic, site, and individual tree variation in the wood properties of loblolly pine. Silvae Genet 9: 149–158
Zobel BJ, van Buijtenen JP 1989 Wood variation, its causes and control. Springer, Berlin Heidelberg New York, 363 pp
Zobel BJ, van Wyk G, Stahl P 1987 Growing exotic forests. Wiley, New York, 505 pp
Zoerb MH, Remeta G 1968 Increment borer drive assembly. Woodlands Res Notes Union Camp Corp Res Note Southern Forest Tree Improvement Conference, Houston TX 19, 4 pp

Species Index

Conifers

Hardwoods

Subject Index

Springer-Verlag and the Environment

We at Springer-Verlag firmly believe that an international science publisher has a special obligation to the environment, and our corporate policies consistently reflect this conviction.

We also expect our business partners – paper mills, printers, packaging manufacturers, etc. – to commit themselves to using environmentally friendly materials and production processes.

The paper in this book is made from low- or no-chlorine pulp and is acid free, in conformance with international standards for paper permanency.

MIX
Papier aus verantwortungsvollen Quellen
Paper from responsible sources
FSC® C105338

If you have any concerns about our products,
you can contact us on
ProductSafety@springernature.com

In case Publisher is established outside the EU,
the EU authorized representative is:
Springer Nature Customer Service Center GmbH
Europaplatz 3, 69115 Heidelberg, Germany

Printed by Libri Plureos GmbH
in Hamburg, Germany